MIMO Antenna Systems for 5G and Beyond

IEEE Press
445 Hoes Lane
Piscataway, NJ 08854

MIMO Antenna Systems for 5G and Beyond

Xiaoming Chen
School of Information and Communications Engineering, Xi'an Jiaotong University, Xi'an, China

Ping Jack Soh
Centre for Wireless Communications, University of Oulu, Oulu, Finland

Mohammad S. Sharawi
Blue Origin LLC, Affiliate Professor with the Electrical and Computer Engineering Department, University of Washington, Seattle, WA, USA

IEEE Press
WILEY

Published by John Wiley & Sons, Inc., Hoboken, New Jersey.
Published simultaneously in Canada.

Library of Congress Cataloging-in-Publication Data:

Names: Chen, Xiaoming, author. | Soh, Ping Jack, author. | Sharawi, Mohammad S., author.
Title: MIMO antenna systems for 5G and beyond / Xiaoming Chen, Ping Jack Soh, Mohammad S. Sharawi.
Description: Hoboken, New Jersey : Wiley, [2025] | Includes index.
Identifiers: LCCN 2024017354 (print) | LCCN 2024017355 (ebook) | ISBN 9781119932512 (hardback) | ISBN 9781119932529 (adobe pdf) | ISBN 9781119932536 (epub)
Subjects: LCSH: Antennas (Electronics) | MIMO systems. | 5G mobile communication systems.
Classification: LCC TK7871.6 .C455 2024 (print) | LCC TK7871.6 (ebook) | DDC 621.384–dc23/eng/20240503
LC record available at https://lccn.loc.gov/2024017354
LC ebook record available at https://lccn.loc.gov/2024017355

Cover Design: Wiley
Cover Images: © metamorworks/Shutterstock

Set in 9.5/12.5pt STIXTwoText by Straive, Chennai, India

To my family for their generous and ever-lasting supports.

Xiaoming Chen

To my family, who fill my life with happiness and love I could not repay: Nora, Jasper and Julien; Mum and Dad.

Ping Jack Soh

To my amazing wife, RANA; my three growing angels, BASIMA, RAND, and IBRAHIM; and my beloved parents, Dr. SAID SHARAWI and BASIMA SHARAWI.

Mohammad S. Sharawi

Contents

About the Authors *xv*
Preface *xvii*
Acknowledgments *xix*

1 Introduction *1*
1.1 Wireless Technology Evolution *3*
1.2 Benefits of MIMO Technology *6*
1.3 Future Technology Trends *7*
1.4 Book Organization *9*
References *11*

2 MIMO Antenna Fundamentals and Characterization Methods *13*
2.1 Performance Metrics for Conventional Antennas *13*
2.1.1 Gain *13*
2.1.2 Efficiency *14*
2.1.3 Polarization *14*
2.2 Performance Metrics of MIMO Antenna Systems *15*
2.2.1 MIMO Channel *15*
2.2.2 MIMO Capacity *16*
2.2.2.1 Capacity of the Frequency Flat MIMO Channel *16*
2.2.2.2 Capacity of the Frequency-Selective MIMO Channels *17*
2.2.3 Isolation *17*
2.2.4 Correlation Coefficient *18*
2.2.4.1 Correlation Matrix *19*
2.2.4.2 Correlation Coefficient *19*
2.2.5 Total Active Reflection Coefficient *20*
2.2.5.1 Generalized Scattering Matrix *21*
2.2.5.2 TARC Calculation *23*

2.2.6 Mean Effective Gain and Diversity Gain *23*
2.2.6.1 Mean Effective Gain *23*
2.2.7 Diversity Combining Methods *25*
2.2.7.1 Diversity Gain *25*
2.2.7.2 System Model of Receiver Diversity *26*
2.2.7.3 Selection Combining *26*
2.2.7.4 Threshold Combining *26*
2.2.7.5 Equal-Gain Combining *27*
2.2.7.6 Maximal-Ratio Combining *27*
2.3 MIMO Antenna Characterization Methods *28*
2.3.1 Assessment Using the *S*-Parameters *28*
2.3.1.1 Normalized Coupling *29*
2.3.1.2 Correlation and Diversity Gain *29*
2.3.1.3 Mean MIMO Capacity in Rayleigh Environment *30*
2.3.2 Anechoic Chamber Characterization *30*
2.3.2.1 MIMO Capacity *31*
2.3.2.2 MRC Diversity Gain *33*
2.3.3 Reverberation Chamber Characterization *33*
2.3.3.1 Reverberation Chamber Versus Anechoic Chamber – on MIMO Performance Characterization *33*
2.3.3.2 MRC Diversity Gain *34*
2.3.3.3 MIMO Capacity and Correlation Coefficient *36*
2.4 Summary *36*
References *37*

3 Feed Network Design and Analysis for MIMO Antenna Systems *41*
3.1 Impedance-Matching Networks *41*
3.1.1 Single-Stub Impedance Matching *42*
3.1.2 Double-Stub Impedance Matching *44*
3.1.3 Matching Networks Using Lumped Elements *46*
3.1.3.1 Resistive L-Section Matching Circuits *47*
3.1.3.2 Reactive L-Section Matching Circuits *48*
3.2 Passive Feeding Networks *50*
3.2.1 Microstrip Line Feed *50*
3.2.2 Coaxial Probe Feed *51*
3.2.3 Aperture Coupling Feed *52*
3.2.4 Proximity (Electromagnetically) Coupled Feed *52*
3.2.5 Coplanar Waveguide Feed *53*
3.3 Active Feed Networks *56*
3.3.1 Open-Loop Tuning *57*

3.3.2 Closed-Loop Tuning *58*
3.3.3 PIN Diode Switches *60*
3.3.4 Varactor-Based Switching *62*
3.3.5 RF MEMS-Based Switching *63*
3.4 Beamforming Networks *65*
3.4.1 Analog, Digital, and Hybrid Beamforming Networks *66*
3.4.2 Passive and Active Beamforming Networks *70*
3.4.3 Architecture Example: Transmit Phase Array Beamformer *72*
3.4.3.1 Broadband Wilkinson Divider Network and Single-Ended to Differential Stage *72*
3.4.3.2 Ultrawideband Active Phase Shifter and Variable Gain Amplifier *73*
3.4.3.3 T-Coil with Resistive Feedback Power Amplifier *74*
3.4.4 Architecture Example: Ultrawideband Programmable Digital Beamforming Receiver *75*
3.4.4.1 Reconfigurable Input Switching Network *78*
3.4.4.2 Reconfigurable Output Combining Network *79*
3.5 Summary *79*
References *80*

4 Passive MIMO Antenna Systems *85*
4.1 Single-Band Implementations *87*
4.1.1 MIMO Antenna for the FR1 Band *88*
4.1.1.1 Dipole-Based Antennas *88*
4.1.1.2 Loop-Based Antennas *91*
4.1.1.3 Inverted-F-Based Antennas *94*
4.1.2 MIMO Antennas in the FR2 Band *96*
4.1.2.1 Dielectric Resonator-Based Antenna *96*
4.1.2.2 Microstrip-Based Antenna *99*
4.2 Multi-Band MIMO Antennas *102*
4.2.1 Higher-Order Resonances *102*
4.2.2 Resonant Trap *102*
4.2.3 Multiple Resonant Structures *103*
4.2.4 Parasitic Resonators *103*
4.2.5 Example Implementations of Multiband MIMO Antennas *104*
4.3 Wideband and Ultra-Wide Band (UWB) MIMO Antennas *113*
4.3.1 UWB MIMO Antenna for the Sub-6 GHz Band *114*
4.3.1.1 Vivaldi-Based UWB Antennas *114*
4.3.1.2 Monopole-Based UWB Antennas *120*
4.3.1.3 IFA-Based MIMO Antennas *124*
4.3.1.4 Dielectric Resonator-Based UWB Antennas *127*
4.3.1.5 Dipole-Based Antennas *131*

4.4 Practical Implementations *136*
4.4.1 Single-Band 5 GHz Linearly Polarized High Gain Patch Module Antenna *136*
4.4.2 Multiband Antenna *137*
4.4.3 Wideband/UWB Antenna *138*
4.5 Design Examples *140*
4.5.1 Antenna for the Sub-6 GHz Band *140*
4.5.2 Antenna for the Millimeter Wave Band *141*
4.6 Summary *142*
References *143*

5 Active Reconfigurable MIMO Antenna Systems *147*
5.1 Frequency-Reconfigurable MIMO Antennas *147*
5.2 Pattern Reconfigurable MIMO Antennas *155*
5.3 Polarization Reconfigurable MIMO Antennas *161*
5.4 Challenges and Practical Considerations *165*
5.5 CAD Modeling Example *166*
5.6 Summary/Conclusion *168*
References *169*

6 Massive-MIMO (m-MIMO) Antenna Systems *173*
6.1 Advantages of m-MIMO *173*
6.1.1 Spectrum Efficiency Improvement *173*
6.1.2 Energy Efficiency Improvement *174*
6.1.3 Air Interface Access Delay Reduction *174*
6.1.4 Signal Processing Complexity Reduction *175*
6.1.5 Expansion of System Coverage *175*
6.1.6 Insensitive to Non-Ideal Performance of Single Hardware *175*
6.1.7 System Reliability Improvement *175*
6.2 m-MIMO Antenna Implementations *176*
6.2.1 Implementation of the Antenna Unit *176*
6.2.1.1 Loading Parasitic Structures *177*
6.2.1.2 Change the Structure of the Antenna *177*
6.2.1.3 Using High Dielectric Constant Dielectric *178*
6.2.1.4 Metamaterial *178*
6.2.2 Implementation of Antenna Array *179*
6.2.2.1 In-Band Antenna Array *180*
6.2.2.2 Cross-Band Antenna Array *181*
6.3 5G Massive-MIMO Array Example (Practical/Industry Related) *185*
6.3.1 Huawei's 5G Blade Active Antenna Unit (AAU) Base Station *185*
6.3.2 Ericsson's AIR 6488 5G m-MIMO Base Station *186*

6.3.3 Ericsson's Street Macro 6701 5G m-MIMO Base Station *186*
6.4 Challenges in m-MIMO Antenna Designs *187*
6.5 Conclusion *188*
References *189*

7 Recent Topics in MIMO Antenna Systems *191*
7.1 On-Chip and On-Package MIMO Antenna Systems *192*
7.1.1 Antenna On-Chip *192*
7.1.2 Codesign of Antennas and Circuits *193*
7.1.3 Benefits of On-Chip Antennas *195*
7.1.3.1 Non-50 Ω Co-Design *195*
7.1.3.2 Fabrication *197*
7.1.3.3 Size and Features *198*
7.2 Millimeter-Wave MIMO Antenna Systems *199*
7.3 Sub-THz Antennas *204*
7.4 MTS *217*
7.4.1 MTS in 5G Communication Systems and Beyond *218*
7.4.1.1 Beam Manipulation *218*
7.4.1.2 Time-Modulated MTS *225*
7.4.1.3 Non-Reciprocal Transmit-Receive MTS *231*
7.5 Reflective Intelligent Surfaces *239*
7.5.1 Introduction *239*
7.5.2 System Architecture *240*
7.5.3 RIS-Based Wireless Communication Prototypes *242*
7.5.3.1 Electronic-Based RIS Prototypes *243*
7.5.3.2 Phase Change Material-Based RIS Prototypes *247*
7.5.3.3 Microfluidic-Based RIS Prototypes *255*
7.6 MIMO OAM *258*
7.7 MIMO Antennas for Space Applications *263*
References *271*

8 Decoupling Techniques for MIMO Antenna Systems *285*
8.1 Mutual Coupling and Its Impact on MIMO *285*
8.1.1 Generation and Characterization of Mutual Coupling *285*
8.1.1.1 Transmitting Mode *285*
8.1.1.2 Receiving Mode *286*
8.1.1.3 Coupling Path *287*
8.1.2 Coupling Effects on MIMO Antenna Systems *289*
8.1.2.1 Coupling Effects on Antenna Characteristics *291*
8.1.2.2 Coupling on Correlation *294*
8.1.2.3 Coupling on Axial Ratio *297*

8.1.2.4 Coupling Effects on the Nonlinearity of Power Amplifiers *297*
8.2 Decoupling Techniques *298*
8.2.1 Linearly Polarized Arrays *299*
8.2.1.1 Decoupling Networks *299*
8.2.1.2 Defected Ground Structure *300*
8.2.1.3 Neutralization Line *301*
8.2.1.4 Metamaterial Structure *302*
8.2.2 Dual Polarized Array *305*
8.2.2.1 Array Antenna Decoupling Surface *305*
8.2.2.2 Decoupling Ground *307*
8.2.2.3 Dielectric Superstrate *308*
8.2.2.4 SRR-Loaded Baffles *309*
8.2.2.5 Hybrid Decoupling Structures *309*
8.2.3 Circularly Polarized Array *311*
8.3 Challenges and Practical Considerations *315*
8.4 Conclusions *316*
References *316*

9 Measurement Methods for MIMO Antenna Systems *323*
9.1 Over-the-Air Tests in Reverberation Chamber *323*
9.1.1 Total Radiated Power *324*
9.1.2 Total Isotropic Sensitivity *328*
9.1.3 Throughput Testing in RC *329*
9.2 Multi-Probe Anechoic Chamber *332*
9.2.1 Algorithms for Channel Emulation in MPAC *333*
9.2.1.1 Target Channel Model *333*
9.2.1.2 PFS Method *335*
9.2.1.3 PWS Method *336*
9.2.2 Testing of UE in MPAC *338*
9.2.2.1 Comparison of Channel Emulation Methods *338*
9.2.2.2 Channel Emulation Accuracy *341*
9.2.2.3 Requirements of UE Test System Design *342*
9.2.2.4 3D Channel Model Emulation in MPAC *343*
9.2.3 Testing of Massive BS in MPAC *343*
9.2.3.1 Comparison of Channel Emulation Methods *344*
9.2.3.2 Channel Emulation Accuracy *346*
9.2.3.3 Requirements of 3D Massive BS Test System Design *347*
9.3 Radiated Two-Stage Method *348*
9.3.1 Principle of Radiated Two-Stage Method *348*
9.3.1.1 Conducted Two-Stage Method *348*
9.3.1.2 RTS Method for MIMO OTA Test *349*

9.3.2 Solution Method of Inverse Matrix *352*
9.4 Other Testing Methods *356*
9.4.1 Principle of Wireless Cable Method *357*
9.4.1.1 Signal Model *357*
9.4.2 Pros and Cons of Wireless Cable Method *364*
References *365*

Index *371*

About the Authors

Xiaoming Chen received the PhD degree in electrical engineering from Chalmers University of Technology, Gothenburg, Sweden, in 2012. From 2013 to 2014, he was a postdoctoral researcher at Chalmers. From 2014 to 2017, he was a research engineer at Qamcom Research & Technology AB, Gothenburg, Sweden. Since 2017, he has been a professor at Xi'an Jiaotong University, Xi'an, China. His research areas include MIMO antennas, over-the-air testing, and reverberation chambers. He has published more than 220 journal articles on these topics. Professor Chen currently serves as a track editor for IEEE Antennas and Wireless Propagation Letters, and an associate editor for IEEE Transactions on Antennas and Propagation. He was the general chair of the IEEE ICEICT in 2021. He received the IEEE outstanding associate editor awards six times, 2018–2023, and URSI (International Union of Radio Science) Young Scientist Award 2017 and 2018.

Ping Jack Soh is an associate professor in the University of Oulu, Finland, and an academy research fellow with the Research Council of Finland. He received the PhD degree in electrical engineering from KU Leuven, Belgium, in 2013. He was a postdoctoral fellow (2013–2014) in KU Leuven, and then a senior lecturer (2014–2017) and associate professor (2017–2021) in Universiti Malaysia Perlis before joining UOulu. In Oulu, he coordinated the "Devices and Circuit Technology" strategic research area within Finland's 6G Flagship (2022–2023). His research interests include antennas for wearables/WBAN; metasurfaces and reflectors; 5G/6G communications; compact satellites; and wireless healthcare, resulting in over 350 publications (160+ in journals) and four book chapters. Dr. Soh was the recipient of the URSI Young Scientist Award (2015), the IEEE MTT-S graduate fellowship (2013), and the IEEE AP-S Doctoral Research Award (2012). He serves as an associate editor for *IEEE Antennas and Wireless Propagation Letters* and the *International Journal of Numerical Modelling* (Wiley); and was the TPC co-chair of IEEE IMBioC 2023. He also volunteers in the MTT-S Education Committee.

Mohammad S. Sharawi was a full-tenured professor of electrical engineering at Polytechnique Montréal, Montréal, Québec, Canada, 2019–2023. He was with King Fahd University of Petroleum and Minerals (KFUPM), Saudi Arabia, between 2009 and 2018, where he founded and directed the Antennas and Microwave Structure Design Laboratory (AMSDL). He was a visiting professor at the University of Calgary, Canada, and Oakland University, USA. Currently, he is a lead research and principal development engineer with Blue Origin LLC, USA, and an Affiliate full Professor with the Electrical and Computer Engineering Department at the University of Washington, Seattle, USA. Professor Sharawi's areas of research include Multiband Printed Multiple-Input-Multiple-Output (MIMO) Antenna systems, Reconfigurable and Active integrated Antennas, Applied Electromagnetics, Millimeter-wave MIMO antennas, and Integrated 4G/5G antennas for wireless handsets and access points. He has more than 400 papers published in refereed journals and international conferences, 11 book chapters, 3 books, and 28 issued/granted patents in the US Patent Office. Professor Sharawi served as area editor and associate editor for various IEEE, IET, and Wiley Journals, and served on the technical committees of all major Antennas and Propagation conferences. He is a fellow of the IEEE and a fellow of the IET.

Preface

Wireless technology has been evolving and advancing at an astonishing rate over the past four decades, and the need for more data and connectivity has been increasing exponentially every year. There is no doubt that radio frequency (RF) front-ends and antenna systems play a major role in the development and advancement of wireless technology. The design and development of antenna systems is a combination of science and engineering art. While single-antenna wireless systems were adopted in the first three wireless generations, the fourth generation introduced a new paradigm of wireless communications in the form of multiple-input-multiple-output (MIMO) system. The MIMO system can improve the data rate and reliability of wireless communications without increasing power or bandwidth, but via the use of multiple antenna elements (channels). Thus, it thrives in the fifth generation (5G) as well. However, the introduction of multiple antennas in compact devices poses serious challenges to live up to the MIMO's full potential. This book focuses on the design and engineering of MIMO antenna systems for current and future wireless technologies. While the antenna concepts and examples in this book focus on the current 5G requirements and use cases, they are equally applicable to any other wireless standards, as the fundamental aspects of antennas are to be satisfied regardless of the targeted technology. MIMO antenna systems in particular will be given special attention in addition to millimeter-wave (mm-wave) solutions for 5G and beyond.

The book includes the latest designs of MIMO antenna systems for 5G wireless applications as well as a comprehensive review of important MIMO antenna system designs and configurations. It provides (1) a comprehensive guide to the latest methods of developing MIMO antenna systems for various wireless communication devices covering a wide spectrum of standards; (2) a complete design procedures for the various elements of MIMO antenna systems, such as active circuits for reconfigurable antennas, power dividers and splitters, active feeding networks at mm-wave bands, as well as active integration procedures and methods (amplifier based); (3) a single reference for the latest designs and performance metrics for MIMO antenna system designs for all applications and wireless standards for practicing researchers, scholars, and engineers; (4) design examples to the presented concepts with complete evaluation of their performance metrics to aid understanding and to provide a quick start to his or her first design. It hopes to serve as a single reference for practicing engineers in the consumer electronics and wireless communications industry, who are responsible for antenna and radio frequency front-end designs. This book also offers a comprehensive guide to current design practices and performance metrics of MIMO antenna systems that will be useful for senior-level undergraduate students focusing on antenna design for wireless applications and first-year graduate students in electrical engineering with a focus on antennas and wireless systems.

Dec. 2023

Xiaoming Chen
Ping Jack Soh
Mohammad S. Sharawi

Acknowledgments

The authors start by thanking GOD (Allah) for helping them complete this big endeavor. Special thanks go to our current and previous graduate students who made ideas into working prototypes through their ingenuity and hard work, we learned a lot from working with you. We thank the following people in particular for providing some raw materials that helped to the completion of this book: Bo Liu, Yipeng Wang, Xiaoyu Huang, Huang Shan, Boju Chen, Bingyi Qian, Xin Cheng, Rifaqat Hussain, Muhammad Afifi bin Mohd Shukri, and Rizwan Khan. We also thank Nandhini Karuppiah, Sarah Lemore, Jayashree Saishankar, Becky Cowan, and Brett Kurzman from Wiley for their support during the book preparation and in helping us meet the deadlines. We thank our institutions for providing the environment and resources to conduct our research work that led to some of the content in this book.

Finally, we thank our families for their patience and support during our sleepless nights, besides sacrificing a part of the family activities during the course of writing this book, thank you, from the bottom of our hearts.

1

Introduction

Wireless technology has been evolving and advancing at an astonishing rate over the past four decades. The amount of development and accomplishments that the wireless industry has achieved is beyond imagination, and no one 40 years ago would have predicted the level of advancement we are experiencing nowadays. What is more, new frontiers and developments are underway to even go beyond some of the fundamental laws that have predicted the technology trends thus far. Wireless technology has relied heavily on advancements in electronics, algorithms, and RF front-ends and antenna systems. Moore's law has been driving the electronics and chip industry for several decades but recently has been overcome by the rapid development of computing power. The need for multi-band, and multi-functional RF front-ends and antenna systems has become a reality nowadays with smaller footprints and higher efficiencies. Our cell phones today have more computing, connectivity, and functionality than complete computers of the past decade.

This rapid technological advancement has been noticeable in every aspect of our daily lives. We are living in smart homes, driving smart cars, using smart phones, and living in smart cities, and these are keeping us aware and connected to our loved ones, assets, and work at any time. Internet coverage is covering more ground with satellite services that are now providing services to remote areas. This has increased the safety aspects of our lives and improved our decision-making and selection capabilities. We can always track our kids, follow up with coworkers, and perform our duties in a much more efficient way that can save a lot of time and effort. The need for more data and connectivity has been increasing exponentially every year (see Figure 1.1), thus pushing the need for rapid improvements on the device and network levels.

While a wireless system depends on RF connectivity, fast and efficient hardware and software components, displays, and control devices, the RF front-ends and antenna systems also play a major role in providing a reliable connection with the network. The very first component of the RF front-end is the antenna system.

MIMO Antenna Systems for 5G and Beyond, First Edition.
Xiaoming Chen, Ping Jack Soh, and Mohammad S. Sharawi.

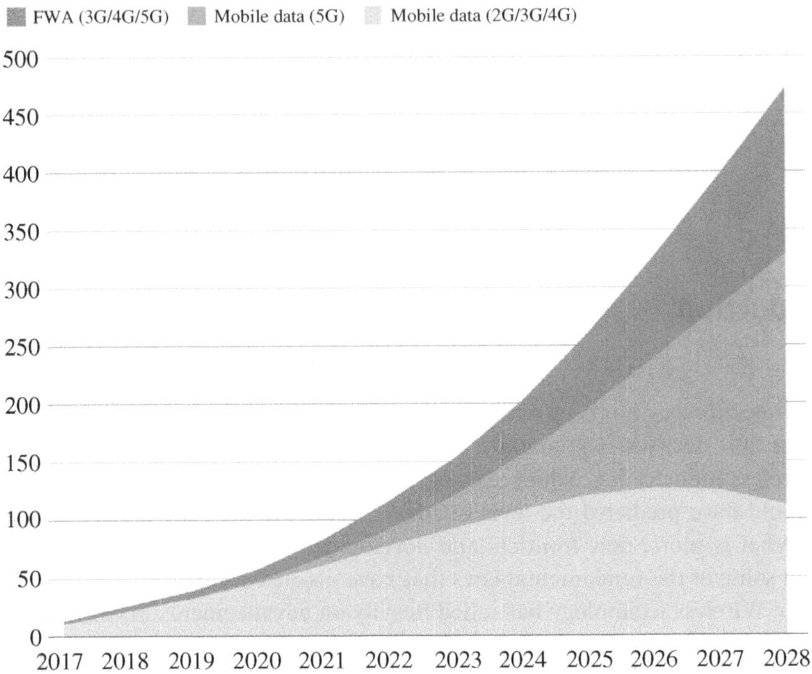

Figure 1.1 Global mobile network traffic in Exabytes (EB) per month. Source: Adapted from Ref. [1].

Antennas are the signal converters that convert the electrical signal from the electronics of your device (i.e., your data, whether you are browsing a web page or making a phone call) into an electromagnetic waveform that can travel in the air from your device to the receiving serving station (i.e., a cell phone tower, a wireless local area network access point, or your car radio via Bluetooth). The antenna can transmit your data to the base station, or receive data from the base station (i.e., it is a reciprocal element). Proper design of such antenna systems is essential to having a proper connection in any wireless system.

Antenna systems' design and development is a combination of science and engineering art. Their characteristics are affected by various parameters such as size, material, shape, surroundings, and types, among many others, and have to be carefully designed and characterized. Careful integration with other system components is critical after the initial design and verification phase is completed to fulfil the expected performance and metrics. While single antenna wireless systems were adopted in the first three wireless generations (i.e., first, second, and third), the fourth generation introduced a new paradigm of wireless communications. It is based on sending multiple data streams to improve the system capacity (data rates) and combat the degradation from multipath signals that exist in all modern environments. This new paradigm is called MIMO communications, which require the existence of multiple antenna systems at the receiver and the transmitter sides. The introduction of multiple antennas in small form factor devices poses serious challenges to make such systems a complete success and achieve the desired high capacity.

This book focuses on the design and engineering of antenna systems for current and future wireless technologies and standards. While the antenna concepts and examples in this book focus on the current 5G requirements and use cases, they are equally applicable to any other wireless standards, as the fundamental aspects of antennas are to be satisfied regardless of the targeted technology. MIMO antenna systems in particular will be given special attention in addition to millimeter-wave (mm-wave) solutions for 5G and beyond.

1.1 Wireless Technology Evolution

While wireless technology evolution is usually associated with different cellular telephony generations, several other factors have driven wireless technology evolution and fast advancement, especially in the past two decades. Among others, additive manufacturing, ultra-scale integration, automation and cost reduction, and fast computing power via the dense integration of processor cores and fabricating nanoscale transistors have contributed heavily toward the level of the technology we have today.

First-generation (1G) cellular technology appeared in the 1980s and relied on the advanced mobile phone system (AMPS) that provided voice-only services to mobile terminals. This was a purely analog technology. The antennas used were whip structures such as helical- or monopole-based ones that protruded from the device. A decade later, technology started to advance more in the digital domain and the global system mobile (GSM) was introduced, resulting in new mobile terminals introduced to the consumer market. It is based on digital techniques that allow the transmission of voice and data (mainly text) with data rates up to 270 kbps. This was considered the second generation of cellular technology (2G). GSM systems were based on time division multiple access (TDMA). Integrated printed antennas started to appear in some models in 2G mobile terminals, at the same time with smaller form factor devices. This is due to the large-scale integration of functionality in the components used.

The addition of multimedia features and incorporation of diversity techniques appeared in the third generation of wireless cellular technology (3G) in the early 2000. Wideband code division multiple access (WCDMA) technology was incorporated and provided wider bandwidths that allowed data transfers in the tens of Mbps. Internet browsing and online gaming became a reality for the first time. International roaming and more secure protocols were utilized. Antenna systems that are or printed fashion were used and multiple functionalities were supported by dedicated antennas, i.e., global positioning system (GPS), wireless local area networks (WLAN), etc.

The introduction of MIMO technology was the driver for 4G. While MIMO was the main focus of 4G, several advancements in modulation and coding and access techniques provided a noticeable capability improvement over 3G that allowed for ultra-fast data transfers in excess of 80 Mbps, making it 20 times faster than its predecessor. MIMO takes advantage of the multipath phenomena that were not very favorable in 3G systems. It creates multiple signal paths that can significantly improve the amount of data transferred within the same frequency band and transmission power levels. Long-term evolution (LTE) networks were built around MIMO technology where multiple antennas on the mobile terminal and multiple antennas in the base station are used to support a communication link. Placing multiple antennas on the mobile terminal is very challenging and needs careful attention to changes in antenna parameters such as isolation and field correlation. These aspects will be discussed in detail in the coming chapters.

The fifth-generation (5G) standard came to provide another leap in connection speeds reaching Gbps levels, with a new paradigm that provides a much broader connectivity framework with three vertices – one focusing on enhanced mobile broadband (eMBB), second providing ultra-reliable low-latency communications (URLLC) that consider fast decision-making features (in autonomous cars for example), and third is the massive machine-type communications (mMTC). For the eMBB which is used for cellular networks, two bands were selected to support 5G requirements; the sub-6 GHz band, also known as frequency range 1 (FR-1), and millimeter-wave band (mm-wave) also called frequency range 2 (FR-2). As there are multiple frequency bands with a large frequency ratio, two sets of antenna systems are expected to support each, in addition to several copies of each to support MIMO. With this, we can imagine how the antenna systems will be packed within a mobile terminal with limited space, and thus this becomes a very challenging problem. This book provides several solutions to 5G-enabled designs. Also, mm-wave bands need antenna arrays with higher gains to overcome the free space path loss (FSPL). Thus, careful integration and array

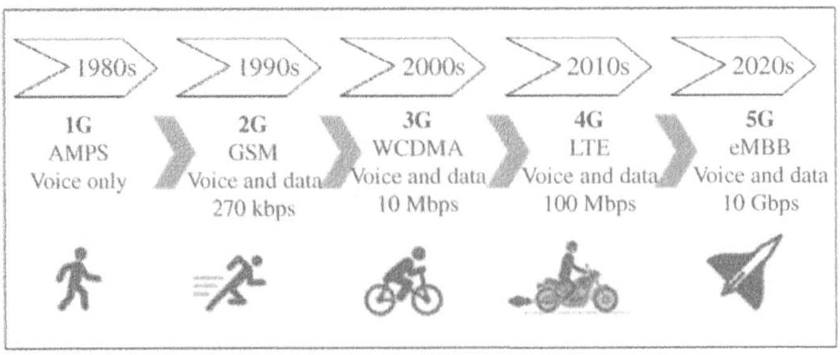

Figure 1.2 Wireless standard evolution from 1G to 5G and their main features.

designs with beam steering/switching are required. A summarizing figure that compares the differences in the services and speeds of various wireless cellular generations is shown in Figure 1.2

Currently, we are thinking beyond 5G or the sixth-generation (6G) wireless standard and its underlying technologies. Key features that are being considered are the use of sub-THz bands for extremely high data transfers of 100 Gbps at short distances, virtual reality (VR) and augmented reality (AR), the use of visible light communications (VLC), utilization of smart programmable radio environments such as reconfigurable intelligent surfaces (RIS), the use of non-terrestrial networks (NTNs), and pervasive artificial intelligence (AI) techniques to make networks smarter and faster. This framework includes so many new technologies that need seamless integration and interconnection to be able to communicate with each other, and that is why this is a hot area of current research and development. Figure 1.3 shows some of the use cases beyond 5G or 6G use cases and high-level architecture.

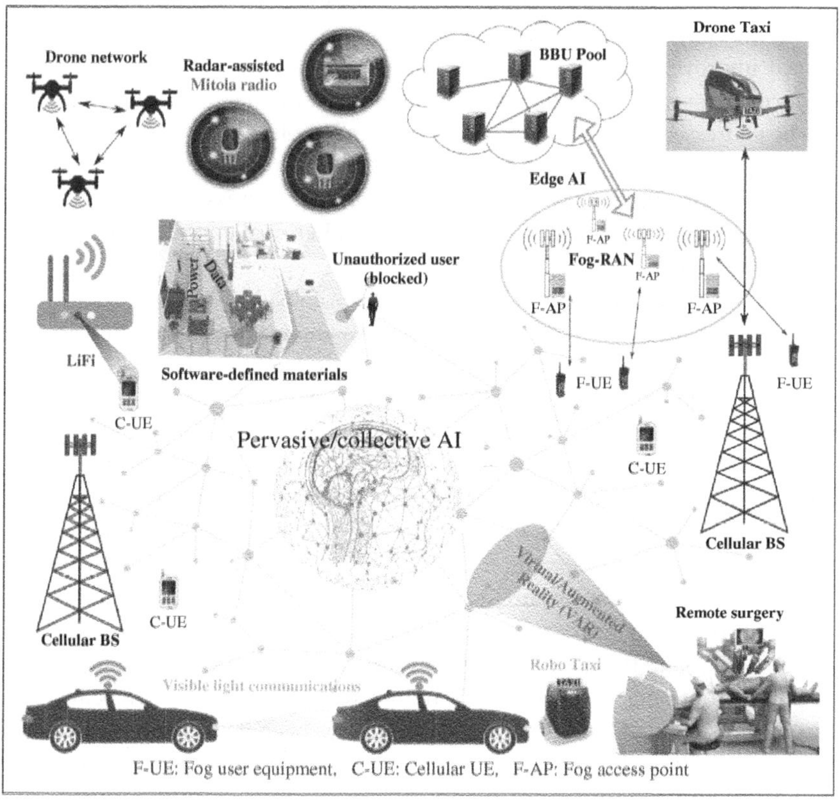

Figure 1.3 Beyond 5G (aka 6G) use cases and high-level architecture. Source: Reproduced with permission from Ref. [2]/IEEE.

1.2 Benefits of MIMO Technology

The fundamental relationship that relates the available bandwidth, signal power, channel conditions, and number of transmitters and receivers, and the achievable channel capacity (i.e., data rate) is given by:

$$C = \mathrm{BW} \log_2 \left[\det \left(I_R + \frac{\rho}{N_T} H H^T\right)\right] \tag{1.1}$$

where C is the channel capacity in bits, BW is the bandwidth in Hz, I_R is the number of receiving channels (i.e., receive antennas), ρ is the signal-to-noise ratio (SNR), N_T is the number of transmitting elements (i.e. antennas) and H is the channel matrix. The major advantage of MIMO is the ability to increase C by increasing the independent channels used (i.e., the rank of H and value of I_R) without any increase in the BW or power. This can multiply the data rates by the increase in the number of independent channels [3].

To create multiple data streams from wireless terminals, multiple antennas are needed. Each antenna will be supporting an independent stream or channel, and based on the channel characteristics and conditions, the multiple streams can significantly improve the data rates achieved. The design of multiple antennas within small form factor wireless terminals is a challenging task that has been and continues to be addressed in past and future designs and applications. The introduction of new frequency bands, and standard coexistence requirements (i.e., LTE, WLAN, etc.) impose more challenges to make sure close-by transmissions do not disturb the others. Field interference and coupling are other challenges that need to be addressed as well.

Due to the clear advantages of MIMO, it has been considered a major technology in 4G wireless standards. All 4G wireless devices and terminals have at least two antennas operating simultaneously within the same frequency band. MIMO continues to be a major cornerstone in 5G wireless standard in both FR-1 and FR-2 bands. Several 5G-enabled devices are using four antennas to support MIMO and massive-MIMO (m-MIMO) applications. The use of m-MIMO will be considered at the base station side and not the mobile terminal side due to the need for tens to hundreds of antennas at the base station side. The use of a large number of antennas and channels at the base station side can improve the channel capacity due to the use of narrower beams toward the users, thus improving the SNR and capacity. It will also provide higher energy efficiency via the use of linear codes and directive beams in addition to the use of smaller lower power RF frontends as compared to high-power front-ends in conventional architectures. Finally, directed beams will overcome jamming and interference. M-MIMO has been deployed at FR-1 bands but also is considered for FR-2 [4]. Figure 1.4 shows a comparison between legacy and m-MIMO systems.

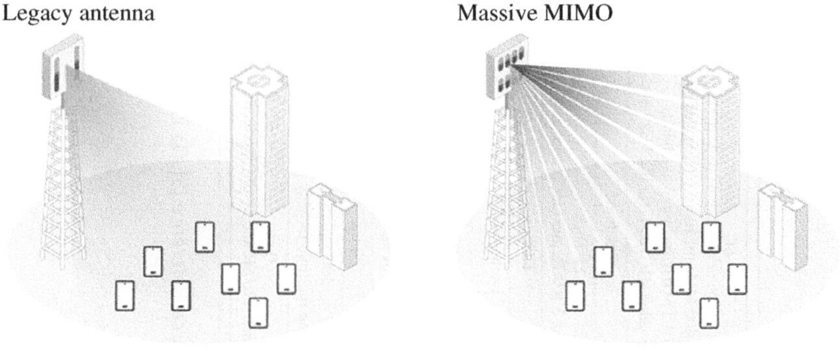

Figure 1.4 Beam coverage of legacy base stations and m-MIMO base stations.

1.3 Future Technology Trends

As we have seen, nowadays, 5G networks are still being deployed all over the world, mainly focusing on the FR-1 bands with few supporting FR-2. We should expect a stable increase in the infrastructure supporting 5G bands, and the deployment of the various applications and supporting networks to allow for seamless connectivity between the three vertices of the 5G triangle (i.e., eMBB, URLLC, mMTC) during the rest of the 2020s. While that is taking place, the research and development of what comes next is underway, i.e., 6G. Figure 1.5 shows a nice summary of the key performance indicators (KPIs) of the 4G, 5G, and 6G, focusing on the main features. As can be seen, 6G will introduce new wideband frequencies in the hundreds of GHz to allow for speeds of 100 Gbps to Tbps, opening a new domain of applications and use cases.

Holographic and multi-sensing communication systems are the new frontiers in wireless communications. The post-COVID area has shown the importance of video conferencing and instant multi-location world collaboration but lacked the feel of interaction and presence. Holograms will allow people to give talks, attend meetings, and give the feel that they are nearby while they are thousands of kilometers away. This will also open the doors of multi-sensing communications where interactions via AR and VR give the feel that someone is immersed into the environment that exists in the metaverse or at another location worldwide. To make these holograms a real-time reality, Tbps transmissions are required.

While 5G came with a framework of seamless connectivity between the three vertices mentioned earlier, the next-generation standard and networks will expand the scale of connectivity to include NTNs consisting of satellites in various orbits, high-altitude platforms (HAPs), and many others. At the same time, terrestrial networks can be integrated with new AI techniques that will

KPI	4G	5G	6G
Operating bandwidth	Up to 400 MHz (band dependent)	Up to 400 MHz for sub-6 GHz bands (band dependent) Up to 3.25 GHz for mmWave bands	Up to 400 MHz for sub-6 GHz bands Up to 3.25 GHz for mmWave bands Indicative value: 10–100 GHz for THz bands
Carrier bandwidth	20 MHz	400 MHz	To be defined
Peak data rate	300 Mbps with 4×4 arrays 150 Mbps with 2×2 antenna arrays	20 Gbps	≥1 Tbps (holographic, VR/AR, and tactile applications)
User experience rate	10 Mbps (shared over UEs)	100 Mbps	1 Gbps
Average spectral efficiency	25 Mbps with 2×2 antenna arrays 40–45 Mbps with 4×4 antenna arrays	7.8 bps/Hz (DL) and 5.4 bps 5.4 bps/Hz (UL)	1× that of 5G
Connection density	N/A	10^6 devices/km^2	10^7 devices/km^2
User plane latency	50 ms	4 ms (eMBB) and 1 ms (uRLLC)	25 μs to 1 ms (holographic, VR/AR and tactile applications)
Control plane latency	50 ms	20 ms	20 ms
Mobility	350 km/h	500 km/h	1000 km/h Handling multiple moving platforms
Mobility interruption time	N/A	0 ms (uRLLC)	0 ms (Holographic, VR/AR and tactile applications)

Figure 1.5 Key performance indicators (KPIs) between current wireless standards and upcoming 6G. Source: Reference [5]. (© 2021 IEEE. Reprinted with permission).

support the massive amount of information being handled in large data and information centers. The computing power should be increased multi-fold, and when considering the RF front-end, much faster and more reliable architectures will be required. Optical and RF-based communications will be integrated seamlessly with a massive amount of hotspot deployment to provide Tbps links with user terminals. The emergence of non-IP-based Internet protocols with AI engines will be the key to achieving this ultra-scale evolution of the network infrastructure.

New antenna designs will always be key to satisfying such wireless links using radio waves. As seen in Figure 1.5, antenna systems will cover extremely wide frequency ratios, from several hundreds of MHz to several hundreds of GHz. Switched beam antennas and phased arrays will be key for mm-wave and sub-THz applications, while reconfigurable and low-profile ones will be needed for sub-6 GHz bands. The aim of this book is to shine some light and give detailed examples and explanations on the design of such antenna systems for beyond 5G applications and frequency bands.

1.4 Book Organization

This book focuses on potential antenna technologies and solutions for 5G and beyond. Wireless terminals will become more environment-aware and will be supporting various standards and functions. Thus, it is inevitable to have multiple antenna systems within a single terminal. The need to support sub-6 GHz bands as well as millimeter-wave ones is essential and currently, the norm in 5G-enabled wireless terminals. This book revisits the fundamentals of some of the enabling technologies and provides various examples and implementation techniques to realize these concepts. More importantly, contents of this book also provide some practical solutions to the reader, be it in the research and development sector or the industrial sector.

Chapter 2 of the book starts by presenting some fundamental antenna metrics followed by a detailed description of MIMO technology. The definition of the performance metrics for MIMO such as the correlation coefficient, the total active reflection coefficient, port and field isolation, and capacity are explained in detail. Various diversity-combining techniques are discussed along with their advantages and disadvantages. The chapter ends with a discussion on MIMO antenna characterization using reverberation and anechoic chambers and the differences and capabilities of each.

The fundamental topic of feed-network design for passive and active antenna systems is presented in Chapter 3. The chapter starts with a revision on impedance-matching techniques using stubs and lumped elements. Various antenna-feeding mechanisms are presented such as microstrip feeding, aperture coupling, coaxial, and coplanar waveguide feeding. Open-loop tuning and closed-loop tuning techniques are discussed, followed by active switching methods using diodes for antenna selection control and frequency tuning. Active beamforming networks for arrays and MIMO antennas are then presented with several practical examples. A comparison between the advantages and limitations of digital and analog beamforming is discussed. Finally, the chapter ends by presenting wideband and reconfigurable feeding techniques.

MIMO technology has been implemented in the 4G wireless standards and will continue in all upcoming generations. Chapter 4 presents a comprehensive discussion of passive MIMO antenna solutions for 5G and beyond. The chapter starts with single-band implementations of sub-6 GHz (Frequency range 1–FR1) antennas with various types and characteristics (dipoles, monopoles, loops, inverted-F). Then, it presents antenna solutions and analysis for single band antenna solutions covering FR2 at mm-waves bands, which include dielectric resonator-based antennas (DRAs) and microstrip-based ones. Multi-band antenna design techniques are presented next with several practical examples covering both FR-1 and FR-2 bands. The chapter ends with ultra-wide-band (UWB) MIMO antenna solutions covering FR-1 and FR-2 bands using microstrip and dielectric resonator-based solutions. Several practical design solutions are presented and discussed to put the reader on track to develop such antenna systems.

Active MIMO antenna system design is discussed in Chapter 5. Reconfigurable antenna systems in frequency, pattern, and polarization are discussed in detail with several practical examples in guiding the design process for active diode-biasing networks and such control mechanisms. Various antenna types (microstrip, DRA, slots) and covered frequency bands are presented. The application of reconfigurable antennas to MIMO configuration is discussed for frequency, pattern, and polarization techniques. The challenges of designing such antenna systems are highlighted and discussed with practical considerations. The chapter ends with a detailed computer-aided design (CAD) example of modeling a reconfigurable MIMO antenna system and methods for biasing circuits with the diodes. The design file provided was made using ANSYS HFSS and is provided in the accompanying CD.

Massive MIMO (M-MIMO) architecture and antenna solutions are presented in Chapter 6. M-MIMO advantages such as improvements in capacity, latency, efficiency, and complexity are presented first. The design of the single antenna element is discussed with several implementation candidates covering various use cases and features (i.e., loading, use of metamaterials). Several examples of different unit element implementations followed by m-MIMO arrays for various frequency bands are presented next. A detailed practical example (i.e., AIR 6488) of an M-MIMO antenna system is discussed, followed by the challenges faced when considering such large and complex antenna systems.

Chapter 7 discussed the most recent developments and topics in MIMO antenna systems in various applications. The chapter considers MIMO developments in on-package and on-chip radios and systems operating primarily at mm-waves. Then some recent advancements in sub-THz applications and antenna solutions with MIMO capability are presented for beyond 5G use cases. Several recent technologies are highlighted and presented. The need for MIMO in metasurface implementations to increase the throughput is discussed with several examples. Reflective intelligent surfaces (RIS) and the need for MIMO/M-MIMO are presented and discussed with several examples from recent works. MIMO in orbital angular momentum (OAM) is gaining importance due to its potential ability to increase the channel capacity with the use of the simultaneous transmission of orthogonal beams. Finally, MIMO solutions for space applications are presented and discussed.

Due to the close proximity of antenna elements in small form factor devices (i.e. mobile handsets), the need for port decoupling becomes essential to achieve the benefits of MIMO technology and implementations. This is the focus of Chapter 8; decoupling techniques for MIMO antenna systems. The impact of mutual coupling is first discussed in detail, prior to the assessments of these effects on the main MIMO parameters such as correlation and antenna characteristics. Decoupling techniques then follow, with several examples of each method highlighting its pros and cons. Polarization-specific decoupling methods are presented as well.

Finally, Chapter 9 focuses on characterization and measurement methods for MIMO antenna systems which is an essential part of the design process. Over-the-air (OTA) tests using reverberation chambers and their capabilities are discussed in detail with some examples. The definition of the total radiated power, and total isotropic sensitivity as well as throughput testing are demonstrated. The details of multi-probe anechoic chambers and their operating algorithms are presented. Characterization and testing of user equipment (UE), i.e., mobile terminal, as well as base station arrays are shown. Other methods of testing and characterization from the industry are presented and discussed with the advantages and limitations of each.

References

1 P. Jonsson et al. (2023). Ericsson Mobility Report. www.ericsson.com/mobility-report.

2 F. Tariq, M. Khandaker, K.-K. Wong, et al. "A speculative study on 6G", *IEEE Wireless Communications Magazine*, Vol. 27, No. 4, p. 118-125, 2020.

3 Sharawi, M.S. (2014). *Printed MIMO Antenna Engineering*. Artech House.
4 Mumtaz, S., Rodriguez, J., and Dai, L. (2017). *mmWave Massive MIMO*. Academic Press.
5 Tataria, H., Shafi, M., Molisch, A.F. et al. (2021). 6G wireless systems: vision, requirements, challenges, insights and opportunities. *Proceedings of the IEEE* 109 (7): 1166–1199.

2

MIMO Antenna Fundamentals and Characterization Methods

In today's telecommunications systems, multiple-input multiple-output (MIMO) techniques are employed widely. MIMO technology is a key enabler for the fifth-generation (5G) and next-generation communications. MIMO antennas benefit from multipath scenarios. Thus, only using conventional antenna metrics is insufficient to characterize its performance. This chapter first briefly introduces performance metrics for conventional antennas, such as gain, efficiency, and polarization. Then, diversity gain and MIMO capacity – two widely used performance metrics for describing MIMO antennas – are introduced. Additionally, several metrics related to MIMO capacity and diversity gain – including isolation, correlation coefficient, total active reflection coefficient (TARC), mean effective gain (MEG), and power-combining techniques – are covered in detail. Finally, we discuss S-parameter-based and port-level MIMO antenna characterization techniques. Furthermore, methods for evaluating MIMO capacity and diversity gain using anechoic and reverberation chambers are also presented.

2.1 Performance Metrics for Conventional Antennas

For a better understanding of MIMO antenna performance, electromagnetic definitions of key conventional antenna characteristics will be discussed separately in what follows.

2.1.1 Gain

Antenna gain is a metric for describing an antenna's performance since it accounts for its efficiency and directional properties. It is defined as "the ratio of the intensity, in a given direction, to the radiation intensity that would be obtained if the power accepted by the antenna were radiated isotopically. The radiation intensity corresponding to the isotopically radiated power equals the power

MIMO Antenna Systems for 5G and Beyond, First Edition.
Xiaoming Chen, Ping Jack Soh, and Mohammad S. Sharawi.

accepted (input) by the antenna divided by 4π [1].” If no direction is specified, the gain typically represents the performance in the direction of maximum radiation, which can be expressed as [1]

$$G = 4\pi \frac{\text{radiation intensity}}{\text{total input (accepted) power}} = 4\pi \frac{U(\theta, \phi)}{P_{\text{in}}}. \tag{2.1}$$

Additionally, there are certain specific definitions of gain that are easy to apply in different situations; two of these definitions are listed below:

- *Relative gain*: It is defined as “the ratio of the power gain in a given direction to the power gain of a reference antenna in its referenced direction [1].” In most cases, the reference antenna is a lossless isotropic source. Thus

$$G_{\text{relative}} = 4\pi \frac{U(\theta, \phi)}{P_{\text{in}}(\text{lossless isotropic source})}. \tag{2.2}$$

- *Absolute gain (realized gain)*: It considers the reflection/mismatch losses due to the connection of the antenna element to the transmission line. In other words, considering the reflection (mismatch) efficiency e_{r}, the absolute gain can be represented as [1]

$$G_{\text{abs}}(\theta, \phi) = e_{\text{r}} G(\theta, \phi) = e_{\text{r}} \left[4\pi \frac{U(\theta, \phi)}{P_{\text{in}}} \right]. \tag{2.3}$$

2.1.2 Efficiency

The overall antenna efficiency e_0 is employed to account for losses at the input terminals and within the antenna structure. In general, the overall efficiency e_0 can be written as [1]

$$e_0 = e_{\text{r}} e_{\text{c}} e_{\text{d}} \tag{2.4}$$

where e_{r}, e_{c}, and e_{d} denote reflection (mismatch with the feeding line or previous stages) efficiency, conduction efficiency (due to conduction/metallic losses), and dielectric efficiency (due to the losses within the substrate or dielectric medium) of the antenna, respectively.

2.1.3 Polarization

The antenna polarization in a given direction is defined as “the polarization of the wave transmitted (radiated) by the antenna. When no direction is specific, the polarization is taken to be in the direction of maximum gain [1].” Different parts of the pattern may have different polarizations because, in practice, the polarization of the radiated energy varies with the direction with respect to the center of the antenna. The field is said to be linearly polarized if the vector that characterizes the electric field at a location in space as a function of time is always directed along a line. Similarly, if the electric field traces an ellipse, then the field is described as elliptically polarized. Linear and circular polarizations are special cases of elliptical polarizations, which are created when the ellipse is transformed into a straight line or a circle, respectively [1]. Most terrestrial-based systems use linearly polarized antennas and wave transmissions while satellite and space links use circularly polarized waves and antennas.

2.2 Performance Metrics of MIMO Antenna Systems

This section considers typical spatial-temporal (ST) wireless systems that utilize multiple antennas at the transmitter and receiver, i.e., MIMO systems. We will start by defining the MIMO channel, and then we will discuss various MIMO performance metrics and their formulations.

2.2.1 MIMO Channel

Consider a MIMO system with M_{T} transmit antennas and M_{R} receive antennas, as shown in Figure 2.1, and ignore additive noise at the receiver. The signals $s_1, \ldots, s_j$ are transmitted by the transmit antennas and received as $y_1, \ldots, y_i$ by the receive antennas. Denoting the impulse response between the jth ($j = 1, 2, \ldots, M_{\mathrm{T}}$) transmit antenna and the ith ($i = 1, 2, \ldots, M_{\mathrm{R}}$) receive antenna by $h_{i,j}(\tau, t)$, the MIMO channel is given by the $M_{\mathrm{R}} \times M_{\mathrm{T}}$ matrix as [2]

$$\mathbf{H}(\tau, t) = \begin{bmatrix} h_{1,1}(\tau, t) & \cdots & h_{1,M_{\mathrm{T}}}(\tau, t) \\ \vdots & \ddots & \vdots \\ h_{M_{\mathrm{R}},1}(\tau, t) & \cdots & h_{M_{\mathrm{R}},M_{\mathrm{T}}}(\tau, t) \end{bmatrix}. \tag{2.5}$$

The vector $[h_{1,j}(\tau, t)\, h_{2,j}(\tau, t) \cdots h_{M_R,j}(\tau, t)]^{\mathrm{T}}$ (where the superscript ^{T} denotes the transpose operator) is the ST signature or channel induced by the jth transmit antenna across the receive antenna array. Further, given that the signal $s_j(t)$ is launched from the jth transmit antenna, the signal received at the ith receive antenna, $y_i(t)$, is given by

$$y_i(t) = \sum_{j=1}^{M_{\mathrm{T}}} h_{i,j}(\tau, t) \star s_j(t), i = 1, 2, \ldots M_{\mathrm{R}} \tag{2.6}$$

where $\star$ denotes the convolution operator. In matrix notation, the input-output relationship for the MIMO channel can be written as

$$\boldsymbol{y}(t) = \boldsymbol{H}(\tau, t) \star \boldsymbol{s}(t) \tag{2.7}$$

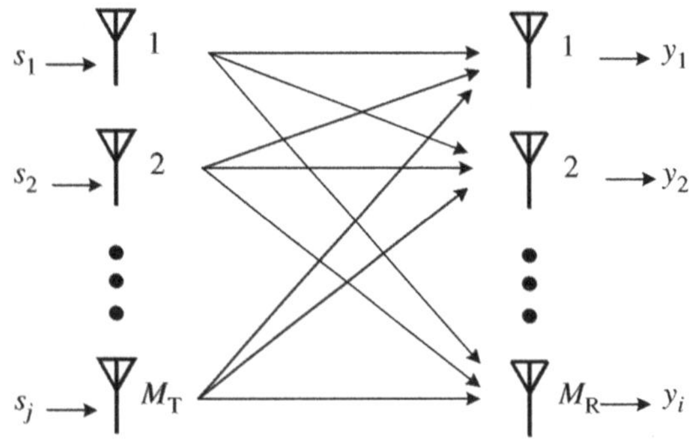

Figure 2.1 Illustration of MIMO system with M_T transmit antennas and M_R receive antennas.

where $\boldsymbol{s}(t) = [s_1(t)\, s_2(t) \ldots s_{M_T}(t)]^T$ is an $M_T \times 1$ vector and $\boldsymbol{y}(t) = [y_1(t)\, y_2(t) \ldots y_{M_R}(t)]^T$ is a vector of dimension $M_R \times 1$. Note that the convolution-based input-output relationship (2.7) is generally valid for both single- and multi-antenna systems in the time domain [2]. In the frequency domain though, the convolution becomes multiplication, i.e., $\tilde{\boldsymbol{y}}(f) = \tilde{\boldsymbol{H}}(f)\tilde{\boldsymbol{s}}(f)$; or in a narrowband system, the convolution can be approximated by multiplication, i.e., $\boldsymbol{y}(t) = \boldsymbol{H}(t)\boldsymbol{s}(t)$.

2.2.2 MIMO Capacity

The maximum error-free data rate a channel can support is called the channel capacity. Claude Shannon first derived the channel capacity for additive white Gaussian noise (AWGN) channels in 1948 in his celebrated paper "A Mathematical Theory of Communication" [3]. However, the MIMO channel functions differently from the scalar AWGN channel due to fading and encompassing a spatial dimension. Additionally, MIMO capacity can be expressed differently from frequency-flat to frequency-selective MIMO channels [2]. Frequency-flat channels have constant frequency responses within the communication bandwidth, whereas frequency-selective channels' frequency responses vary rapidly within the communication bandwidth.

2.2.2.1 Capacity of the Frequency Flat MIMO Channel

Considering a MIMO channel with M_T transmit antennas and M_R receive antennas, and assuming that the channel is frequency flat, the input-output relation for the MIMO channel is given by

$$\boldsymbol{y} = \sqrt{\frac{E_s}{M_T}}\boldsymbol{H}\boldsymbol{s} + \boldsymbol{n} \tag{2.8}$$

where $\mathbf{H}$ is the narrowband MIMO channel matrix, $\mathbf{y}$ is the $M_R \times 1$ received signal vector, $\mathbf{s}$ is the $M_T \times 1$ transmitted signal vector, $\mathbf{n}$ is the zero mean

Figure 2.2 The capacity of a frequency-selective MIMO channel is the sum of the capacities of frequency flat sub-channels.

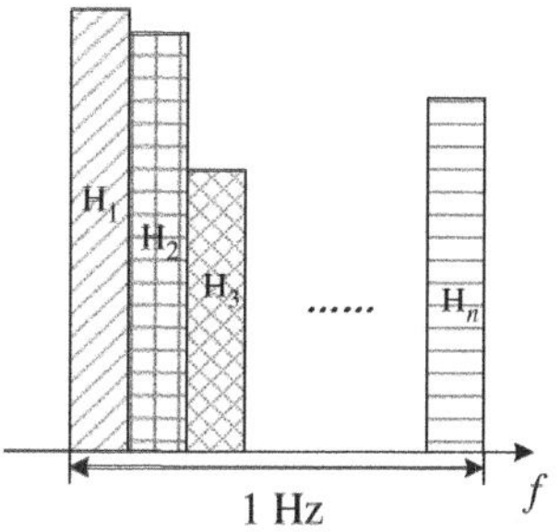

circularly symmetric complex Gaussian (ZMCSCG) noise with covariance matrix $\mathcal{E}\{\mathbf{n}\mathbf{n}^H\} = N_0\mathbf{I}_{M_R}$, and E_s is the total average energy available at the transmitter over a symbol period. It equals the total average transmit power if the symbol period is assumed to be 1s. The covariance matrix of $\mathbf{s}$, $\mathbf{R}_{ss} = \mathcal{E}\{\mathbf{s}\mathbf{s}^H\}$ ($\mathbf{s}$ is assumed to have zero mean) must satisfy $\mathrm{Tr}(R_{ss}) = M_T$ in order to constrain the total average energy transmitted over a symbol period. Thus, the capacity of the MIMO channel is given by [2]

$$C = \max_{\mathrm{Tr}(\mathbf{R}_{ss})=M_T} \log_2 \det\left(\mathbf{I}_{M_R} + \frac{E_s}{M_T N_0}\mathbf{H}\mathbf{R}_{ss}\mathbf{H}^H\right) \text{ bps/Hz}. \quad (2.9)$$

The capacity C in Eq. (2.9) is also often referred to as the error-free spectral efficiency, or the data rate per unit bandwidth that can be sustained reliably over the MIMO link. Thus, the highest data rate that can be achieved over a bandwidth of W Hz using a MIMO channel is WC bps.

2.2.2.2 Capacity of the Frequency-Selective MIMO Channels

As illustrated in Figure 2.2, the capacity of a frequency-selective MIMO channel can be determined by splitting the frequency band of interest (say 1 Hz) into N narrower sub-channels, each of which has a bandwidth of $1/N$ Hz. The capacity of the frequency-selective MIMO channel equals [2]

$$C = \frac{1}{N}\max_{\mathrm{Tr}(\mathbf{R}_{ss})=NM_T} \log_2 \det\left(\mathbf{I}_{M_R} + \frac{E_s}{M_T N_0}\mathcal{H}\mathbf{R}_{ss}\mathcal{H}^H\right) \text{ bps/Hz} \quad (2.10)$$

where $\mathcal{H}$ is an $NM_R \times NM_T$ block diagonal matrix with $\mathbf{H}_l$ as the block diagonal elements, and the covariance matrix of the signals $\mathcal{S}$ with $\mathbf{R}_{ss} = \mathcal{E}\{\mathcal{S}\mathcal{S}^H\}$ is constrained so that Tr $(\mathbf{R}_{ss}) = NM_T$. Note that $\mathcal{S} = [\mathbf{s}_1^T\ \mathbf{s}_2^T \ldots \mathbf{s}_N^T]$ is a vector of $NM_T \times 1$ which contains transmitted signal information of each sub-channel.

2.2.3 Isolation

In some modern communication transceivers, such as mobile terminals and base stations (BSs), compact MIMO antennas are required due to limited space and aesthetic reasons. However, (electromagnetic) mutual coupling becomes unavoidable when antenna elements are close to each other. Mutual coupling describes the energy absorbed by a nearby antenna when one antenna is operating. Free space radiations, surface currents, and surface waves all contribute to mutual coupling in MIMO antennas. While the first two apply to all arrays, the third is more frequent with microstrip antennas [4]. The mutual coupling can seriously degrade the signal-to-interference-noise ratio (SINR) of an adaptive array and the convergence of array signal processing algorithms [5, 6], the estimations of carrier frequency offset [7], channel estimation [8], angle of arrival [9], and the active reflection coefficient [10] of a MIMO array. Moreover, multiple power amplifiers (PAs) in the presence of mutual coupling can cause significant out-of-band (OOB) emission [11], creating interferences to communication systems at adjacent channels.

Isolation is basically equivalent to mutual coupling and thus can be used to measure the mutual coupling level. Specifically, the dB-valued isolation between the mth and nth elements in a MIMO array is usually quantified using the corresponding S-parameter,

$$\text{Isolation} = -20\log 10(|S_{mn}|). \tag{2.11}$$

The isolation requirement differs for different MIMO applications. For example, MIMO antennas in BSs are usually required to have more than 25-dB isolation [12, 13]. Nevertheless, for some special micro-cell scenarios, where less than 0.4-wavelength (instead of the common half-wavelength) inter-element spacing is required, the isolation requirement can be loosened to 20 dB as a compromise of the more compact MIMO array [14]. On the other hand, 10-dB isolation is usually regarded as sufficient for MIMO antennas in smartphones [15, 16]. In general, decoupling for BS arrays is more challenging than that for smartphone antennas. This is because the former dictates undistorted radiation patterns to ensure good cellular coverage, whereas the latter imposes almost no limit on the pattern distortion incurred by the decoupling.

2.2.4 Correlation Coefficient

In some real-world cases, inadequate antenna spacing or scattering increases spatial correlation, which has a significant impact on the MIMO channel ($\mathbf{H}$), causing it to deviate from the spatially white MIMO channel, thus deteriorating its capacity [2]. In this section, we address the correlation matrix and its entries (i.e., correlation coefficients). For convenience and better illustration, we assume no delay spread (i.e., a frequency-flat channel), drop the τ dependence, and refer to the MIMO channel $\mathbf{H}$ as a $M_R \times M_T$ matrix.

2.2.4.1 Correlation Matrix

The MIMO channel **H** can be approximated using the full-correlation-based channel model [2]

$$\mathrm{vec}(\mathbf{H}) = \mathbf{R}^{\frac{1}{2}}\mathrm{vec}(\mathbf{H}_{\mathrm{w}}) \tag{2.12}$$

where $\mathbf{H}_{\mathrm{w}}$ is the spatially white $M_{\mathrm{R}} \times M_T$ MIMO channel, the superscript $^{1/2}$ denotes the Hermitian square root of a matrix, and **R** is a $M_{\mathrm{T}}M_{\mathrm{R}} \times M_{\mathrm{T}}M_{\mathrm{R}}$,positive semi-definite Hermitian matrix, defined as

$$\mathbf{R} = \mathcal{E}\{\mathrm{vec}(\mathbf{H})\mathrm{vec}(\mathbf{H})^{\mathrm{H}}\}. \tag{2.13}$$

Note that if $\mathbf{R} = \mathbf{I}_{M_{\mathrm{T}}M_{\mathrm{R}}}$, then $\mathbf{H} = \mathbf{H}_{\mathrm{w}}$. Although the model described above is capable of capturing the overall correlation effects between the elements of **H**, a less generalized yet more popular model is often used instead. It is the so-called Kronecker channel model [2]

$$\mathbf{H} = \mathbf{R}_{\mathrm{r}}^{1/2}\mathbf{H}_{\mathrm{w}}\mathbf{R}_{\mathrm{t}}^{1/2} \tag{2.14}$$

where $\mathbf{R}_{\mathrm{t}}$ is the $M_{\mathrm{T}} \times M_{\mathrm{T}}$ transmit correlation matrix and $\mathbf{R}_{\mathrm{r}}$ is the $M_{\mathrm{R}} \times M_{\mathrm{R}}$ receive correlation matrix. Both $\mathbf{R}_{\mathrm{t}}$ and $\mathbf{R}_{\mathrm{r}}$ are positive semi-definite Hermitian matrices. Equation (2.13) implies that the receive antenna correlation matrix $\mathbf{R}_{\mathrm{r}}$ is equal to the covariance of the $M_{\mathrm{R}} \times 1$ receive vector channel when excited by any transmit antenna, and is, therefore, the same for all transmit antennas. The model can occur when the angle spectra of the scatterers at the receive array for signals arriving from any transmit antenna are identical. This condition arises if all the transmit antennas are closely located and have identical radiation patterns. These remarks also carry over to the transmit antenna correlation $\mathbf{R}_{\mathrm{t}}$. From (2.12) and (2.13), we can show that **R**, $\mathbf{R}_{\mathrm{t}}$, and $\mathbf{R}_{\mathrm{r}}$ are related by

$$\mathbf{R} = \mathbf{R}_{\mathrm{t}}^{\mathrm{T}} \otimes \mathbf{R}_{\mathrm{r}} \tag{2.15}$$

where $\otimes$ stands for the Kronecker product.

Note also that $\mathbf{H}_{\mathrm{w}}$ is a full-rank matrix with probability of one. In the presence of receive or transmit correlation, the rank of **H** is constrained by $\min(r((\boldsymbol{R})_{\mathrm{r}}), r((\boldsymbol{R})_{\mathrm{t}}))$, where $r(\mathbf{A})$ denotes the rank of **A**.

2.2.4.2 Correlation Coefficient

The correlation coefficients, i.e., the entries of the correlation matrix shown in (2.13), between the mth and nth antennas in a MIMO array, can be explicitly expressed using the conventional antenna metrics

$$r_{mn} = \frac{\oint | G_{12}(\Omega) | \, \mathrm{d}\Omega}{\sqrt{\oint | G_{11}(\Omega) | \, \mathrm{d}\Omega}\sqrt{\oint | G_{22}(\Omega) | \, \mathrm{d}\Omega}} \tag{2.16}$$

$$G_{mn}(\Omega) = \kappa E_{\theta m}(\Omega)E_{\theta n}^{*}(\Omega)P_{\theta}(\Omega) + E_{\phi m}(\Omega)E_{\phi n}^{*}(\Omega)P_{\phi}(\Omega)$$

where κ represents the cross-polarization discrimination (XPD) of the propagation channel, which is the ratio between the vertically polarized and horizontally polarized components. The superscript * denotes the complex conjugate operator, $E_{\theta n}$ and $E_{\phi n}$ are the θ and ϕ polarized components of the radiation pattern of the nth array element; $P_\theta(\Omega)$ and $P_\phi(\Omega)$ are the power angular spectra (PAS) of θ, and ϕ polarized propagation waves from the solid angle Ω, respectively. These parameters are all linked to the propagation channel. The correlation coefficients at the transmitting and receiving array can be determined in the same way. It is clear from the correlation expression (2.16) that the correlation coefficients depend on the radiation patterns and the propagation channel. That is, in general, it is incomplete to calculate the correlation coefficients using antenna patterns alone.

The correlation expression (2.16) is mainly helpful for theoretical analysis. To calculate the correlation coefficients from simulated or measured multipath channels [17], the following equation is usually used instead:

$$r_{mn} = \frac{E\left[s_m s_n^*\right]}{\sqrt{E\left[|s_m|^2\right] E\left[|s_n|^2\right]}} \tag{2.17}$$

where s_m is the complex received signal field at the mth array element in the multipath environment and E denotes the mathematical expectation operator. As long as the angular distribution of the propagation wave is accurately described by $P_\theta(\Omega)$ and $P_\phi(\Omega)$ with known XPD, (2.16) and (2.17) essentially yield the same result [18].

The (complex-valued) correlation coefficients of complex random signals are often used in the correlation-based channel model, whereas for antenna engineers, the (real-valued) envelope correlation coefficient (ECC), i.e., correlation between signal magnitudes, is usually preferred. The ECC equals the squared magnitude of the complex correlation coefficient [19]. Usually, an implicit assumption of ECC is isotropic scattering environment so that it only depends on the antenna patterns. Because of its simplicity, the ECC is heavily used by antenna engineers[1]. Nevertheless, as the ECC loses the phase information of the complex correlation coefficient, it contains incomplete information about the channel's spatial structure or the eigenvalues of the correlation matrix for MIMO antennas with more than two ports. Interested readers can refer to [2, 20] for in-depth discussions on this matter.

2.2.5 Total Active Reflection Coefficient

To describe the frequency bandwidth and radiation capabilities of an N-port antenna, the TARC was introduced in [21]. For a desired port excitation, the

1 Despite of its popularity, there is no specific ECC requirement in the antenna industry except for the common belief that the less the better.

incident power is defined as the total sum of all available powers at all excitation ports, the transmitted power is defined as radiation power, and the reflected power is defined as the difference between these two. The square root of the ratio of reflected power to incident power is defined as the TARC. The frequency bandwidth must be calculated using the TARC values, not the individual port scattering parameter or the active reflection coefficient (ARC) [22, 23]. Multiport antenna engineers can better comprehend the antenna bandwidth independent of the beam-forming feeding network by evaluating the true antenna bandwidth for a desired port excitation in this manner.

2.2.5.1 Generalized Scattering Matrix

A MIMO antenna generally can be seen as a microwave network with two radiation ports representing the radiated fields' vertical and horizontal polarization states in the far-field region at coordinates θ and ϕ and one or more excitation ports (see Figure 2.3). The generalized scattering matrix of an N-port antenna is an $(N+2)\times(N+2)$ matrix that can be expressed as

$$\begin{bmatrix} b_h \\ b_v \\ b_1 \\ \vdots \\ b_N \end{bmatrix} = \begin{bmatrix} S_{hh} & S_{hv} & S_{h1} & \cdots & S_{hN} \\ S_{vh} & S_{vv} & S_{v1} & \cdots & S_{vN} \\ S_{1h} & S_{1v} & S_{11} & \cdots & S_{1N} \\ \vdots & \vdots & \vdots & \vdots & \vdots \\ S_{Nh} & S_{Nv} & S_{N1} & \cdots & S_{NN} \end{bmatrix} \cdot \begin{bmatrix} a_h \\ a_v \\ a_1 \\ \vdots \\ a_N \end{bmatrix} \tag{2.18}$$

where the parameters a_h and a_v are horizontally and vertically polarized received waves while b_h and b_v are horizontally and vertically polarized transmitted waves by the antenna for that given point. $a_1 \cdots a_N$ and $b_1 \cdots b_N$ are N forward and N backward waves at excitation ports, respectively. The generalized scattering matrix has four submatrices as described below:

$$\begin{bmatrix} b_h \\ b_v \\ b_1 \\ \vdots \\ b_N \end{bmatrix} = \begin{bmatrix} \mathbf{S}_S & \mathbf{S}_{SP} \\ \mathbf{S}_{PS} & \mathbf{S}_P \end{bmatrix} \cdot \begin{bmatrix} a_h \\ a_p \\ a_1 \\ \vdots \\ a_N \end{bmatrix}. \tag{2.19}$$

The structural antenna scattering characteristics are described by the submatrix $\mathbf{S}_S$. When all excitation ports are terminated by matched loads, this 2×2 submatrix relates the scattered waves for both polarizations at a certain location caused by the incident plane wave. The relationships between the excitation and radiation ports (i.e., the receive and transmit characteristics of the MIMO antenna) are described in the submatrices $\mathbf{S}_{SP}$ and $\mathbf{S}_{PS}$. For example, submatrix $\mathbf{S}_{SP}$ describes the contributions of input signals $(a_1 \cdots a_N)$ at excitation ports on horizontally and vertically polarized transmitted waves, i.e., b_h and b_v. Similarly, submatrix $\mathbf{S}_{PS}$ shows the contributions on reflected signals $(b_1 \cdots b_N)$ at excitation

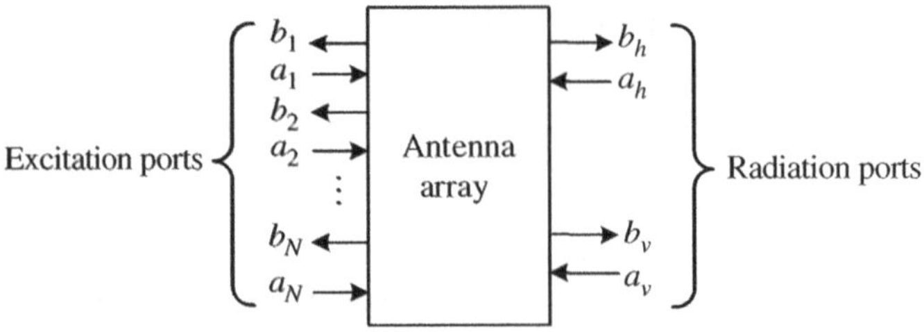

Figure 2.3 N-port antenna with dual-polarization radiation ports. $a_1 \cdots a_N$ and $b_1 \cdots b_N$ denote input and reflected signals at each port, respectively. a_h and a_v are horizontally and vertically polarized received waves while b_h and b_v are the horizontally and vertically polarized transmitted waves by the antenna.

ports from horizontally and vertically polarized received waves, i.e., a_h and a_v. The relationship between the excitation ports when there is no received signal ($a_v = a_h = 0$) is contained in the submatrix $\mathbf{S}_P$.

The generalized scattering matrix of the antenna is frequency-dependent, just like other scattering matrices. Additionally, the three submatrices $\mathbf{S}_S$, $\mathbf{S}_{PS}$, and $\mathbf{S}_{SP}$, which are associated with the radiation characteristics of the MIMO antenna, are functions of θ and ϕ (i.e., the coordinates of the observation point).

All the far-field components in the generalized scattering matrix are directional, and their directions are identified by their subscripts. For instance, when all excitation ports are terminated to matched loads and the MIMO antenna is illuminated by a horizontally polarized wave, the notation s_{vh} describes the vertical component of the scattered wave at the observation point. The radiated power associated with the ith port, when the other ports are terminated by matched loads, is calculated by:

$$p_{ri} = \oint_S \mathbf{E}_i \times \mathbf{H}_i^* \cdot \mathrm{d}s = |a_i|^2 \oint_S \left(s_{hi}s_{hi}^* + s_{vi}s_{vi}^*\right) \cdot \mathrm{d}s \tag{2.20}$$

where $\mathbf{E}_i$ and $\mathbf{H}_i$ are the radiated electric and magnetic fields of the antenna and a_i is the forward signal at the ith port. The horizontal and vertical components of the radiated field associated with the ith port in the generalized scattering matrix are given by s_{hi} and s_{vi}, respectively. This is based on the assumption that there is no loss in the antenna and that all of the energy produced by the excitations either radiates or goes through the other ports.

For a lossless MIMO antenna, the entries of its generalized scattering matrix are related as

$$\begin{cases} \sum_{j=1}^{N} s_{ij}s_{ij}^* + \oint_S \left(s_{iv}s_{iv}^* + s_{ih}s_{ih}^*\right) \mathrm{d}s = 1 \\ \sum_{i=1}^{N} s_{ij}s_{ij}^* + \oint_S \left(s_{vj}s_{vj}^* + s_{hj}s_{hj}^*\right) \mathrm{d}s = 1 \end{cases}, i,j = 1, \ldots, N. \tag{2.21}$$

The surface integrals are conducted on a closed surface in the far-field region. In this case, the radiated power associated with each port normalized to the excitation can be defined as radiation loss, which can be calculated using the elements in $\mathbf{S}_p$ as

$$p_{rj} = \oint_S \left(s_{vj} s_{vj}^* + s_{hj} s_{hj}^* \right) \cdot ds = 1 - \sum_{i=1}^{N} s_{ij} s_{ij}^*, j = 1, \ldots, N. \tag{2.22}$$

Submatrix $\mathbf{S}_P$ can be treated as the scattering matrix of a lossy N-port network. Equation (2.22) also shows that $\mathbf{S}_P$ can be employed to determine the radiation performance of the antenna.

2.2.5.2 TARC Calculation

For a desired port excitation, the TARC of the antenna is defined as the square root of the available power generated by all excitations minus radiated power, divided by the available power as follows:

$$\Gamma_a^t = \sqrt{\frac{\text{available power} - \text{radiated power}}{\text{available power}}} = \sqrt{\frac{p_a - p_r}{p_a}}. \tag{2.23}$$

For instance, if an N-port antenna is excited at the ith port and the other ports are connected to matched loads, the TARC can be calculated by:

$$\Gamma_{ai}^t = \sqrt{1 - p_{ri}} = \sqrt{\sum_{j=1}^{N} |s_{ij}|^2}, i = 1, \ldots, N. \tag{2.24}$$

For a lossless antenna, the TARC can be calculated using $\mathbf{S}_P$. For a given excitation $[a]$, the Γ_a^t as a function of frequency is

$$\Gamma_a^t = \frac{\sqrt{\sum_{i=1}^{N} |b_i|^2}}{\sqrt{\sum_{i=1}^{N} |a_i|^2}} \tag{2.25}$$

where

$$[b] = [S_P] \cdot [a]. \tag{2.26}$$

Note that the TARC is a real number between zero and one. When the value of the TARC is equal to zero, all the delivered power is radiated, and when it is equal to one, all the power is either reflected or goes to the other ports.

2.2.6 Mean Effective Gain and Diversity Gain

2.2.6.1 Mean Effective Gain

The MEG, which is a scalar quantity describing the impact of the antenna while accounting for the multipath environment, has emerged as a way to characterize the communication performance of the antenna in real propagation environments. The concept of MEG was introduced in [24], and the authors in [25] provided the derivation of the MEG in Rician channels.

In spherical coordinates, Ω defines the angle of arrivals (AoA), θ and ϕ are the elevation and azimuth angles, respectively. $p_\theta(\Omega)$ and $p_\phi(\Omega)$ denote the weighted PAS (also known as the weighted probability density function of the AoA) of the stochastic components in the θ-polarization and ϕ-polarization, respectively. Moreover, the available powers of the stochastic components in the θ-polarization and ϕ-polarization are denoted by p_θ and p_ϕ, respectively. In a multipath environment characterized by a mixed field with both uncorrelated random, unpolarized, component, and one deterministic, polarized, component, the MEG of an antenna is given by [25]

$$\begin{aligned}
G_e &= \frac{1}{1+\chi}\int \frac{\chi G_\theta(\Omega)p_\theta(\Omega)}{1+K_\theta} + \frac{G_\phi(\Omega)p_\phi(\Omega)}{1+K_\phi} d \\
&\quad + \frac{1}{1+\chi}\left(\sqrt{\frac{\chi K_\theta G_\theta(\Omega_0)}{1+K_\theta}} + \sqrt{\frac{K_\phi G_\phi(\Omega_0)}{1+K_\phi}}\right) \\
&= G_e^{NLOS} + G_e^{LOS}
\end{aligned} \tag{2.27}$$

where K_θ and K_ϕ are the Ricean K-factors of the vertical and horizontal polarization components, respectively, defined as

$$K_\theta = \frac{P_{0\theta}}{P_\theta}, K_\phi = \frac{P_{0\phi}}{P_\phi} \tag{2.28}$$

where $P_{0\theta}$ and $P_{0\phi}$ denote the powers of the deterministic field in θ-polarization and ϕ-polarization, respectively. In (2.27), χ denotes the XPR, which can be computed in Rician channels as

$$\chi = \frac{P_{0\theta} + P_\theta}{P_{0\phi} + P_\phi} = \chi_{\text{unpol}}\frac{1+K_\theta}{1+K_\phi} \tag{2.29}$$

where χ_{unpol} is the corresponding XPR of the stochastic (unpolarized) component.

In (2.27), the MEG is basically the sum of the MEGs due to the non-line of sight (NLOS) (unpolarized) component and the LOS (polarized) component of the incident field. When K_θ and K_ϕ both are zero, the MEG is completely defined by the stochastic, unpolarized NLOS component (i.e., in a Rayleigh-fading environment) as

$$G_e = \int \frac{\chi}{\chi+1}G_\theta(\Omega)p_\theta(\Omega) + \frac{1}{\chi+1}G_\phi(\Omega)p_\phi(\Omega)\mathrm{d}\Omega. \tag{2.30}$$

On the other hand, when K_θ and K_ϕ both tend to infinity, the MEG is completely defined by the deterministic, polarized LOS component as

$$G_e = \frac{1}{1+\chi}(\sqrt{\chi G_\theta(\Omega_0)} + \sqrt{G_\phi(\Omega_0)})^2. \tag{2.31}$$

2.2.7 Diversity Combining Methods

2.2.7.1 Diversity Gain

In a wireless system, uncorrelated fading channels can be realized in a variety of methods. Using MIMO antennas to transmit or receive, where the elements of the array are spaced apart, is one way to achieve diversity. This type of diversity is referred to as space diversity. A diversity gain can be achieved with receiver space diversity without an increase in transmit signal power or bandwidth. Furthermore, compared to the SNR that would be obtained with just a single receive antenna, coherent combining of the diversity signals improves the signal-to-noise ratio (SNR) at the receiver. This SNR increase, called *array gain*, can also be obtained with transmitter space diversity by appropriately weighting the transmit antennas relative to the complex channel gains. In addition to the array gain, space diversity also provides *diversity gain*, defined as the change in the slope of the error probability resulting from the diversity combining [26]. The maximum diversity gain for either transmitter or receiver space diversity typically requires the separation between antennas so that the fading channels corresponding to the antennas are uncorrelated. For example, in a uniform scattering environment with omnidirectional transmit and receive antennas, the minimum antenna separation required for uncorrelated fading on each antenna is approximately one half-wavelength [27].

Note that the above description of the diversity gain is rooted in the communication society. In the antenna and propagation society, another definition of diversity gain that involves realistic antennas is often preferred. In practice, diversity gain can be achieved by utilizing orthogonal polarizations or antenna patterns in addition to the spatial domain. The diversity gain attributed to the combination effect of antenna spacing, orthogonal polarizations, and orthogonal patterns can be collectively referred to as the antenna diversity gain. We introduce the antenna diversity gain next. (Note that without specification, we refer to the antenna diversity gain simply as diversity gain in the sequel.)

The diversity gain is defined as the SNR improvement of a multi-port antenna w.r.t. that of a single ideal antenna at 1% outage probability level [28, 29]. The actual diversity gain also depends on the diversity combining schemes, as described next.

Multipath fading has a considerable negative impact on the performance of the single-antenna system. One of the best techniques to mitigate this fading effect is diversity combining of the uncorrelated fading signals using multi-antenna systems. Diversity combining uses the low probability that two or more independent signal pathways may simultaneously experience deep fades. Thus, the idea behind diversity is to send the same data over uncorrelated fading channels. These uncorrelated channels are combined in such a way that the output SNR after diversity combining is enhanced significantly [26].

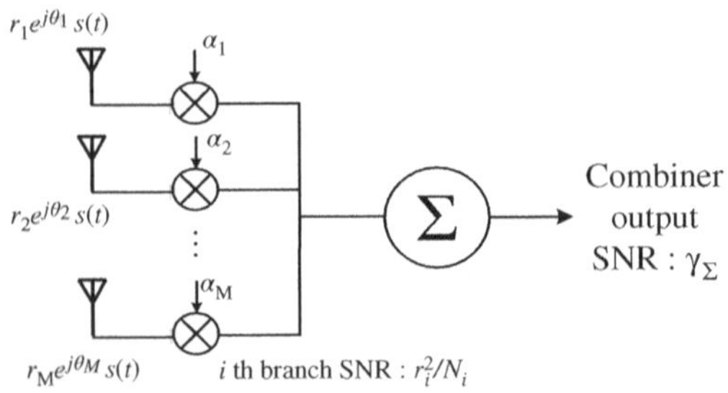

Figure 2.4 Diversity antenna system.

2.2.7.2 System Model of Receiver Diversity

In a diversity receiver, the independent fading channels associated with multiple receive antennas are combined to obtain a signal. The combining can be done in several ways, which vary in complexity and overall performance. Most combining techniques are linear: the output of the combiner is just a weighted sum of the different fading branches, as shown in Figure 2.4 for an M-branch diversity antenna system. The main purpose of the diversity antenna is to coherently combine the independent fading branches so that the fading effects are mitigated. The signal output from the combiner equals the originally transmitted signal $s(t)$ multiplied by a random complex amplitude term $\alpha_\Sigma = \sum_i a_i r_i$. This complex amplitude term results in a random $\text{SNR}_{\gamma\Sigma}$ at the combiner output, where the distribution of γ_Σ is a function of the number of diversity branches, the fading distribution on each branch, and the combining technique.

2.2.7.3 Selection Combining

In selection combining (SC), the combiner selects the signal on the branch with the highest SNR r_i^2/N_i, as shown in Figure 2.4. This is equivalent to choosing the branch with the highest $r_i^2 + N_i$ if the noise powers are the same on all branches, i.e., $Ni = N$. Because only one branch is used at a time, SC often requires just one receiver that is switched into the active antenna branch. However, a dedicated receiver on each antenna branch may be needed for systems that transmit continuously in order to monitor SNR on each branch simultaneously. With SC, the diversity output from the combiner has an SNR equal to the maximum SNR of all the branches.

2.2.7.4 Threshold Combining

SC for systems that transmit continuously may require a dedicated receiver on each branch to continuously monitor branch SNR. A simpler type of diversity combining scheme, called threshold combining (TC), avoids the need for a dedicated receiver on each branch by scanning each of the branches in sequential order and sending the first signal whose SNR is above a given threshold γ_T to the output.

Once a branch is chosen, this combiner's signal output remains as long as the SNR on that branch is above a predefined threshold. If the SNR on the selected branch falls below the threshold, the combiner switches to another branch.

2.2.7.5 Equal-Gain Combining

A slightly simpler technique (compared to the MRC) is equal-gain combining (EGC), which coherently combines the signals on all the branches with equal amplitude weighting, i.e., $\alpha_i = e^{-\theta_i}$. The output SNR of the EGC, assuming equal noise variance $N_0/2$ in each branch, is then given by [26]

$$\gamma_\Sigma = \frac{1}{N_0 M}\left(\sum_{i=1}^{M} r_i\right)^2. \tag{2.32}$$

2.2.7.6 Maximal-Ratio Combining

In SC and TC, the output of the combiner equals the signal on one of the branches. They enjoy lower complexity and cost but are clearly suboptimal. To achieve optimal diversity performance, the Maximal-Ratio Combining (MRC) scheme has been proposed. In an MRC receiver, the output is a weighted sum of all branches, so α_i in Figure 2.4 are all nonzero, and $\alpha_i = a_i e^{-j\theta_i}$, where a_i and θ_i are the amplitude and phase of the incoming signal on the ith branch, respectively. Thus, the envelope of the combiner output will be $r = \sum_{i=1}^{M} a_i r_i$. Assuming the same noise variance $N_0/2$ in each branch yields a total noise variance at the combiner output $N_{tot}/2 = \sum_{i=1}^{M} a_i^2 N_0/2$, thus, the output SNR of the combiner is [26]

$$\gamma_\Sigma = \frac{r^2}{N_{tot}} = \frac{1}{N_0}\frac{\left(\sum_{i=1}^{M} a_i r_i\right)^2}{\sum_{i=1}^{M} a_i^2}. \tag{2.33}$$

It is easy to show that the maximum of γ_Σ is achieved when a_i is proportional to the amplitude of the ith branch and the output SNR of an MRC receiver equals the sum of the SNRs of all the branches.

To compare the performances of different diversity combining schemes, we assume an ideal uncorrelated two-port diversity antenna in a Rayleigh fading environment and simulated its cumulative distribution functions (CDFs) with SC, EGC, and MRC. The corresponding results are shown in Figure 2.5. As expected, the MRC has the best performance. Note that even though MRC has slightly higher complexity compared with SC and EGC, it is far more practical than the other two schemes and is actually implemented in the 4G long-term evolution system, as demonstrated in the experiments of [30].

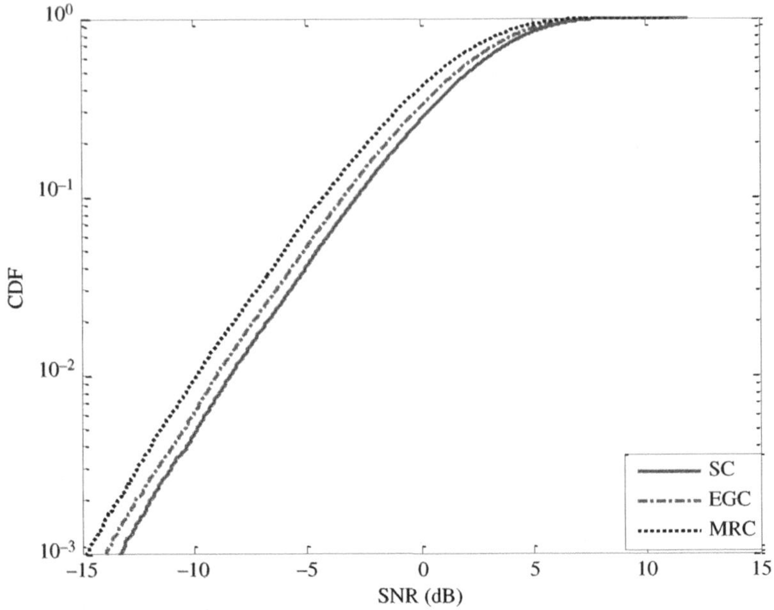

Figure 2.5 Diversity performances of SC, EGC, and MRC for an ideal uncorrelated two-port diversity antenna in Rayleigh fading environment.

2.3 MIMO Antenna Characterization Methods

2.3.1 Assessment Using the *S*-Parameters

The *S*-parameters are often used for MIMO antennas to evaluate the antenna port-matching and coupling between ports. In addition, in some special cases, *S*-parameters can be used to evaluate MIMO antennas' correlation and channel capacity. A scattering matrix can describe the properties of an N-port network, which can be thought of as a MIMO antenna with N ports, as shown in Figure 2.6. The scattering matrix can relate the voltage wave input to the network port and

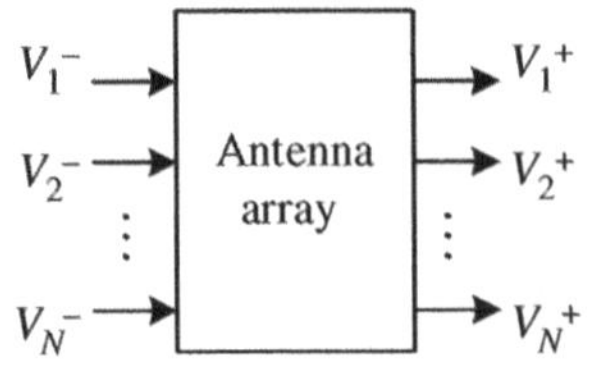

Figure 2.6 N-port network of an N-port MIMO antenna.

the reflected wave from the port as

$$\begin{bmatrix} V_1^- \\ V_2^- \\ \vdots \\ V_N^- \end{bmatrix} = \begin{bmatrix} S_{11} & \cdots & S_{1N} \\ \vdots & \ddots & \vdots \\ S_{N1} & \cdots & S_{NN} \end{bmatrix} \begin{bmatrix} V_1^+ \\ V_2^+ \\ \vdots \\ V_N^+ \end{bmatrix} \tag{2.34}$$

or in matrix form as $\mathbf{V}^- = \mathbf{S}\mathbf{V}^+$. The entries of matrix $\mathbf{S}$ can be expressed as

$$S_{ij} = \left.\frac{V_i^-}{V_j^+}\right|_{V_k^+=0,k\neq j}. \tag{2.35}$$

This indicates that S_{ij} can be obtained by measuring the reflected voltage V_i at port i when exciting at port j with V_j, provided that the remaining ports are terminated with matching loads (except for ports i and j). Therefore, the elements on the diagonal of the scattering matrix $\mathbf{S}$ can evaluate the matching of the MIMO antenna ports, and the non-diagonal elements indicate the coupling between the corresponding ports.

2.3.1.1 Normalized Coupling

Sometimes, the matching condition (i.e., S_{ii} shown in (2.34), $i = 1\ldots N$) of different antenna ports is unequal. In this case, if the antenna loss is relatively low, the normalized S-parameters can be used to compare the coupling fairly. For the ith port and jth port, the normalized S_{ij} can be expressed as

$$S_{ij}^{\text{Normalized}} = \frac{S_{ij}}{\sqrt{\left(1-|S_{ii}|^2\right)\left(1-|S_{jj}|^2\right)}}. \tag{2.36}$$

2.3.1.2 Correlation and Diversity Gain

In addition, the ECC ρ_{env} and diversity gain (DG) of two antenna ports with SC can be expressed using entries in the scattering matrix as[2]

$$\rho_{\text{env}} = \frac{|S_{11} * S_{12} + S_{21} * S_{22}|^2}{\left(1-\left(|S_{11}|^2+|S_{21}|^2\right)\right)\left(1-\left(|S_{22}|^2+|S_{12}|^2\right)\right)}. \tag{2.37}$$

$$\text{DG} = 10 \times \sqrt{1-\rho_{\text{env}}}. \tag{2.38}$$

2 Caution should be practiced when using S-parameter-based approximations in (2.37) and (2.38) as they are based on ideal antenna configurations that are not usually satisfied in most of practical applications. Readers are recommended to refer to [4, 31] for more discussion on this subject. It is best to rely on the field-based equations in the calculation of ECC.

2.3.1.3 Mean MIMO Capacity in Rayleigh Environment

MIMO antenna designs and analyses are typically challenging and complex because they require a plethora of information about the complex radiation patterns and propagation channels. However, the authors in [32] proposed a simple method to calculate the mean MIMO channel capacity for lossless MIMO antennas based only on S-parameters. In a 3-D isotropic random environment, where scatters randomly exist in all directions, the mean MIMO channel capacity for a MIMO system containing two arrays and each with M element antennas can be approximated as [32][3]

$$C = \log_2 \left\{ \left| \left(\mathbf{I} - \mathbf{S}_\mathrm{R}^H \mathbf{S}_\mathrm{R}\right) \left(\mathbf{I} - \mathbf{S}_\mathrm{T}^H \mathbf{S}_\mathrm{T}\right) \frac{\gamma}{M} \right| M! \right\} -1.6\log_{10} M - 0.7 \tag{2.39}$$

where $\mathbf{I}$ is an $M \times M$ identity matrix, γ denotes the SNR, and ! denotes the factorial operator. The matrices $\mathbf{S}_\mathrm{T}$ and $\mathbf{S}_\mathrm{R}$ represent S-parameters of the transmitting and receiving arrays, respectively.

2.3.2 Anechoic Chamber Characterization

An anechoic chamber is a shielded room covered with an electromagnetic absorber to mimic the free-space environment. In a well-designed microwave anechoic chamber, the equipment under test can receive radio frequency (RF) signals from a signal source without being influenced by external or internal reflected interferences. Most anechoic chambers are enclosed in faraday cages (sealed metallic enclosures) to minimize external electromagnetic interference. The inside of an anechoic chamber is lined with large foam pyramids called radar absorbent material (RAM) to nullify reflections. A photo of a multi-probe anechoic chamber is shown in Figure 2.7. Note that the vertical probe ring is used for spherical near-field measurement of passive antennas [33] or radiated measurements of active devices [34], while the horizontal probe ring is reserved for emulating 2-D spatial MIMO channel for, e.g., throughput measurements of MIMO terminals [35].

Conventionally, anechoic chambers are widely used to evaluate the performance of antennas in a free-space scenario. Several metrics for conventional antennas can be characterized using the anechoic chamber, such as gain, directivity, radiation pattern, beam width, and sidelobes. Nevertheless, evaluating MIMO antennas' performance (i.e., capacity, diversity gain) is not straightforward using an anechoic chamber. As MIMO antennas are intended for use in multipath environments with

---

3 It should be noted that (2.39) is an approximation in an extreme (i.e., 3-D isotropic) scattering environment. For more realistic (non-isotropic) multipath environment, one should resort to the general capacity formula (2.9).

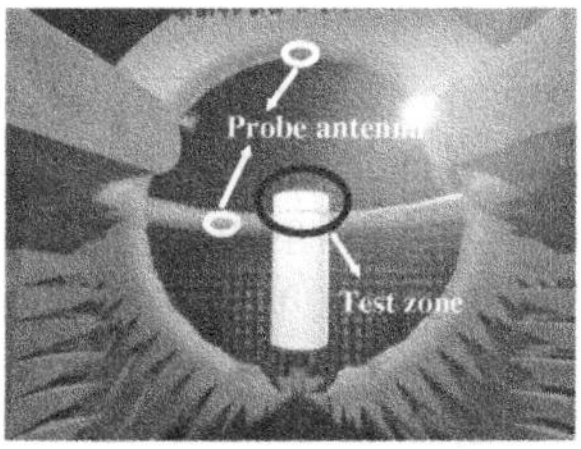

Figure 2.7 Photo of anechoic chamber.

noticeable time-delay spread, the wireless systems have now been provided with digital MIMO and orthogonal frequency domain multiplexing (OFDM) technology to improve the system performance, e.g., the throughput. These technologies cannot be tested in the conventional anechoic chamber. Nevertheless, a multipath environment with specific parameters, such as power delay profile, power Doppler spectrum, cross-polarization ratio (XPR), and spatial correlation at the Rx side, can be achieved by combining the multi-probe anechoic chamber with channel emulator [35]. The spatial radio propagation environment can be reproduced using the prefaded signal synthesis (PFS) method [36], which maps the output of the channel emulator to the probes in the anechoic chamber. Interested readers may refer to [35, 36] for more detailed spatial channel reconstructions and throughput tests. In this section, we will focus on a simple method that evaluates MIMO capacity using the embedded far-field functions and radiation efficiency measured in conventional anechoic chambers [29].

2.3.2.1 MIMO Capacity

Assuming that the transmit power is distributed equally among the transmit branches and that the receive MIMO antennas under test have complete channel state information, the ergodic capacity of the MIMO system is [29]

$$C_{N_r \times N_t} = E\left\{ \log_2 \left[\det \left(\mathbf{I}_{N_t} + \frac{\gamma}{N_t} \mathbf{H}_{N_r \times N_t} \mathbf{H}^H_{N_r \times N_t} \right) \right] \right\} \tag{2.40}$$

where $\mathbf{H}_{N_r \times N_t}$ is the MIMO channel matrix, N_t and N_r are numbers of transmit and receive antennas, respectively. Moreover, to focus on the characterization of multi-port antennas, assuming that the multi-port antenna being tested is on the receive side and that the transmit antennas are ideal (i.e., with unity efficiency and no correlation), the ergodic capacity, including overall antenna effects, can be expressed as

$$\begin{aligned} C &= E\left\{ \log_2 \left[\det \left(\mathbf{I} + \frac{\gamma}{N_t} \left(\mathbf{R}^{\frac{1}{2}} \mathbf{H}_w \right) \left(\mathbf{R}^{\frac{1}{2}} \mathbf{H}_w \right)^H \right) \right] \right\} \\ &= E\left\{ \log_2 \left[\det \left(\mathbf{I} + \frac{\gamma}{N_t} \mathbf{R} \mathbf{H}_w \mathbf{H}_w^H \right) \right] \right\} \end{aligned} \tag{2.41}$$

where $\mathbf{H}_{\mathrm{w}}$ denotes the normalized spatially white MIMO channel, whose Frobenius norm satisfies $E\left[\|\mathbf{H}_{\mathrm{w}}\|_{\mathrm{F}}^{2}\right] = N_{\mathrm{t}}N_{\mathrm{r}}$. The physical meaning of this normalization is that every sub-channel (i.e., entry of the channel matrix) should have unity average channel gain.

The covariance matrix **R** can be constructed by using embedded radiation efficiencies at each antenna port and complex correlations between all the antenna ports. The embedded radiation efficiency can be measured in an anechoic chamber, while the complex correlation between mth and nth ports is determined from measured embedded far-field functions at the corresponding ports [29], as shown in

$$[\mathbf{\Phi}]_{mn} = \frac{\iint_{4\pi} \mathbf{g}_m^H(\Omega)\mathbf{P}_{\mathrm{inc}}(\Omega)\mathbf{g}_n(\Omega)\mathrm{d}\Omega}{\sqrt{\iint_{4\pi} \mathbf{g}_m^H(\Omega)\mathbf{P}_{inc}(\Omega)\mathbf{g}_m(\Omega)\mathrm{d}\Omega \cdot \iint_{4\pi} \mathbf{g}_n^H(\Omega)\mathbf{P}_{\mathrm{inc}}(\Omega)\mathbf{g}_n(\Omega)\mathrm{d}\Omega}} \tag{2.42}$$

where Ω is the solid angle of arrival, $\mathbf{g}_i$ $(i=1, \ldots, N_{\mathrm{r}})$ is the embedded far-field function (a column vector with elements representing different polarization components) at the ith antenna port, and $\mathbf{P}_{\mathrm{inc}}$ is dyadic PAS of the incident waves. Note that in polarization-balanced isotropic scattering environments, e.g., reverberation chambers, $\mathbf{P}_{\mathrm{inc}}(\Omega) = \mathbf{I}$. Denote the efficiency vector consisting of all the total embedded radiation efficiencies and the correlation matrix as

$$\mathbf{e} = \left[e_{\mathrm{emb1}}\ e_{\mathrm{emb2}} \cdots e_{\mathrm{embN_r}}\right]^{\mathrm{T}}$$

$$\mathbf{\Phi} = \begin{bmatrix} 1 & \rho_{12} & \cdots & \rho_{1N_{\mathrm{r}}} \\ \rho_{12}^{*} & 1 & \cdots & \rho_{2N_{\mathrm{r}}} \\ \vdots & \vdots & \ddots & \vdots \\ \rho_{1N_{\mathrm{r}}}^{*} & \rho_{2N_{\mathrm{r}}}^{*} & \cdots & 1 \end{bmatrix}. \tag{2.43}$$

The covariance matrix based on anechoic chamber measurements can be expressed as

$$\mathbf{R} = \mathbf{\Xi} \circ \mathbf{\Phi} \tag{2.44}$$

where $\mathbf{\Xi} = \sqrt{\mathbf{e}}\sqrt{\mathbf{e}^{\mathrm{T}}}$, $\circ$ denotes entry-wise product, the superscript ^{T} is the transpose operator, and $\sqrt{\mathbf{e}}$ is the entry-wise square root of matrix **e**. The matrix $\mathbf{\Xi}$ can be expressed as

$$\mathbf{\Xi} = \begin{bmatrix} e_{\mathrm{emb1}} & \sqrt{e_{\mathrm{emb1}}e_{\mathrm{emb2}}} & \cdots & \sqrt{e_{\mathrm{emb1}}e_{\mathrm{embN_r}}} \\ \sqrt{e_{\mathrm{emb1}}e_{\mathrm{emb2}}} & e_{\mathrm{emb2}} & \cdots & \sqrt{e_{\mathrm{emb2}}e_{\mathrm{embN_r}}} \\ \vdots & \vdots & \ddots & \vdots \\ \sqrt{e_{\mathrm{emb1}}e_{\mathrm{embN_r}}} & \sqrt{e_{\mathrm{emb2}}e_{\mathrm{embN_r}}} & \cdots & e_{\mathrm{embN_r}} \end{bmatrix}. \tag{2.45}$$

Since the embedded radiation efficiencies obtained from anechoic chamber measurements are based on measured embedded far-field functions as well, we call this method embedded far-field function method. For power-balanced multi-port antenna with a scalar embedded radiation efficiency of e_{emb}, $\mathbf{R}$ becomes $e_{\text{emb}}\mathbf{\Phi}$, and (2.41) reduces to [37]

$$C = E\left\{\log_2\left[\det\left(\mathbf{I} + \frac{\gamma e_{\text{emb}}}{N_t}\mathbf{\Phi}\mathbf{H}_{\text{w}}\mathbf{H}_{\text{w}}^H\right)\right]\right\} \tag{2.46}$$

which is a special case of (2.41). Based on the generalized method, MIMO capacity of any multi-port antenna can be calculated based on measurements in an anechoic chamber.

2.3.2.2 MRC Diversity Gain

For MRC diversity calculations, the covariance matrix (2.44) only needs to be constructed using the embedded radiation patterns measured in the anechoic chamber. Then, the *covariance-eigenvalue approach* [29] is applied directly, which will be presented in detail in **MRC Diversity** in the next section (i.e., Section 2.3.3).

2.3.3 Reverberation Chamber Characterization

2.3.3.1 Reverberation Chamber Versus Anechoic Chamber – on MIMO Performance Characterization

The reverberation chamber is a large metal cavity provided with mode stirrers and (optional) turn-table platform. It has been used for more than 30 years for electromagnetic compatibility (EMC) measurements of electromagnetic susceptibility and undesired emissions, as described in an overview article in [38]. Its basic theories have also been documented in [39–42].

Like the anechoic chamber that emulates free space as a well-accepted reference environment for characterizing antennas, the reverberation chamber is an efficient instrument for over-the-air (OTA) characterization of desired radiation of antennas and wireless devices, particularly when the antennas and wireless devices are intended for use in dense urban or indoor environments. A typical reverberation chamber for OTA measurements is shown in Figure 2.8. The reverberation chamber emulates Rayleigh fading when the modes are well stirred [39] by mechanical movement of translating plates or rotating paddles. In [43], it is shown that the cavity modes represent plane waves with an omnidirectional distribution of angles of arrival (AoA). Thus, the reverberation chamber corresponds to an isotropic multipath environment.

The main advantage of the reverberation chamber over the anechoic chamber is that it is smaller, cheaper, and highly efficient for measuring antenna efficiency [44], total radiated power (TRP) or total isotropic sensitivity (TIS) [45]. Traditionally, the reverberation chamber is believed to be not suitable for antenna pattern measurement. Previous antenna pattern measurement methods either

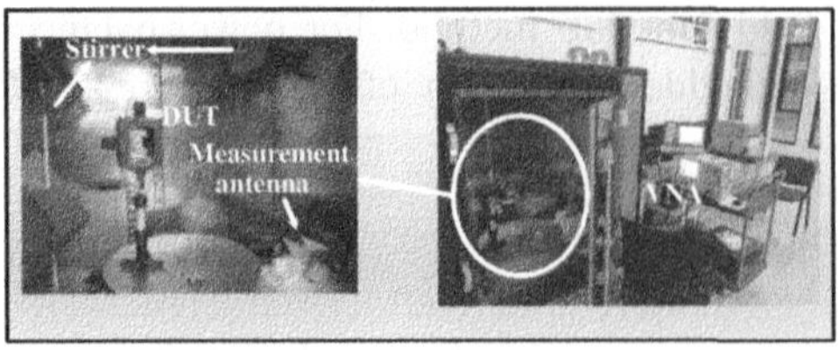

(a) (b)

Figure 2.8 Photos of the reverberation chamber: (a) zoomed in and (b) overview.

are limited to broadband antennas or suffer from a tremendous increase in measurement time as compared with the anechoic chamber measurement. Recently, the authors in [46] proposed a new method of antenna pattern measurements by expressing the antenna correlation coefficients in terms of the spherical wave coefficients of the antenna. The method is valid for most practical antennas and is much more efficient. However, the method only yields normalized radiation patterns (i.e., without antenna gain information) and its performance degrades drastically in loaded reverberation chambers. These drawbacks are solved using the improved method proposed in [47].

2.3.3.2 MRC Diversity Gain

Diversity technology offers effective leverage to mitigate fading in a wireless multipath environment. There are mainly three types of diversity techniques: frequency diversity, time diversity, and antenna diversity [25]. For antenna diversity, there are three main diversity-combining techniques: SC, EGC, and MRC [25, 28]. MRC offers the best performance at the expense of complexity. As the technology has developed over the past decades, MRC has become more feasible and popular. This section will focus on characterizing the MRC diversity gain of MIMO antennas using a reverberation chamber. Assuming an N-port diversity antenna in a Rayleigh-fading environment (e.g., a reverberation chamber), its covariance matrix is expressed as [29]

$$\mathbf{R} = E[\mathbf{h}\mathbf{h}^H] \tag{2.47}$$

where $\mathbf{h}$ is the complex baseband column-vector fading channel including overall antenna effects,[4] the superscript ^{H} is Hermitian operator, E is mathematical expectation, and we denote the ith eigenvalues of $\mathbf{R}$ as λ_i. The effective diversity gain (EDG) is defined as the output of a diversity antenna improvement compared with that of a single ideal antenna at a certain outage probability level, e.g., 1%.

---

4 Note that the channel transfer function (i.e., the channel's frequency response) between a transmit-receive antenna pair can be conveniently measured using a vector network analyzer in the form of S_{21}. The interested reader can refer to [27] for more details.

The MRC EDG is expressed as [29]

$$\text{EDG} = \left.\frac{F^{-1}(\gamma)}{F_{\text{ideal}}^{-1}(\gamma)}\right|_{1\%} \tag{2.48}$$

where γ denotes the instantaneous SNR and denotes functional inversion. $F_{\text{ideal}}(\gamma)$ is the CDF of the output SNR of a single ideal antenna, and for Rayleigh fading it can be expressed as $F_{\text{ideal}}(\gamma) = 1 - \exp(-\gamma)$. The CDF of the MRC output SNR $F(\gamma)$ in Rayleigh fading is known for two cases:

- When all eigenvalues are different from each other, $F(\gamma)$ is given by

$$F(\gamma) = 1 - \sum_{i=1}^{N} \frac{\lambda_i^{N-1} \exp\left(-\frac{\gamma}{\lambda_i}\right)}{\prod_{k \neq i}^{N}(\lambda_i - \lambda_k)}. \tag{2.49}$$

- When all eigenvalues are equal, i.e., $\lambda_i = \lambda (i = 1 \ldots M)$, $F(\gamma)$ is given by

$$F(\gamma) = 1 - \exp\left(-\frac{\gamma}{\lambda}\right) \sum_{i=1}^{N} \frac{(\gamma/\lambda)^{i-1}}{(i-1)!}. \tag{2.50}$$

Since the EDG is determined from the CDF consisting of eigenvalues of the covariance matrix, we refer to the method of EDG calculation shown above as the covariance eigenvalue approach. Compared with the traditional way of diversity measurement [48], the covariance-eigenvalue approach offers better accuracy with the same finite number of samples. As seen from (2.49), it has an apparent singularity when any two eigenvalues of the covariance matrix are equal. Therefore, it is usually believed that (2.49) will result in a large numerical error when two eigenvalues are close to each other. However, the authors in [25] have proved that (2.49) converges to the true CDF when the eigenvalues approach each other when the covariance matrix is estimated by the sample mean,

$$\hat{\mathbf{R}} = \frac{1}{M} \sum_{m=1}^{M} \mathbf{h}_m \mathbf{h}_m^H \tag{2.51}$$

where $\mathbf{h}_m$ is the mth realization of random channel vector $\mathbf{h}$, and M is the number of realizations (or samples). We therefore refer to (2.44) as sample covariance matrix and its eigenvalues, $\hat{\lambda}_i (i = 1 \ldots N)$, as sample eigenvalues. The probe antennas (mounted on three orthogonal walls with different polarizations inside the chamber) are generally regarded as independent transmit antennas in the reverberation chamber. The AUT functions as the receive antennas to form a MIMO system together with the probe antennas. The covariance matrix at the receive side can be estimated using (2.44). Once the covariance matrix is calculated, the MRC diversity gain can be readily obtained using the covariance eigenvalue approach presented above.

2.3.3.3 MIMO Capacity and Correlation Coefficient

The reverberation chamber is a metal cavity with many excited and stirred modes to create an isotropic Rayleigh-fading environment. Assuming that the receiver has perfect channel state information and the transmitted power is equally allocated among the transmit branches, the ergodic capacity of the MIMO system is [25]

$$C_{N_r \times N_t} = E\left\{ \log_2 \left[\det \left(\mathbf{I}_{N_t} + \frac{\gamma}{N_t} \mathbf{H}_{N_r \times N_t} \mathbf{H}^H_{N_r \times N_t} \right) \right] \right\} \tag{2.52}$$

where $\mathbf{H}_{N_r \times N_t}$ is the MIMO channel matrix, N_t and N_r are number of transmit and receive antennas, respectively. In the reverberation chamber, the ergodic MIMO capacity can be easily calculated using (2.52) with the (normalized) measured MIMO channel samples.

As to the correlation matrix, for convenience, the normalized measured channel matrix can be expressed as [37]

$$\mathbf{H}_{\text{meas}} = \frac{\mathbf{H}_{N_r \times N_t}}{\sqrt{P_{\text{ref}}}} \tag{2.53}$$

where P_{ref} denotes the reference level, which is obtained by dividing the average power level with the total radiation efficiency of the reference antenna. For antenna i and antenna j, the complex correlation coefficient ρ_{ij} in a reverberation chamber can be easily calculated using

$$\begin{aligned} \rho_{ij} &= \frac{\mathbf{R}(i,j)}{\sqrt{\mathbf{R}(i,i)\mathbf{R}(j,j)}}, \\ \mathbf{R} &= E\left\{ (\mathbf{H}_{\text{meas}} - E[\mathbf{H}_{\text{meas}}])(\mathbf{H}_{\text{meas}} - E[\mathbf{H}_{\text{meas}}])^H \right\}. \end{aligned} \tag{2.54}$$

2.4 Summary

In this chapter, we introduced several performance metrics for conventional antennas and MIMO antennas, providing a comprehensive view of MIMO antenna evaluations. Moreover, three MIMO antenna characterization methods with different measurement techniques are discussed. Using the S-parameters of MIMO antennas, we can assess coupling, correlation, diversity gain, and capacity in the 3-D isotropic scattering environment. Furthermore, MIMO capacity and MRC diversity gain can also be estimated by anechoic chamber and reverberation chamber. Compared to the anechoic chamber, the reverberation chamber is smaller, cheaper, and highly efficient for measuring radiation efficiency of passive antenna, and TRP and TIS of active devices. Though more complex, the anechoic chamber is suitable for measurements of conventional antenna metrics as well as MIMO antenna metrics.

References

1 Balanis, C.A. (2015). *Antenna Theory: Analysis and Design*. Wiley.

2 Paulraj, A., Rohit, A.P., Nabar, R., and Gore, D. (2003). *Introduction to Spacetime Wireless Communications*. Cambridge University Press.

3 Shannon, C.E. (1948). A mathematical theory of communication. *The Bell System Technical Journal* 27 (3): 379–423.

4 Sharawi, M.S. (2017). Current misuses and future prospects for printed multiple-input, multiple-output antenna systems [wireless corner]. *IEEE Antennas and Propagation Magazine* 59 (2): 162–170.

5 Yuan, Q., Chen, Q., and Sawaya, K. (2006). Performance of adaptive array antenna with arbitrary geometry in the presence of mutual coupling. *IEEE Transactions on Antennas and Propagation* 54 (7): 1991–1996.

6 Wang, B., Chang, Y., and Sun, Y. (2016). Performance of the large-scale adaptive array antennas in the presence of mutual coupling. *IEEE Transactions on Antennas and Propagation* 64 (6): 2236–2245.

7 Wu, Y., Bergmans, J., and Attallah, S. (2010). Effects of antenna correlation and mutual coupling on the carrier frequency offset estimation in MIMO systems. In: *in 2010 6th International Conference on Wireless Communications Networking and Mobile Computing (WiCOM)*, 1–4. IEEE.

8 Lu, S., Hui, H.T., Bialkowski, M.E. et al. (2007). The effect of antenna mutual coupling on channel estimation of MIMO-OFDM systems. *2007 IEEE Antennas and Propagation Society International Symposium* (09-15 June 2007), Honolulu, HI, IEEE, pp. 2945–2948.

9 Lui, H.S. and Hui, H.T. (2010). Mutual coupling compensation for direction-of- arrival estimations using the receiving-mutual-impedance method. *International Journal of Antennas and Propagation* 2010: 37306.

10 Pozar, D.M. (2003). A relation between the active input impedance and the active element pattern of a phased array. *IEEE Transactions on Antennas and Propagation* 51 (9): 2486–2489.

11 Fager, C., Bland, X., Hausmair, K., Cahuana, J.C., and Eriksson, T. (2014). Prediction of smart antenna transmitter characteristics using a new behavioral modeling approach. *2014 IEEE MTT-S International Microwave Symposium (IMS2014)*, IEEE, pp. 1–4.

12 Da, Y., Chen, X., and Kishk, A.A. (2022). In-band mutual coupling suppression in dual-band shared-aperture base station arrays using dielectric block loading. *IEEE Transactions on Antennas and Propagation* 70 (10): 9270–9281.

13 Wu, K., Wei, C., Mei, X., and Zhang, Z. (2017). Array-antenna decoupling surface. *IEEE Transactions on Antennas and Propagation* 65 (12): 6728–6738.

14 Liu, B., Da, Y., Chen, X., and Kishk, A. (2022). Hybrid decoupling structure based on neutralization and partition schemes for compact large-scale base station arrays. *IEEE Antennas and Wireless Propagation Letters* 21 (2): 267–271.

15 Zhao, A. and Ren, Z. (2019). Wideband MIMO antenna systems based on coupled-loop antenna for 5G N77/N78/N79 applications in mobile terminals. *IEEE Access* 2019: 93761–93771.

16 Sun, L., Li, Y., Zhang, Z. et al. (2020). Self-decoupled MIMO antenna pair with shared radiator for 5G smartphones. *IEEE Transactions on Antennas and Propagation* 68 (5): 3423–3432.

17 Guan, K., Li, G., Kürner, T. et al. (2016). On millimeter wave and the mobile radio channel for smart rail mobility. *IEEE Transactions on Vehicular Technology* 66 (7): 5658–5674.

18 Chen, X., Pei, H., Li, M. et al. (2020). Revisit to mutual coupling effects on multi-antenna systems. *Journal of Communications and Information Networks* 5 (4): 411–422.

19 J. N. Pierce and S. Stein, "Multiple diversity with nonindependent fading," *Proceedings of the IRE*, vol. 48, no. 1, pp. 89-104, Jan. 1960.

20 Y. Wang, X. Chen, H. Pei, W. E. I. Sha, Y. Huang, and A. A. Kishk, "MIMO performance enhancement of MIMO arrays using PCS-based near-field optimization technique," *Science China Information Sciences*, 66, no. 6, 2023, Art. no. 162302.

21 Manteghi, M. and Rahmat-Samii, Y. (2005). Multiport characteristics of a wide-band cavity backed annular patch antenna for multipolarization operations. *IEEE Transactions on Antennas and Propagation* 53 (1): 466–474.

22 Gomez-Tagle, J. and Christodoulou, C. (2000). Broadband characterization of the active reflection coefficient of finite-sized phased array microstrip antennas. *Proceedings 2000 IEEE International Conference on Phased Array Systems and Technology (Cat No. 00TH8510)*, 21-25 May 2000, *Dana Point*, CA, USA, IEEE, pp. 255–258.

23 Deshpande, M.D. and Bailey, M.C. (1989). Analysis of finite phased arrays of circular microstrip patches. *IEEE Transactions on Antennas and Propagation* 37 (11): 1355–1360.

24 Taga, T. (1990). Analysis for mean effective gain of mobile antennas in land mobile radio environments. *IEEE Transactions on Vehicular Technology* 39 (2): 117–131.

25 Glazunov, A.A., Molisch, A.F., and Tufvesson, F. (2009). Mean effective gain of antennas in a wireless channel. *IET Microwaves, Antennas & Propagation* 3 (2): 214–227.

26 Goldsmith, A. (2005). *Wireless Communications*. Cambridge University Press.

27 Teal, P.D., Abhayapala, T.D., and Kennedy, R.A. (2002). Spatial correlation for general distributions of scatterers. *IEEE Signal Processing Letters* 9 (10): 305–308.

28 Schwartz, M., Bennett, W.R., and Stein, S. (1995). *Communication Systems and Techniques*. Wiley.

29 Chen, X., Kildal, P.-S., Carlsson, J., and Yang, J. (2012). MRC diversity and MIMO capacity evaluations of multi-port antennas using reverberation chamber and anechoic chamber. *IEEE Transactions on Antennas and Propagation* 61 (2): 917–926.

30 Chen, X. (2014). Throughput modeling and measurement in an isotropic-scattering reverberation chamber. *IEEE Transactions on Antennas and Propagation* 62 (4): 2130–2139.

31 Sharawi, M. et al. (2017). Correlation coefficient calculations for MIMO antenna systems: a comparative study. *International Journal of Microwave and Wireless Technologies* 9: 1991–2004.

32 Honma, N., Sato, H., Ogawa, K., and Tsunekawa, Y. (2015). Accuracy of MIMO channel capacity equation based only on s-parameters of MIMO antenna. *IEEE Antennas and Wireless Propagation Letters* 14: 1250–1253.

33 Hansen, J.E. (1988). *Spherical Near-Field Antenna Measurements*. London: IET Press.

34 Huang, S., Li, F., and Chen, X. (2022). An improved method for total radiated power tests in anechoic chambers. *IEEE Transactions on Instrumentation and Measurement* 71: 1005909.

35 Pei, H., Chen, X., Huang, X. et al. (2022). Key issues and algorithms of multiple-input-multiple-output over-the-air testing in the multi-probe anechoic chamber setup. *Science China Information Sciences* 65 (3): 1–27.

36 Kyosti, P., Jamsa, T., and Nuutinen, J.-P. (2012). Channel modelling for multiprobe over-the-air MIMO testing. *International Journal of Antennas and Propagation* 2012.

37 Chen, X., Kildal, P.-S., Carlsson, J., and Yang, J. (2011). Comparison of ergodic capacities from wideband MIMO antenna measurements in reverberation chamber and anechoic chamber. *IEEE Antennas and Wireless Propagation Letters* 10: 446–449.

38 Backstrom, M., Lunděn, O., and Kildal, P.-S. (2002). Reverberation chambers for EMC susceptibility and emission analyses. In: *Proc. Review of Radio Science 1999–2002* (ed. W.R. Stone), 429–452. New York: IEEE Press.

39 Kostas, J.G. and Boverie, B. (1991). Statistical model for a mode-stirred chamber. *IEEE Transactions on Electromagnetic Compatibility* 33 (4): 366–370.

40 Hill, D.A., Ma, M.T., Ondrejka, A.R. et al. (1994). Aperture excitation of electrically large, lossy cavities. *IEEE Transactions on Electromagnetic Compatibility* 36 (3): 169–178.

41 Hill, D.A. (1999). Linear dipole response in a reverberation chamber. *IEEE Transactions on Electromagnetic Compatibility* 41 (4): 365–368.

42 Hill, D.A. (1994). Electronic mode stirring for reverberation chambers. *IEEE Transactions on Electromagnetic Compatibility* 36 (4): 294–299.

43 Rosengren, K. and Kildal, P.-S. (2001). Study of distributions of modes and plane waves in reverberation chambers for the characterization of antennas in a multipath environment. *Microwave and Optical Technology Letters* 30 (6): 386–391.

44 Chen, X. (2013). On statistics of the measured antenna efficiency in a reverberation chamber. *IEEE Transactions on Antennas and Propagation* 61 (11): 5417–5424.

45 Xue, W., Li, F., Chen, X. et al. (2020). A unified approach for uncertainty analyses for total radiated power and total isotropic sensitivity measurements in reverberation chamber. *IEEE Transactions on Instrumentation and Measurement* 70: 1–12.

46 Xu, Q., Huang, Y., Xing, L. et al. (2016). 3-d antenna radiation pattern reconstruction in a reverberation chamber using spherical wave decomposition. *IEEE Transactions on Antennas and Propagation* 65 (4): 1728–1739.

47 Zheng, J., Chen, X., Liu, X. et al. (2022). An improved method for reconstructing antenna radiation pattern in a loaded reverberation chamber. *IEEE Transactions on Instrumentation and Measurement* 71: 1–12.

48 Chen, X., Kildal, P.-S., and Carlsson, J. (2011). Fast converging measurement of MRC diversity gain in reverberation chamber using covariance-eigenvalue approach. *IEICE Transactions on Electronics* 94 (10): 1657–1660.

3

Feed Network Design and Analysis for MIMO Antenna Systems

The main function of multiple-input-multiple-output (MIMO) systems is to increase the channel's capacity without the use of additional time or frequency resources. It necessitates a large number of antennas configured in a transceiver with minimized coupling with each other. One of the main design considerations for efficient MIMO antenna systems is the design of their feed networks. This chapter begins by providing a brief overview of passive feeding network methods, including coaxial feeds, microstrip line feeds, aperture-coupled feeds, and electromagnetically coupled feeds. While such feeds are well applicable in both single-input antennas and MIMO systems, examples of feeding multiport MIMO antennas will be presented, considering suppression of mutual coupling. Then, a discussion of active feeding networks is conducted. This includes switching-based techniques implemented in feeding networks and transmission lines, consisting of components such as PIN diodes, varactors, and RF MEMS. Finally, active beamforming networks and circuits, such as analog, digital, and hybrid beamforming, applicable to MIMO antenna systems, are included with several architecture examples.

3.1 Impedance-Matching Networks

One of the most critical requirements in the design of antennas and high-frequency electronic circuits is transferring the maximum possible energy from one sub-component to another, which is satisfied with complex conjugate matching via matching networks. In the case of antennas, methods to minimize reflections due to impedance mismatch from the feedline to the antenna should be small enough to be negligible. Therefore, impedance-matching networks are typically introduced between two sub-components (antennas and feed) with dissimilar impedances by matching an arbitrary load, Z_L to a transmission line, Z_0 to ensure maximum power transfer between them, as shown in Figure 3.1. In this chapter,

MIMO Antenna Systems for 5G and Beyond, First Edition.
Xiaoming Chen, Ping Jack Soh, and Mohammad S. Sharawi.

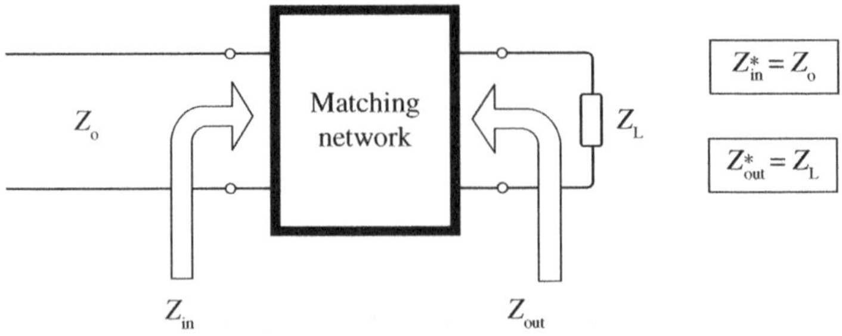

Figure 3.1 Impedance-matching network. Source: Adapted from Ref. [1].

several impedance-matching networks and techniques are presented in the following sub-sections.

3.1.1 Single-Stub Impedance Matching

The most basic impedance-matching technique is the single-stub matching network, which can be used to match any load impedance (e.g. single element antenna) to a transmission line. To do so, a transmission line segment of length L is inserted between the source and load, such that the input impedance looking into the transmission line from the source has a real part equal to the source of resistance. Then, a parallel short-circuit stub or a series short-circuit stub cancel the imaginary part of the input impedance. Both steps are illustrated in Figure 3.2. Two main parameters in designing a single-stub matching network are the location of the stub with reference to the load, d_{stub} and the length of the stub line, L_{stub}. This also implies that when the load impedance changes, the location

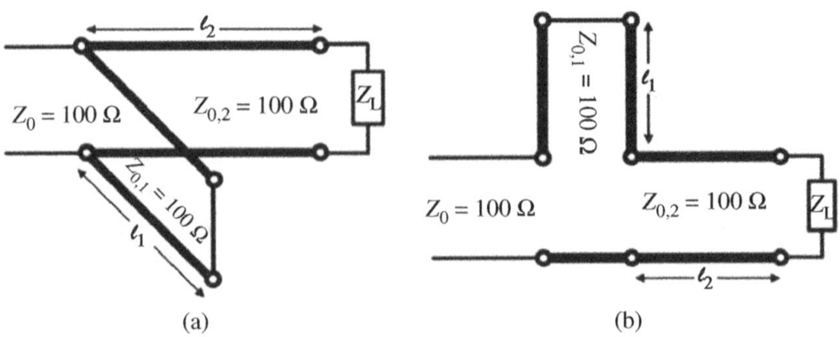

Figure 3.2 Single-stub matching network. (a) Single-stub tuning circuit with parallel short-circuit stub. (b) Single-stub tuning circuit with series short-circuit stub. Source: Reference [2]/with permission of Springer Nature.

of transmission line insertion will need to be moved, prior to the re-matching of the reactance.

The choice of open or short stub depends on several factors. An example is that a short-circuited stub is less prone to leakage of electromagnetic radiation and is easier to realize. On the other hand, an open-circuited stub may be more practical for certain types of transmission lines. A suitable example is microstrip lines, where one would have to drill the insulating substrate to short-circuit the two conductors of the line. When the matching network is based on insertion of a parallel stub, it is more convenient to work with admittances, rather than impedances in Figure 3.3.

For impedance matching:

$$Y_A = Y_{stub} + Y(d_{stub}) = Y_o = \frac{1}{Z_o} \tag{3.1}$$

where Y_{stub} is the input admittance of the stub line and $Y(d_{stub})$ is the line admittance at location d_{stub} before the stub is applied. To complete the design, an appropriate location for the stub needs to be set. Note that the input admittance of a stub is always imaginary (inductance if negative, or capacitance if positive),

$$Y_{stub} = jB_{stub} \tag{3.2}$$

A stub should be placed at a location where the line admittance has real part equal to Y_0,

$$Y(d_{stub}) = Y_0 + jB(d_{stub}) \tag{3.3}$$

For matching, a conjugate matching needs to be found,

$$B_{stub} = -B(d_{stub}) \tag{3.4}$$

An example of simple single-stub matching for a two-port MIMO antenna in portable mobile terminals at 2.4 GHz is presented in [4]. The MIMO antenna is printed on the top layer of an FR4 substrate with a thickness of 1.66 mm, a dielectric constant of 4.4, and a loss tangent of 0.02, as shown in Figure 3.4a. A good isolation is obtained for the MIMO antenna pair at about 35 dB when two

Figure 3.3 Single-stub impedance matching with admittances. Source: Adapted from Ref. [3].

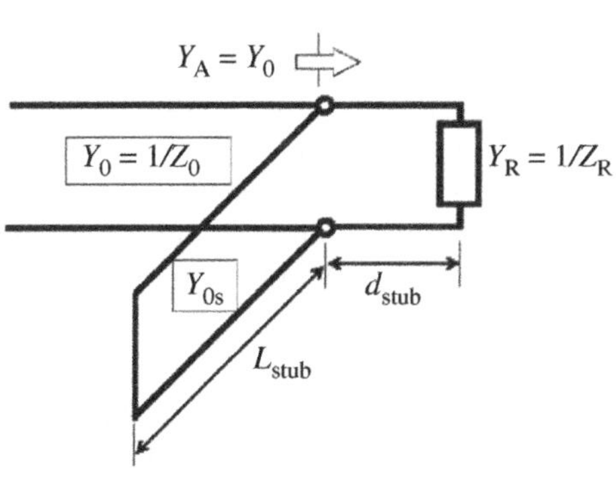

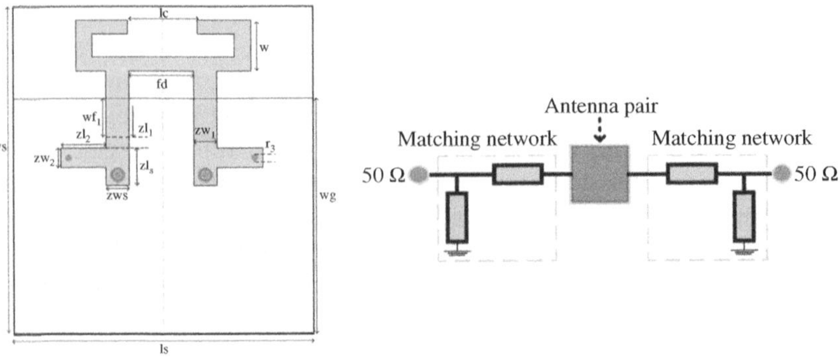

(a) (b)

Figure 3.4 (a) Geometry of the MIMO antenna pair. (b) Sketch of the matching networks. Source: Reproduced with permission from Ref. [4]/IEEE.

matching networks are implemented, as shown in Figure 3.4b. The narrowband matching network is designed from 2.4 to 2.5 GHz.

3.1.2 Double-Stub Impedance Matching

The matching technique presented in Section 3.1.1 requires that a reactive element or stub be placed at a precise distance from the load. This point will shift with the variation of load impedance. Thus, it may not be feasible to match the load using a single reactive element. Another possible technique to match the circuit is by employing two stubs with a fixed separation between them. This device can be inserted at a convenient point before the load. The impedance is matched by adjusting the length of the two stubs. As will be seen later in this section, the separation between the two stubs limits the range of load impedance that can be matched with a given double-stub tuner. In a double-stub matching network, a tuning stub is placed at the location of the load and another one is placed at a distance of $\lambda_g/8$ from the first stub toward the source. Then, the double-stub tuning circuit with parallel short-circuit stubs design is shown in Figure 3.5.

Given that l_1 and l_2 are the lengths for two stubs, as shown in Figure 3.6. The first stub is located at a distance l from the load, $Z_L = R_L + jX_L$ ohms. Separation between the two stubs is d, and the characteristics impedance of every transmission line is Z_0. In double-stub matching, the load impedance Z_L is transformed to the normalized admittance at the location of the first stub. Since the stub is connected in parallel, its normalized susceptance is added to that and the resulting normalized admittance is transferred to the location of the second stub. The matching conditions at this point require that the real part of this normalized

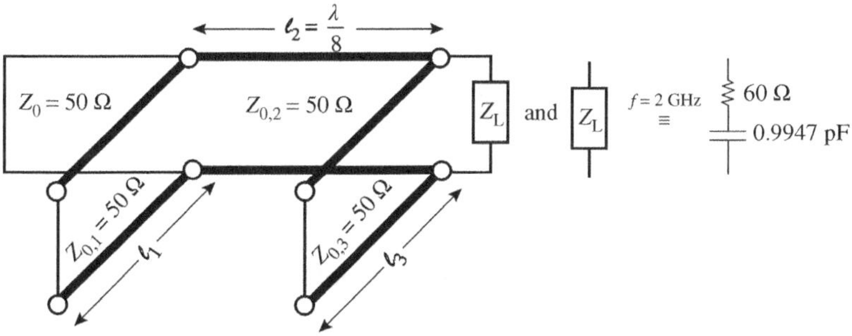

Figure 3.5 Double-stub tuning circuit with parallel short-circuit stubs. Source: Reference [2]/with permission of Springer Nature.

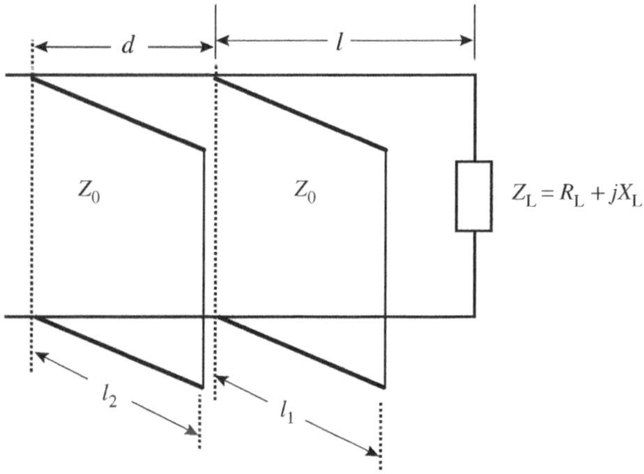

Figure 3.6 Double-stub matching network. Source: Reference [5]/with permission of John Wiley & Sons.

admittance be equal to unity while its imaginary part is canceled by a conjugate susceptance of the second stub.

Two possible susceptances of the first stub, which can match the load are determined using in [5]:

$$\overline{B}_1 = \cot \beta d \left[1 - \overline{B} \tan \beta d \pm \sqrt{\overline{G} \sec^2 \beta d - (\overline{G} \tan \beta d)^2} \right] \tag{3.5}$$

where β is the propagation constant over the line.

The normalized susceptance of the second stub is then calculated as

$$\overline{B}_2 = \frac{\overline{G}^2 \tan \beta d - (\overline{B} + \overline{B}_1 + \tan \beta d)[1 - (\overline{B} + \overline{B}_1) \tan \beta d]}{(\overline{G} \tan \beta d)^2 + [1 - (\overline{B} + \overline{B}_1) \tan \beta d]^2} \tag{3.6}$$

Once the susceptance of a stub is known, its short-circuit length can be determined as follows:

$$l_1 = \frac{1}{\beta}\cot^{-1}(-\overline{B}_1) \tag{3.7}$$

$$l_2 = \frac{1}{\beta}\cot^{-1}(-\overline{B}_2) \tag{3.8}$$

3.1.3 Matching Networks Using Lumped Elements

In this section, L-section matching networks using discrete components are presented. Such matching networks feature simplicity as they consist of only two lumped components: a shunt susceptance, jB, and series reactance, jX. There are two basic configurations of an L-section matching network, depending on the location of the shunt element;

- Τψπε 1: ωηεν $R_L > Z_S'$ ανδ
- Τψπε 2: ωηεν $R_L < Z_S'$.

The shunt element, jB, is placed in parallel with the larger of Z_s or Z_L. As the load and source will be complex, a generalized solution for a complex $Z_L = R_L + jX_L$ and Z_S is shown in Figure 3.7. First, only the resistive parts of Z_L and Z_S are considered, prior to conjugate matching the reactive parts using resulting matching components X and B.

For type 1 matching, the total impedance of the L-section and Z_L should be matched to be equal to Z_s as shown in Figure 3.7a,

$$Z_s = jX + \frac{1}{jB + 1/(R_L + jX_L)} \tag{3.9}$$

Rearranging and separating into real and imaginary parts then provide:

$$B(XR_L - X_L Z_s) = R_L - Z_s \tag{3.10}$$

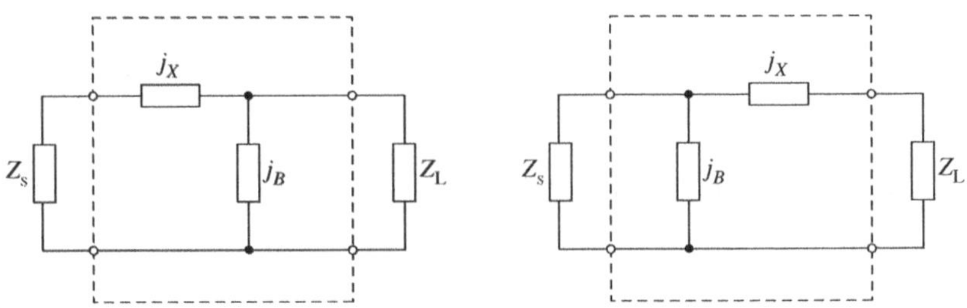

(a) (b)

Figure 3.7 Generalized L-section matching networks. (a) Type 1. (b) Type 2. Source: Adapted from Ref. [6].

$$X(1 - BX_L) = BZ_sR_L - X_L \tag{3.11}$$

Solving for X from simultaneous equations (3.10) and (3.11) and substituting X in Eq. (3.12) for B, the following can be obtained:

$$B = \frac{X_L \pm \sqrt{R_L/Z_s}\sqrt{R_L^2 + X_L^2 - Z_sR_L}}{R_L^2 + X_L^2} \tag{3.12}$$

where positive X represents capacitance and negative values represent inductance.

Since $R_L > Z_s$, argument of the second root is always positive, the series reactance can be found as

$$X = \frac{1}{B} + \frac{X_LZ_0}{R_L} - \frac{Z_0}{BR_L} \tag{3.13}$$

where positive X represents capacitance and negative values represent inductance. Note that two solutions for B are possible, either positive or negative.

For type 2 matching as shown in Figure 3.7b, the total impedance of L-section plus Z_L should be equal to $1/Z_s$,

$$\frac{1}{Z_s} = jB + \frac{1}{R_L + j(X + jX_L)} \tag{3.14}$$

Rearranging and separating into real and imaginary parts then result in Eq. (3.14)

$$BZ_s(X + X_L) = Z_s - R_L \tag{3.15}$$

$$(X + X_L) = BZ_sR_L \tag{3.16}$$

Solving for X and B from simultaneous equations (3.15) and (3.16), the following equations are then obtained

$$X = \pm\sqrt{R_L(Z_s - R_L)} - X_L \tag{3.17}$$

$$B = \pm\frac{\sqrt{(Z_s - R_L)/R_L}}{Z_s} \tag{3.18}$$

Since $R_L < Z_s$, the argument of the square roots is always positive, and again, two solutions for X and B are possible, either positive or negative.

3.1.3.1 Resistive L-Section Matching Circuits

A signal generator with internal resistance R_s that feeds a load resistance R_L is considered, as illustrated in Figure 3.8. Since the source resistance is different from the load, a part of the signal is reflected. Assuming that R_s is larger than R_L and there can be a resistive L-section introduced between the two. Furthermore, voltages at its input and output ports are assumed to be V_{in} and V_{out}, respectively. If this circuit is matched at both its ports, the following two conditions need to be fulfilled:

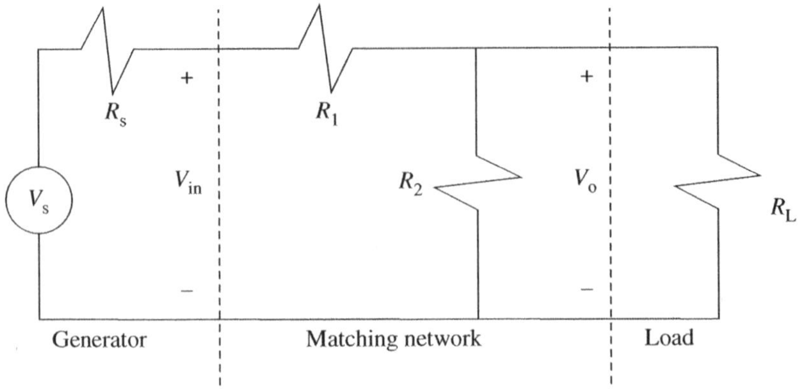

Figure 3.8 Resistive L-section matching circuit. Source: Adapted from Ref. [6].

a) With R_L connected, resistance looking into the input port must be R_s.
b) With R_s terminating the input port, resistance looking into the output port must be R_L.

From the first condition,

$$R_s = R_1 + \frac{R_2 R_L}{R_2 + R_L} = \frac{R_1 R_2 + R_1 R_L + R_2 R_L}{R_2 + R_L} \tag{3.19}$$

and from the second,

$$R_L = \frac{R_2(R_1 + R_s)}{R_2 + R_1 + R_s} = \frac{R_1 R_2 + R_2 R_s}{R_2 + R_1 + R_s} \tag{3.20}$$

Equations (3.19) and (3.20) can be solved for R_1 and R_2 as follows:

$$R_1 = \sqrt{R_s(R_s - R_L)} \tag{3.21}$$

$$R_2 = \sqrt{\frac{R_L^2 R_s}{R_s - R_L}} \tag{3.22}$$

Note that R_1 and R_2 are real only when R_s is greater than R_L. If this condition is not satisfied (i.e., $R_s < R_L$), the circuit shown in Figure 3.8 will require modification. Flipping the L-section the other way around (i.e., R_1 connected in series with R_L and R_2 across the input) will match the circuit.

3.1.3.2 Reactive L-Section Matching Circuits

As mentioned before, resistive matching circuits are frequency insensitive but dissipate part of the signal power that affects the signal-to-noise ratio adversely. Here an alternative design using reactive components is considered, resulting in an ideal zero power dissipation, but frequency-dependent matching. Consider the

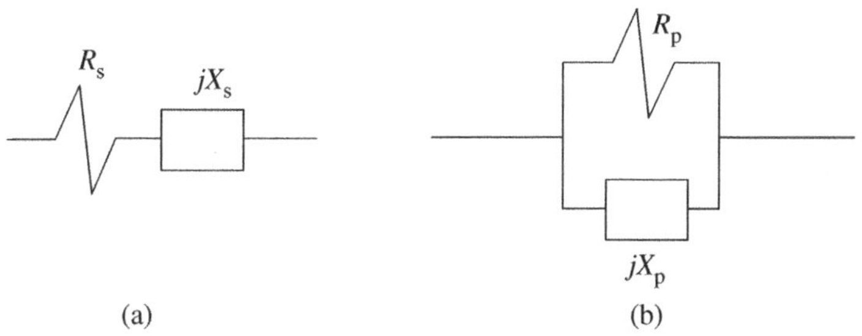

Figure 3.9 (a) Series. (b) parallel connected impedance circuits. Source: Adapted from Ref. [6].

two circuits shown in Figure 3.9. In one of these circuits, resistor R_s is connected in series with a reactance X_s as shown in Figure 3.9a, while in the other, resistor R_p is connected in parallel with a reactance X_p as shown in Figure 3.9b.

If impedance of one circuit is a complex conjugate of the other, then:

$$|R_s + jX_s| = \left|\frac{jX_pR_p}{R_p + jX_p}\right| \tag{3.23}$$

or

$$\sqrt{R_s^2 + X_s^2} = \frac{X_pR_p}{\sqrt{R_p^2 + X_p^2}} \tag{3.24}$$

However, in multiple antenna settings, energy that is coupled to an adjacent antenna when an antenna is operating introduces coupling to adjacent elements, thus causing mutual coupling [7]. In general, this can be overcome by increasing inter-element spacings. On the contrary, insufficient inter-element spacing increases the interactions between multiple antenna elements (assuming identical elements). Required inter-element spacing should be at least half a wavelength to minimize coupling. Moreover, with any specified spacing, different elements, which are excited with different amplitudes and phases for beamsteering purposes consequently result in impedance mismatches due to the changing current distributions.

A simple L-section matching circuit is added to each port to retune the input impedance matching [8]. The network gives a dual-resonant decoupling performance achieving isolation levels of 20 dB across 250 MHz bandwidth at 2.4 GHz. The even and odd mode admittances are equal while simultaneously improving the matching, thereby achieving isolation improvement at the center frequency with limited isolation bandwidth. A T-shaped circuit is used to improve the isolation without considering the effect on input matching and it is visible to both the even and odd modes. Ten lumped components with six different values

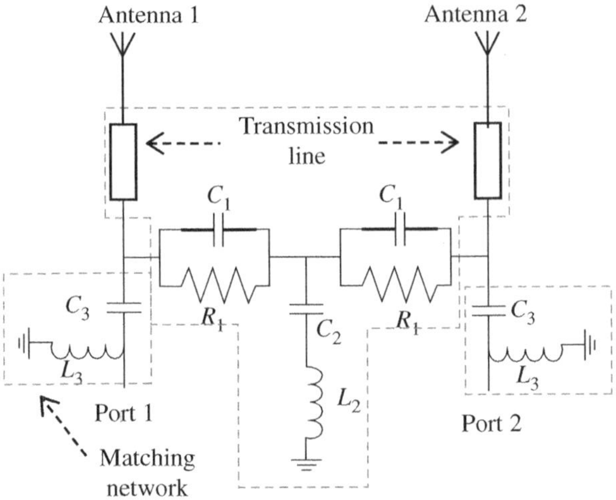

Figure 3.10 L-section matching network to retune the input impedance matching. Source: Reproduced with permission from Ref. [8]/IEEE.

are adopted in this circuit, as shown in Figure 3.10. The use of a single element connected in between the antenna elements affects only the odd mode input admittance and does not affect the even mode impedance. This introduces more design freedom for matching, which results in wideband isolation improvement.

3.2 Passive Feeding Networks

Feeding is the most important part of the antenna to excite the radiating element. There are four basic methods of feeding techniques, which are microstrip line feed, coaxial feeding, aperture coupling, and proximity coupling. Feeding can be done either directly from the feedline to the element (coaxial probe and microstrip line feed) or indirectly/capacitively (aperture coupling and proximity coupling) [9]. Important factors in the choice of feeding methods are the required bandwidth, fabrication requirements, and structural complexity. A variety of planar and non-planar feeds, which will be discussed next include the microstrip line feed, coaxial-probe feed, aperture-coupled feed, proximity-coupled feed, and coplanar waveguide (CPW) feed.

3.2.1 Microstrip Line Feed

The simplest way to feed a microstrip patch is to connect a microstrip line directly to the edge of the patch with both elements located on the same substrate layer, as shown in Figure 3.11. While this eases fabrication, the impedance of the patch at its radiating edge is very high compared to the typical 50 Ω impedance of the

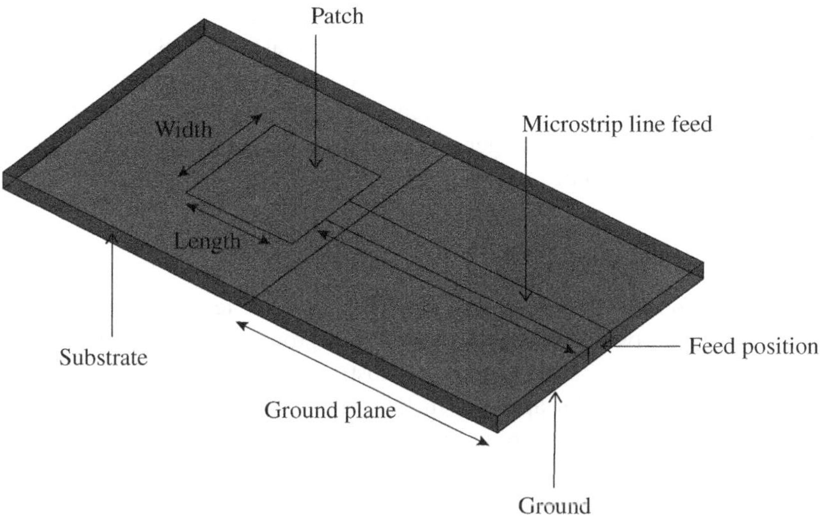

Figure 3.11 Microstrip line feed. Source: Reference [10]/with permission of IEEE.

feed line. The microstrip line can be connected to the patch within an inset in the patch [11]. This modifies matching of the input impedance to the transmission line impedance without additional matching elements by only adjusting the depth of the inset. While microstrip feed is relatively simple to design, fabricate, and implement, it is most effective to be used for single-layered structures, where space requirements are not stringent.

3.2.2 Coaxial Probe Feed

Feeding of antennas can also be done using coaxial probe feed. It typically consists of an inner conductor of a coaxial line, whereas the outer conductor to this line is connected to the antenna's common ground plane, as shown in Figure 3.12. The

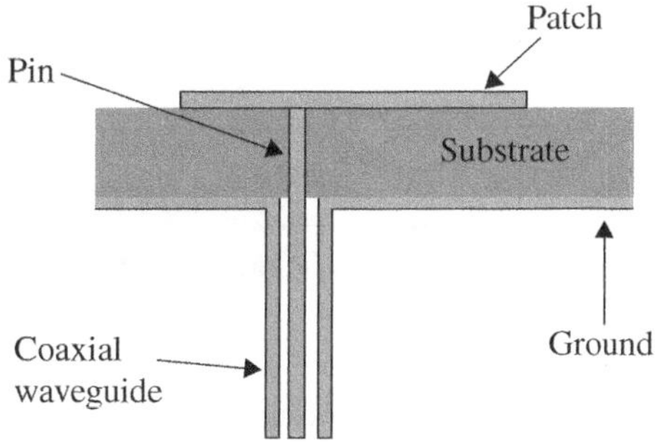

Figure 3.12 A microstrip antenna with a coaxial probe. Source: Reproduced with permission from Ref. [12]/IEEE.

location of the feed point determines the best impedance matching to such coaxial feed, simplifying the design procedure. However, coaxial feeding of a large array or MIMO antenna may require a relatively large number of solder points compared to a typical microstrip array of such antennas, for instance. Besides that, when thicker substrates are needed, longer probes may increase spurious radiation and increase the feed inductance.

3.2.3 Aperture Coupling Feed

The first non-contact feed type, the aperture coupled feed, is illustrated in Figure 3.13. It uses two substrates separated by a ground plane, with power fed onto a microstrip feed line on the lower substrate electromagnetically coupled to the patch through a slot aperture in this ground plane. While the slot can be of any shape or size, matching and bandwidth can be improved by their proper choice. Besides that, the two-layered structure also provides an additional degree of freedom to optimize the feed and radiator more independently. For instance, the substrate for the feed line can be thin and with a high dielectric constant, whereas the substrate for the patch can be thick with a lower dielectric constant. Moreover, due to the shielding effect of the ground plane, radiation from the open end of the feed line does influence the radiation pattern of the patch, improving its polarization purity. Due to the typical symmetry of the coupling aperture, cross-polarization can also be reduced. Finally, this feed-type features improved bandwidth by adjusting the shape and length of the coupling slot, width of the feed line and stub length.

3.2.4 Proximity (Electromagnetically) Coupled Feed

A proximity-coupled feed is shown in Figure 3.14. It consists of two substrate layers, with a microstrip line on the bottom layer and a patch antenna on the top layer. Power fed into the open-ended feed line underneath the patch couples to

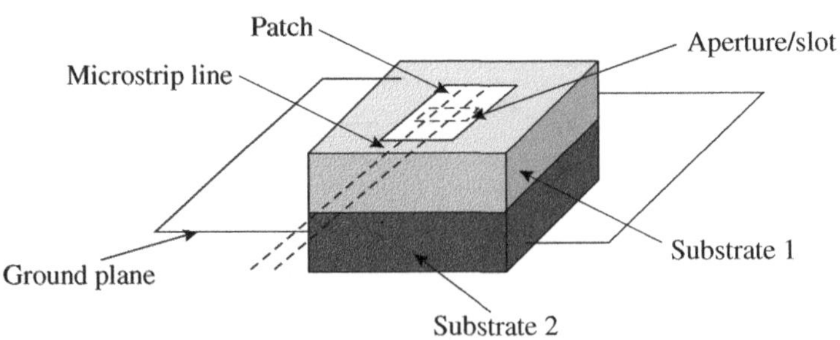

Figure 3.13 Aperture-coupled feed. Source: Reference [13]/CiteSeerX/CC BY 4.0.

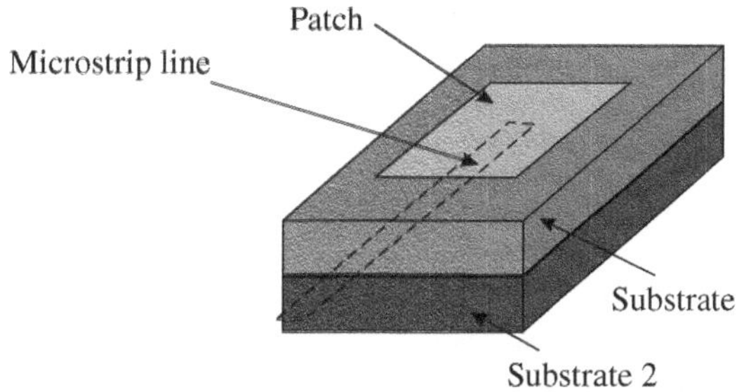

Figure 3.14 Proximity-coupled feed.
Source: Reproduced with permission from Ref. [12]/IEEE.

the patch capacitively. This capacitance/coupling level can typically be optimized for impedance matching of the antenna. To improve bandwidth, the open end of the microstrip line can be terminated in a stub with suitable stub parameters. Besides that, suitable substrate parameters can be selected for the same purpose. For this, a thinner bottom (feed) layer, and a thicker substrate for the radiator can be introduced for a larger bandwidth. As is typical with dual-layered feeding structures, fabrication of such feed is slightly more complex due to the accuracy of alignment between the feed and the patch.

3.2.5 Coplanar Waveguide Feed

Another method to excite patch antennas is by using CPW feeds. In such feeds, the CPW line can be either located on the same (contacting feed) or on the opposite sides of a dielectric substrate (proximity-coupled feed) [14]. For the former, RF power from the CPW feed is directly coupled to the patch antenna, whereas in the latter, it is electromagnetically coupled to the patch antenna from the terminated end of the CPW feed. Termination can either be a modified short circuit or an open circuit as illustrated in Figure 3.15. The short circuit gives rise to excess inductance at the termination and therefore the coupling is termed as inductive. By a similar reasoning, the coupling from an open circuit is termed capacitive. The short-circuit and the open-circuit terminations are located symmetrically above the center of the patch.

The use of CPW feed allows series as well as shunt mounting of components without the use of via holes. Besides, the radiation from the feed structure is negligible because the CPW is excited in the odd mode of the coupling slot line [15]. Due to this mode, the equivalent magnetic currents on both the CPW slots radiate almost out of phase, with negligible contribution to the feed radiation. This CPW feed feature is useful in the design of arrays and MIMO antennas as mutual coupling between adjacent lines is minimized. A comparison of the various types of feed structures is listed in Table 3.1.

An example of a simple and compact MIMO antenna with a coupled feed structure developed for operation in the 0.746–0.787 GHz band is presented

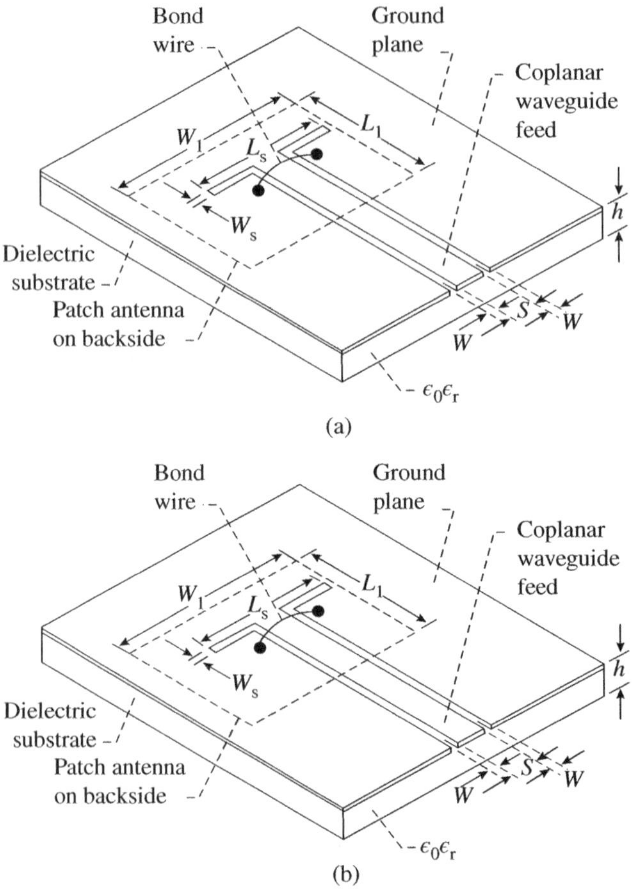

Figure 3.15 Patch antenna electromagnetically coupled CPW: (a) Inductively coupled; (b) capacitively coupled. Source: Adapted from Ref. [14].

in [16]. The antenna uses a Z-shaped capacitive coupled feed strip instead of conventional direct feed structures with the main radiating section to effectively modify the antenna impedance matching and maintain isolation. Two types of antenna arrangements are studied. For type I configuration, the two main radiating elements have a back-to-back arrangement with a feeding point located around the middle portion of the top edge of the ground plane. On the other hand, in the type II arrangement, the two main radiating elements have a face-to-face configuration with a feeding point located around the top corner of the ground plane. The reflection coefficient, S_{11} and isolation characteristics of the antenna with respect to these configurations are shown in Figure 3.16. It is seen that the

Table 3.1 The comparison of various types of feed structures for microstrip patch antennas.

Characteristics	**Inset feed**	**Coaxial-probe**	**Proximity-coupled**	**Aperture feed**	**CPW feed**
Microstrip feed types					
Spurious feed radiation	High	High	High	High	Less
Polarization purity	Poor	Poor	Poor	Excellent	Good
Ease of fabrication	Alignment required	Soldering and drilling needed	Alignment required	Alignment required	Alignment required
Reliability	Better	Can be poor due to soldering	Good	Good	Good

Source: Reference [15]/with permission of Artech House.

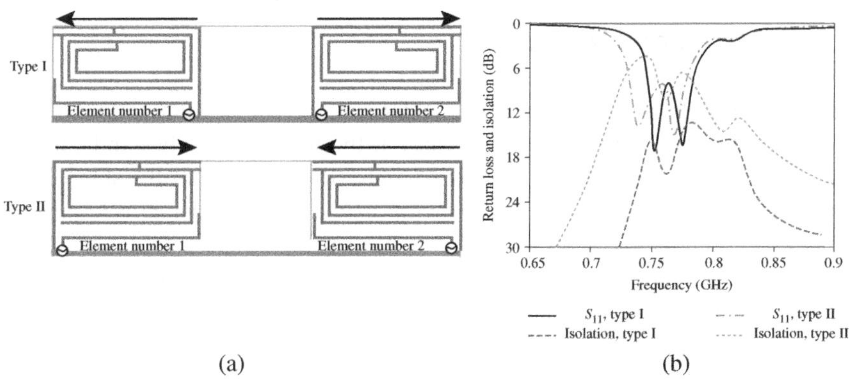

Figure 3.16 Reflection coefficient and isolation characteristics of the antenna with respect to a variation in the antenna element arrangement: (a) types of antenna configuration and (b) reflection coefficient and isolation. Source: Reference [16]/Hindawi Publishing Corporation/CC BY 4.0.

type II configuration showed poorer isolation despite improved bandwidth. The better isolation performance of configuration I is mainly due to the direction of the main radiation from each element, which is placed away from each other. This results in reduced coupling between them than that of type II configuration. The choice of the feeding structure, either using a coupled feed or a direct feed, can affect the level of isolation and matching, as shown in Figure 3.17. A coupled feed can provide better isolation and bandwidth compared to a direct feed structure.

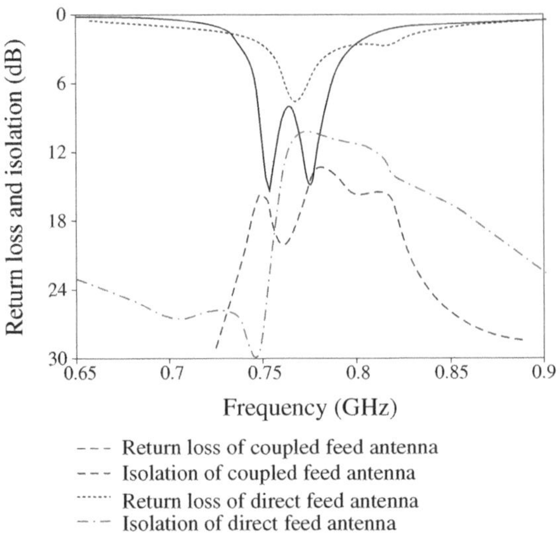

Figure 3.17 Return loss and isolation characteristics of the main folded monopole element with either a coupled feed structure or a direct feed structure. Source: Reference [16]/Hindawi Publishing Corporation/CC BY 4.0.

3.3 Active Feed Networks

Besides passive feeding networks used to feed antennas, feed networks can also be integrated into communication devices in the form of switches and variable impedance tuners. In the microwave bands, they are typically standalone, off-the-shelf components, which are made using semiconductor technologies. Switching components generally come in the form of single-pole double-throw (SPDT), single-pole triple-throw (SP3T), and single-pole quadruple-throw (SPQT), whereas matching circuits can be implemented using capacitance and inductance. Antenna tuning via active feeding networks enables the use of bandwidth more efficiently for potential carrier aggregation while keeping the power consumption low. In MIMO systems, adjustment of resonant frequency for matching and maximum power transfer between ports becomes a necessity. This is to readjust impedance matching when RF performance from the front-end changes under varying conditions such as frequency band, power level, impedance, etc., which changes this parameter. When implementing such active feeding networks, one should pay attention to the potential creation of harmonics and the level of isolation between each port due to additional switching losses. Nonetheless, the progress in materials, design, and manufacturing technologies is gradually improving these factors, besides their costs. In communication devices, two main different types of active feeding methods include open-loop tuning and closed-loop tuning, and they will be explained as follows.

3.3.1 Open-Loop Tuning

The first option to actively feed an antenna is by using an open-loop tuning method, as shown in Figure 3.18. Essentially, in this method, a tuning device can be connected to an integrated circuit (IC) operating at a digital baseband. This is a non-feedback solution controlled by a time-based trigger. The digital baseband IC has a look-up table, which consists of timed band switching information – the switch and variable capacitance are activated by the trigger when the frequency band is changed. The two main methods to achieve active antenna tuning are impedance tuning, illustrated in Figure 3.19 and antenna aperture tuning, shown in Figure 3.20. Despite their similarities, antenna impedance tuning includes a matching circuit, which changes the matching impedance (and operation) of the antenna. These components can either be implemented in a series or shunt configurations.

The antenna aperture tuning is usually achieved by a switch which must have a low loss to avoid degradation of antenna-radiating efficiency. The switches can be of the series type or the shunt type. Shunt type is more commonly used than series type because it has a lower ohmic loss. As a result, great radiation efficiency is attainable. Aperture tuning helps boost isotropic sensitivity and

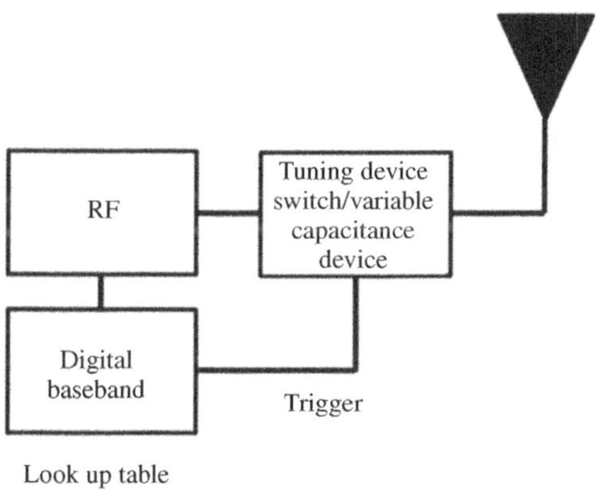

Figure 3.18 Open-loop tuning. Source: Reference [17]/with permission of Artech House.

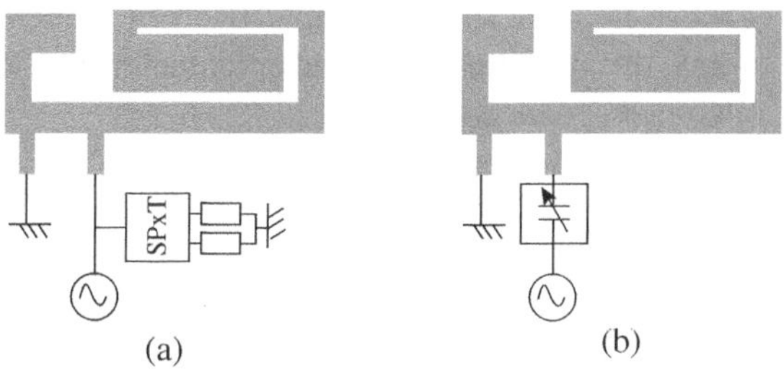

Figure 3.19 Open loop impedance method using (a) switching method and (b) variable capacitance method. Source: Reference [17]/with permission of Artech House.

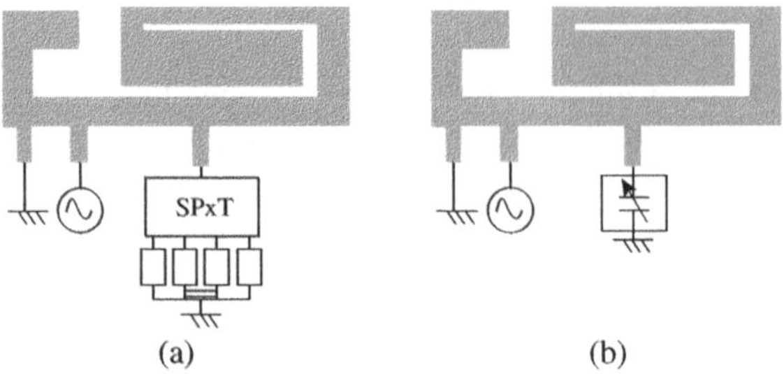

Figure 3.20 Open loop aperture tuning (a) switching method and (b) variable capacitance method. Source: Reference [17]/with permission of Artech House.

radiated power. This is achieved by optimization of efficiencies for transmit and receive frequencies. It requires enough electronic circuitry for external bypassing and does not require filtering. By using aperture tuning, electrical length of antenna ground leg is adjusted to shift resonance of operating band. Meanwhile, the antenna impedance tuning can broaden the antenna bandwidth so that uniform insertion loss and return loss can be achieved at multiple frequencies for both transmit and receive directions. Impedance tuning helps boost power transfer between antenna and RFFE (Radio Frequency Front End) and requires external passive components for bypassing. Impedance tuning improves reflection coefficient and system efficiency accordingly, but it does not improve antenna radiation efficiency as antenna structure has not been changed. Hence, the antenna aperture tuning is needed to change the radiation principle of the antenna without changing the antenna structure.

Despite its simplicity in implementation, the tuning range may be limited, depending on the complexity of the matching circuits. Switches and the antenna form an equivalent RLC circuit to improve matching and system efficiency but have minimal effect in improving the radiation efficiency. This is due to the unchanged fundamental antenna structure. On the other hand, antenna aperture tuning is similar to antenna impedance tuning, except that its main aim is to enable the modification of the electrical length by adjusting the GND location/leg instead of the input impedance to the antenna. It is usually achieved by a low-loss switch, which can be connected in series or shunt forms to avoid degrading radiation efficiency. Shunt-type switches are generally more widely used as they feature lower ohmic losses. Aperture tuning typically enables switching between frequencies either capacitively or inductively, and similar to impedance tuning, the aperture switch and antenna form an RLC circuit to adjust the natural antenna resonance(s).

3.3.2 Closed-Loop Tuning

Closed-loop tuning is shown in Figures 3.21 and 3.22 for impedance and aperture tuning needs, respectively, and offers feedback that automatically adjusts the band

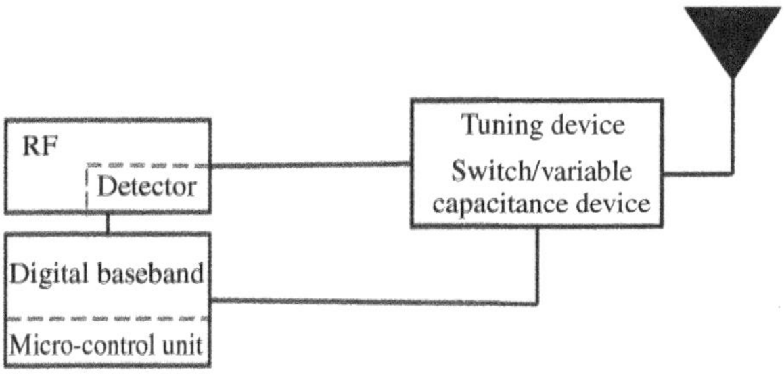

Figure 3.21 Closed loop impedance tuning [17].

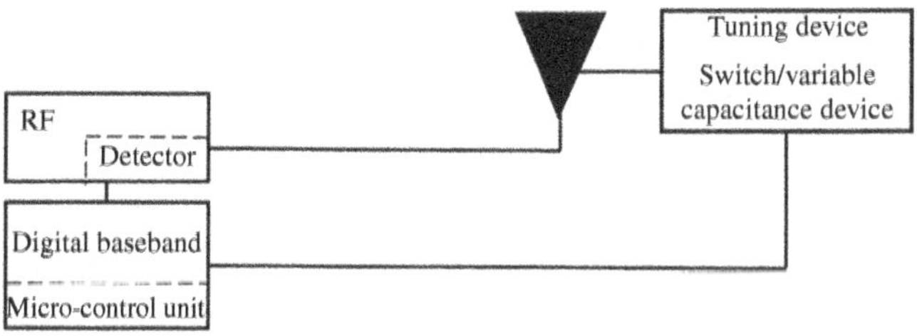

Figure 3.22 Closed loop aperture tuning. Source: Reference [17]/with permission of Artech House.

and uses case variation. This includes frequency detuning caused by external loading effects such as coupling to the users' hands, heads, and other use cases. It relies on the feedback from impedance mismatch sensors or power detectors and includes the need for an algorithm. In recent years, detectors have been implemented in radio frequency (RF) integration circuits (RFIC) and microcontroller units are implemented in digital baseband ICs. Such detectors (e.g., power detector or dual couplers) monitors the change in reflection coefficients. Then, using the same principles explained previously for impedance and aperture tuning, switches and components are activated to provide correction to matching or electrical length to improve the power transfer between the RF front end and the antenna.

Active feed networks are becoming more important in MIMO antenna systems. They provide more flexibility in adjusting the feed power and phases to different feed sub-networks so that more advanced features can be integrated into multiple antenna systems. Most of these active feed networks employ switching-based techniques to expand the working frequency range and improve MIMO operation [18]. They include various reconfiguration features (frequency, radiation pattern, and polarization), which will be explained in more detail in Chapter 5. This sub-section will briefly explain several typical methods used to perform such switching and tuning such as PIN diodes, varactor diodes, and RF microelectromechanical systems (MEMS). The choice of a specific technique depends on the requirements of the application, complexity of the required features, and operating frequency. For example, the switching speed of RF MEMS switches ranges from 1 to 200 μs [19]. On the other hand, PIN diodes offer fast switching with switching rates ranging from 1 to 100 ns [20] and wide impedance bandwidth, depending on the method used. Besides that, combining an antenna with a varactor diode enables the variation of capacitance by changing the bias voltage, which usually results in a large tuning range. This section will explain each of these switchable active feed network options in MIMO antenna systems.

3.3.3 PIN Diode Switches

The PIN diode consists of two semiconductor junctions, each with a p-type and an n-type, separated by a resistive intrinsic region for increased switch speed and smaller package size. For the ON-state operation, the bias current needed is typically in the 10-mA range. Besides, the PIN diode also behaves as a variable resistor as it can be current-controlled. The circuit model is modeled based on resistance (*R*), inductance (*L*), and capacitance (*C*), as shown in Figure 3.23. The series *RL* circuit represents a PIN diode's equivalent circuit in the ON state, in which current can flow to the radiating regions due to its low *R* value. An *RLC* circuit with *L* connected in series is used to simulate the PIN diode in its OFF state, with *C* and *R* connected in parallel. Current is expected to be reduced (ideally blocked) as the impedance of the *RC* part increases in value. As a result, the forward biasing of the PIN diode creates an ideal short circuit, which results in an ON state, whereas the reverse biasing creates an ideal open circuit, which results in an OFF state.

The inductor (L_b) performs as a choke in both ON and OFF modes while maintaining a constant value. In general, the radiation pattern in an antenna can potentially be distorted by the DC biasing circuitry [22]. Besides that, feed network reconfiguration using such PIN-based switching methods may be limited in isolation and matching [23]. Nonetheless, due to their fast-switching times and relatively high current-handling capacity, PIN diodes are packaged in a far smaller form factor compared to mechanical RF switches. An example of the use

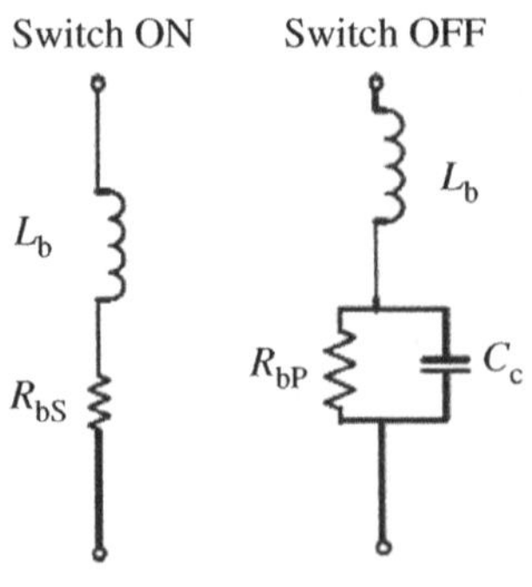

Figure 3.23 RF PIN Diode circuit for ON/OFF states. Source: Reference [21]/with permission of Springer Nature.

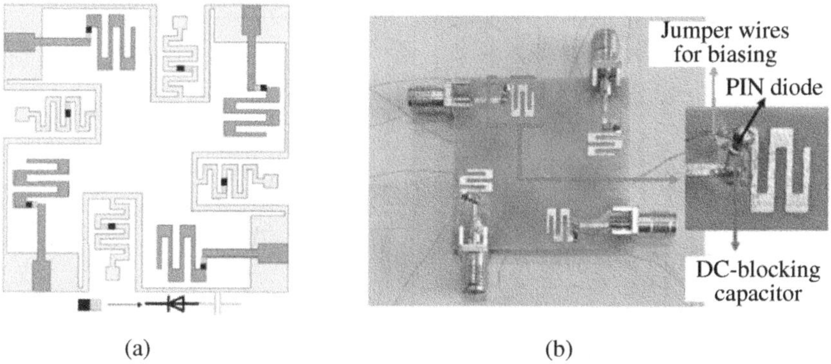

Figure 3.24 PIN diode feedline reconfiguration: (a) layout and (b) fabricated prototype (top view). Source: Reference [24]/with permission of The Institution of Engineering and Technology.

of such switching is demonstrated in a four-element MIMO antenna as shown in Figure 3.24 [24]. Its main purpose is to decouple the feed network to obtain high isolation. Besides that, the frequency and isolation of the structure can be modified due to the additional resonators with PIN diodes and decoupling networks. Capacitors are then located in the resonators to prevent the RF source from leaking into the DC bias voltage.

To calculate the "*R*" value, the biasing circuit shown in Figure 3.25a applies a DC voltage of +9 V making the PIN diode operate in its ON state. It can be observed from the multimeter reading that the DC forward current flowing in the circuit is 99.4 mA, which is close to the desired reading of 100 mA. The used "*R*" value in this circuit was 47 Ω. The total DC resistance of the circuit was 62.12 Ω (47 + 7.56 + 7.56) with a diode forward current of 99.4 mA. During measurements of the antenna, the biasing voltage and resistance "*R*" are kept as +9 V and 47 Ω, respectively, during the PIN diode ON state. Figure 3.25b illustrates the switching

Figure 3.25 (a) Biasing circuit of the antenna and (b) switching circuits for PIN diode ON and OFF states. Source: Reference [24]/with permission of The Institution of Engineering and Technology.

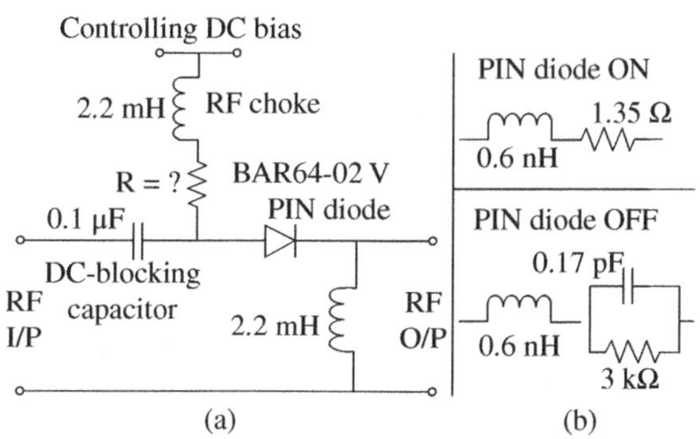

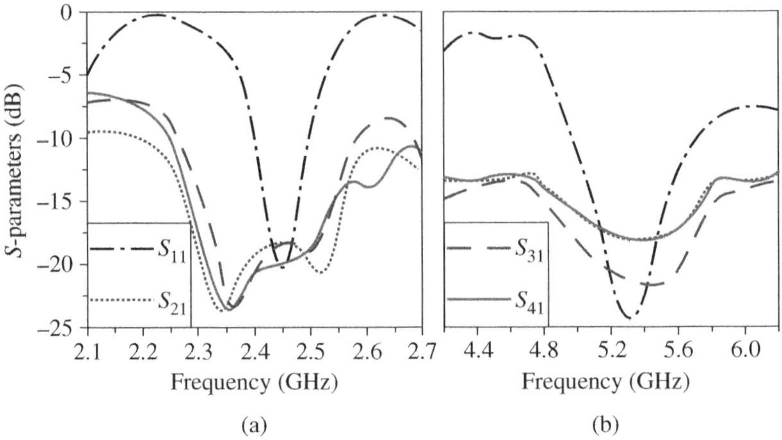

Figure 3.26 Measured S_{11} in the (a) ON state, (b) OFF state. Source: Reference [24]/with permission of The Institution of Engineering and Technology.

between the ON and OFF states. When turned ON, the PIN diode has an extremely low resistance (1.35 Ω). The reverse parallel resistance and diode capacitance values are fixed at 3 kΩ and 0.17 pF, respectively, to emulate the PIN diode being in the OFF state. The same result can be obtained by biasing the DC bias terminal with 0 V. The PIN diode (BAR64-02V) is chosen to avoid any discrepancy between the measured and calculated antenna characteristics.

The antenna system consists of eight PIN diodes connected to eight DC-blocking capacitors. Jumper wires are used to bias the PIN diodes. In the ON state, the device serves as a four-element antenna system operating in the 2.4–2.5 GHz band, whereas in the OFF state, the four-element antenna system becomes active in the 5.1–5.8 GHz band. Figure 3.26 shows the measured S-parameters in the ON and OFF states, respectively. All ECC values in both antenna-operating bands are below 0.2, with a minimum gain of 1.5 dBi and efficiency of at least 70%. The losses from the PIN diodes and capacitors mostly contributed to this decrease in efficiency.

3.3.4 Varactor-Based Switching

There exist two main types of implementations for varactors; they can either be realized using CMOS/P-N junction method or MEMS varactors [25]. The former typically adds to the complexity of circuits, as it requires the use of DC biasing, besides the limited dynamic range and linearity of the varactor [26]. On the other hand, MEMS-based varactors typically feature low-loss and high-Q factors as their main advantages, which will be explained further in the next sub-section.

Table 3.2 Comparisons of the characteristics of CMOS and MEMS varactors.

CMOS varactor	MEMS varactor
Possible current leakages	Small notable leakages
Continuous downscaling decreases the tuning range	Tuning ranges from 5 to 50
Q-factor ranges from 30 to 60	Q-factor ranges from 200 to 300
Relatively more lossy	Relatively low in losses

A comparison between both types of varactors is presented in Table 3.2. An example of the use of tunable varactor diodes is by electrically modifying the size of an antenna element using capacitance variation. This is to enable the increase or decrease of mutual coupling between an active element and a similarly (electrically) sized parasitic element, given that they are designed with suitable spacing. An example of such a concept is implemented to tune the parasitic directors of a planar Yagi Uda antenna to result in a main beam reconfiguration [27]. Two tunable varactors were used on each parasitic director and by varying their capacitance values, the overall antenna can be scanned between $\pm 40°$. Due to the relative size change in comparison to the Yagi-driven element, the beam can either be tuned between $\pm 14.4°$ and $\pm 40°$ due to the coupling with the parasitic element, besides the regular broadside radiation. More methods for radiation pattern reconfigurable antennas will be presented in Chapter 5.

3.3.5 RF MEMS-Based Switching

RF MEMS switches consist of a movable thin metal bridge suspended across the central conductor. Switching speeds of antennas using RF MEMS range from 1 to 200 μs. RF MEMS switches are characterized by low insertion losses, lightweight, compact size, excellent isolation, good linearity, tunability, and power-handling capability [25]. MEMS switches do not pose any matching issues as in their native state, they function like a 50 Ω transmission line. An example of a four-port antenna implementing such an RF MEMS-switching structure is shown in Figure 3.27 [28]. Four slot antennas denoted as $A_1 - A_4$ are first introduced on the four corners of the substrate. Antennas A_2, A_4, and A_1, A_3 are symmetric to one another based on the overall structure's vertical center. Two MEMS switches (denoted as S_1 and S_2) are installed across two slot antennas (A_1 and A_2) with their DC control. The switch control lines are integrated with resistors to minimize the RF effects of the DC control lines. On the other hand, MEMS switches S_3 and S_4 are introduced on the decoupling network. By turning switches S_1 and S_2 ON, the pair of symmetric reconfigurable slot antennas

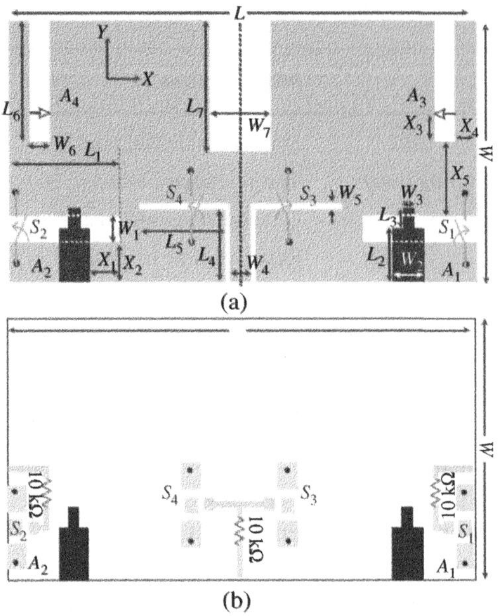

Figure 3.27 Dimensions of the reconfigurable four-port MIMO antenna: (a) top view and (b) bottom view with switches and DC bias lines. Source: Reproduced with permission from Ref. [28]/IEEE.

(A_1 and A_2) can be changed from a $\lambda/4$ open-end slot in the 2.4 GHz band to $\lambda/2$ shorted-end slot in the 5 GHz band. As a result, a two-port antenna operating from 2.4 to 2.5 GHz and a two-port antenna operating from 4.95 to 5.725 GHz are generated when $S_{1-2-3-4}$ are "OFF." Conversely, a four-port antenna working between 4.9 and 5.725 GHz is formed when all switches are turned "ON."

It is also important to ensure that initial simulations during the design stage are performed efficiently. This involves the use of ideal metallic strips to emulate the ON state in such antennas, and the absence of it for the OFF state. Despite this, the final antenna should be integrated and simulated with realistic values of commercially available components. Figure 3.28 compares the reflection coefficient for an ideal copper strip-based hardwired switch and an RF MEMS switch. Due to the difference in parasitic loading from each solution, the slot length for the hardwire solution has been optimized to $L_1 = 17$ mm, whereas the same parameter for the MEMS switch ($L_1 = 12$ mm) to enable antenna resonance at 2.45 and 5.5 GHz. From the surface currents, the E-field is observed to be more concentrated near the opposite end of the $\lambda/4$ open-end slot when one end of the slot is excited. Despite its satisfactory performance for compact devices, efficiency in the antenna efficiency is also observed to be slightly degraded.

Table 3.3 summarizes and compares the different switching components suitable for reconfigurable MIMO antennas. It is seen that PIN diodes or varactors are simple to design and are substantially lower in terms of cost. On the other hand, MEMS switches featured benefits such as good linearity and isolation, low losses,

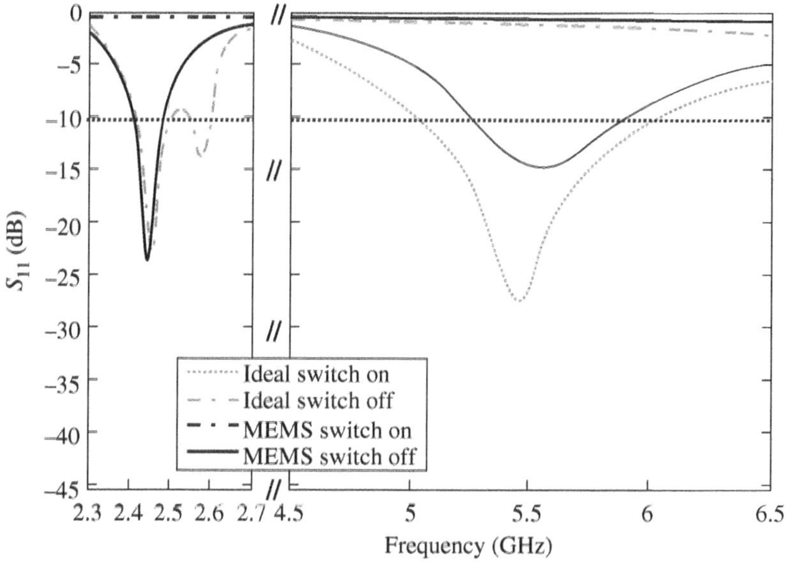

Figure 3.28 Comparison of simulated S_{11} of the single antenna with ideal and MEMS switches. Source: Reproduced with permission from Ref. [28]/IEEE.

Table 3.3 A comparison of different switching components.

Active elements	Advantages	Disadvantages
PIN-diodes	• Low cost • Ease of installation • High-speed switching	• High DC bias in ON-state • Slow response time
Varactors	• Small current requirements • Constant tuning	• Limited dynamic range • Requires advanced bias circuits
RF MEMS	• Wide impedance bandwidth • High linearity • Low losses	• High control voltage • Shorter life span • Slower switching speed

and relatively wider impedance bandwidth. However, such a solution requires large control voltages, is relatively slower in switching speed, has a shorter life cycle, and is relatively costlier.

3.4 Beamforming Networks

Beamforming methods are increasingly being included in MIMO antenna systems. Beamforming is a spatial filtering operation typically using an array of radiators to capture or radiate energy in a specific direction over its aperture. The improvement achieved over omnidirectional transmission/reception is in the transmit/receive gain [29]. In general, beamforming techniques can be either analog, digital, or hybrid. In its most conventional form, beams from antennas can be focused in specific directions from a wave interaction with an array of antennas, where waves can either overlap constructively or destructively to produce beamforming. This same concept has been used to design beamforming feed networks.

3.4.1 Analog, Digital, and Hybrid Beamforming Networks

One main difference between analog and digital beamforming (DBF) is how and where the amplitude and phase variation are applied. In analog beamforming (transmit mode), the amplitude and phase variation are applied after modulation is performed. This signal is then split using power dividers and channeled through the beamformer to change amplitude and phase in each path before arriving at the antenna. As an alternative, analog beamforming can also be done by having a fixed amplitude to each path, and spatial variation of antenna elements can be introduced to modify their phases. As a receiver, analog beamforming is performed via the application of complex weighting (consisting of amplitude and phase) to each signal received at each antenna. These signals are then combined into a single output for a single channel. The block diagrams for the analog beamforming transmitter and receiver are shown in Figure 3.29. This solution is typically chosen when the system requires low power consumption and cost, and when long-range signal coverage is needed.

On the other hand, DBF performs scaling of amplitude, phase, and summing of signals digitally. DBF receivers typically consist of RF downconverters, analog-to-digital converters (ADCs), complex weight multipliers, and summing operations. The RF downconverter first converts RF signals into lower intermediate signals (IF) by mixing signals from a local oscillator (LO) and applying the appropriate filtering (bandpass and lowpass). The IF signals are then digitized using ADCs before being passed to the digital downconverter (DDC) for further downconversion into baseband I/Q components. Complex weights are then applied to these baseband signals and summation of these signals then produces desired radiation patterns. This process is summarized in the block diagram illustrated in Figure 3.30. Conversely, on the transmitter side, DBF is produced by applying amplitude and phase variation to the digital signal after the ADC and DDC conversion at the end of the transmitter. In summary, both DBF receivers and transmitters operate based on the transmit/receive weighting at baseband. The most traditional baseband DBF requires one distinct RF chain per antenna [32], which can easily become complex and incur high power consumption in large arrays.

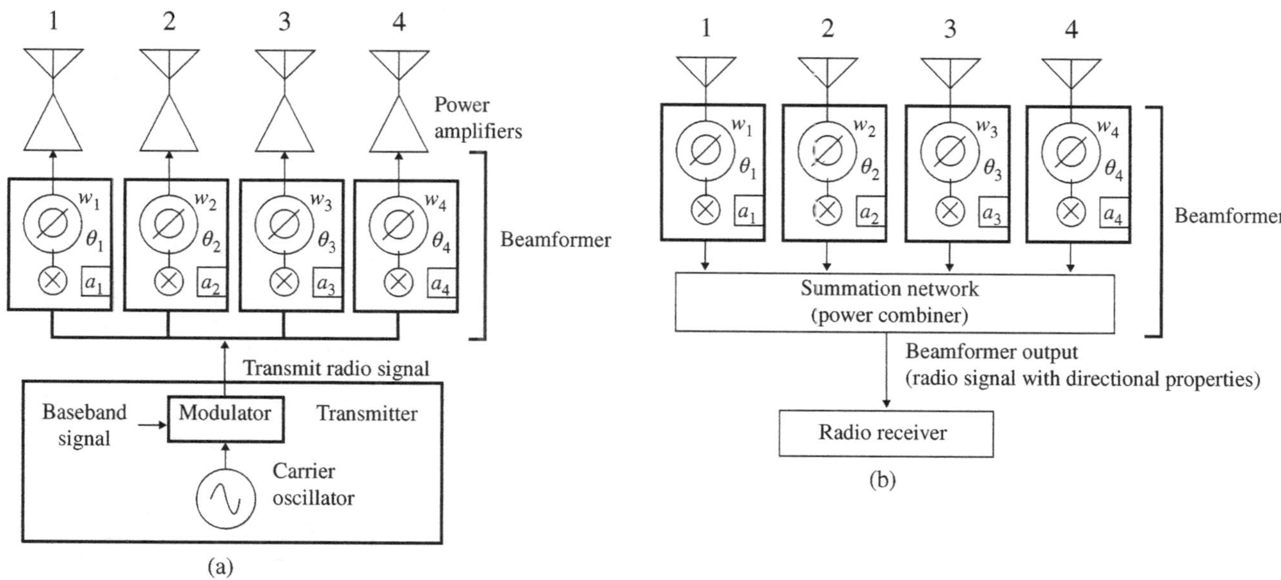

Figure 3.29 Analog beamforming (a) transmitter and (b) receiver. Source: Reproduced with permission from Ref. [30]/IEEE.

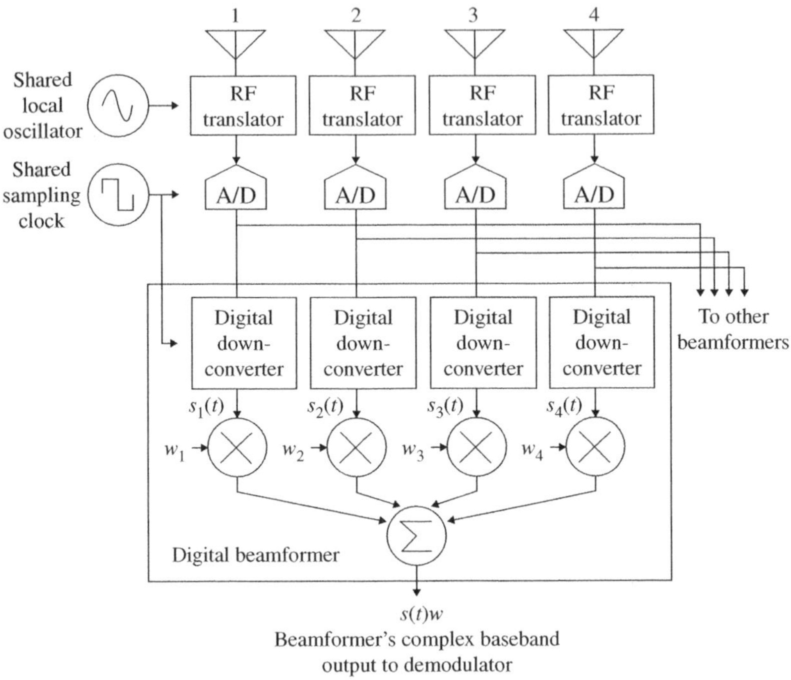

Figure 3.30 Digital beamforming receiver [31].

Therefore, in massive MIMO systems, hybrid beamforming operating in the baseband and analog domains can be considered, including different architectures aiming to reduce the number of RF chains [29]. Hybrid beamforming is a combination of analog and DBF. Hybrid beamforming divides signal processing into analog and digital domains to enable multi-stream (and possibly multi-user) communication. Its architecture consists of a lower number of RF chains (relative to DBF), which enables multi-stream digital baseband processing followed by analog (baseband or RF) processing to realize antenna beamforming gain. A drawback of this architecture is that the lower number of RF chains also reduces the number of simultaneous beams that can be generated. This then limits its scanning ability, limiting angular resolution, reducing beamforming gain, and reducing coverage.

Two main hybrid beamforming architectures are the sub-array hybrid beamforming and full-array hybrid beamforming, as depicted in Figure 3.31. For the former, smaller "sub-arrays" within the large antenna array are separated and connected to RF beamformers of individual RF chains. Fewer RF chains and connected antennas sub-arrays in such architecture causes wide beamwidth, reduced

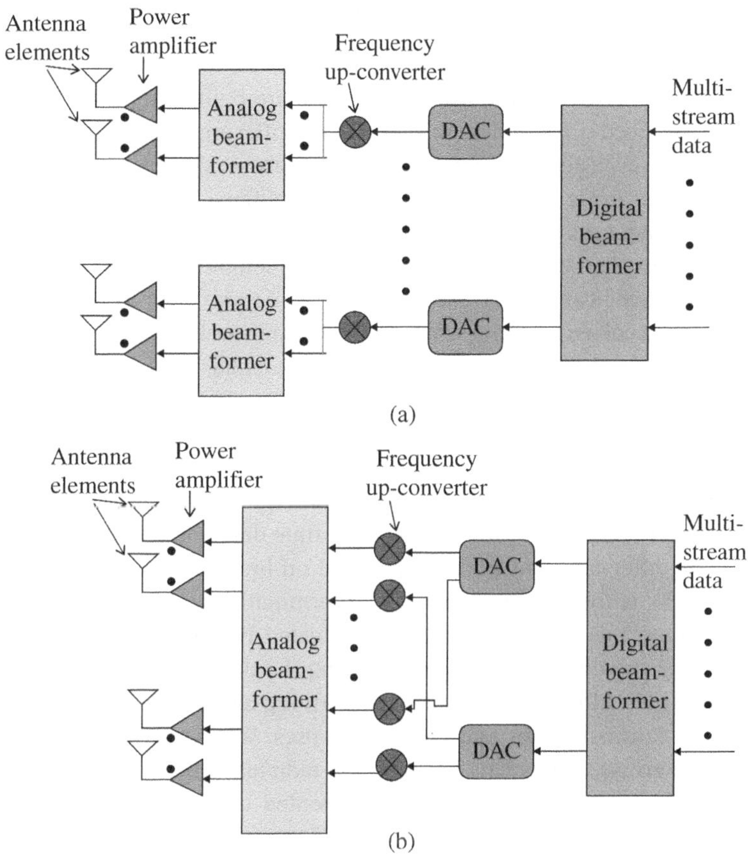

Figure 3.31 Hybrid beamforming array antenna: (a) Sub-array; and (b) Full-array. Source: Reproduced with permission from Ref. [29]/IEEE.

directivity and possibly strong interference from other chains due to inter-beam interference. However, such a hybrid beamforming type reduces circuit complexity and allows relatively more efficient processing. On the other hand, the full array hybrid beamforming uses all antenna elements in the array. In transmit mode, signals from different RF chains are combined before they are transmitted through the antenna. While such architecture is relatively costlier compared to the sub-array architecture, the larger number of RF chains provide higher gain and more directional control over a large angular range. It should be mentioned that full array hybrid beamforming requires calibration and synchronization of RF chains and antenna elements, making its processing more resource-intensive.

3.4.2 Passive and Active Beamforming Networks

Analog beamforming networks can be categorized into either fully passive or active beamformers. Passive beamforming networks are built using a variety of fixed beamforming feed approaches, including the Butler matrix, Blass matrix, and Nolen matrix, as listed in Table 3.4. Besides phase shifters (PSs), such beamformers can be made active by integrating amplifiers. The Butler matrix is a beamforming network consisting of 90° hybrids and PSs, which provides 2^n orthogonal beams that are linear and independent combinations of the array element patterns. Each beam can be used by a dedicated transceiver or by a single transmitter and/or receiver, with the selection made using an RF switch. A Butler matrix can also be used to steer the beam of the antenna via amplitude and phase-weighted input to the ports, followed by variable uniform phase tapering.

A Blass matrix [33] consists of transmission lines and directional couplers, which form beams by introducing progressive time delays and are suitable for broadband operation. One of the ports provides an equal time delay to all elements, producing a broadside beam, whereas the time delays between elements are produced by other adjacent ports to produce off-broadside beams. One of its main drawbacks is the largely resistive load terminations which cause large losses. To overcome this, the Nolen matrix is created by partially cutting the Blass matrix along the diagonal line and replacing the diagonal coupler with a transmission line [34]. Finally, other fixed beamforming methods include the use of lenses such as Rotman lenses and Luneburg lenses. When fed with multiple feeds or multiple arrays, lenses can focus energy radiated by these feeds/arrays with improved directivity. These lenses, implemented either from specialized dielectric materials or space-fed arrays, provide wide scanning angles and ease of use, and are lower in cost [35]. Besides that, it also features large bandwidth and

Table 3.4 Comparison of Butler matrix, Blass matrix, and Nolen matrix.

Parameters	Butler matrix	Blass matrix	Nolen matrix
Power loss	Low	High	Moderate
Degree of freedom	Low	Moderate	High
Components	Crossover, phase shifter, hybrid couplers	Load terminations, couplers, phase shifters	Phase shifters, couplers,
Efficiency	High	Less	Moderate
Design	Simple	Moderate	Complex

real-time delay characteristics, enabling beamsteering irrespective of operating frequency [36].

In DBF, different signal processing algorithms are used, which modify the amplitude or phase to a digital signal before the digital-to-analog (DAC) conversion in a transmitter. The main aim of these algorithms is to optimize the beam toward a desired direction and by forming nulls in the direction of the interfering signals by iteratively calculating the corresponding complex weights for each antenna element in the array. These weights are iteratively determined based on the array output and a reference signal, which is assumed to be identical to the desired signal, with the aim of minimizing the difference between them. To find these optimum weights, several techniques are commonly employed. They include the least mean square (LMS), sample matrix inversion (SMI), and recursive least mean square (RLMS), as summarized in Table 3.5. The SMI algorithm is a time average estimate of the array correlation matrix using k-time samples. SMI has a faster conversion rate as it employs direct inversion of the covariance matrix. The LMS algorithm uses a steepest-descent method to compute the weight vector with the knowledge of the desired signal. This can be done by periodically transmitting a training sequence known to the receiver and converging slowly when the spread of the eigenvector of the received signal is large. However, the LMS-based algorithms tend to suffer from an increase in the noise level. The RLMS scheme is the least sensitive to a noisy reference signal. In most cases, the RLMS scheme achieves a larger enhancement in performance at lower input signal-to-interference to noise ratio (SINR) [37]. Furthermore, it is shown that the RLMS algorithm is also more robust when the reference signal used is noisy. However, this technique is higher in complexity than that of the SMI algorithm as the complexity for the LMS is very low.

Table 3.5 Comparison of algorithms (SMI, LMS, and RLMS).

Parameters	Sample matrix inversion	Least mean square	Recursive least mean square
Quality of service	Poor	Moderate	Good
Complexity	High	Low	Moderate
Logic	Linear algebra	Gradient descent	Gradient descent
Memory	Not required	Required	Required
Application	Fixed beamforming	Adaptive beamforming	Adaptive beamforming
Methodology	Matrix	Eigenvalues	Eigenvalues

3.4.3 Architecture Example: Transmit Phase Array Beamformer

A recent example of a transmit beamformer will be discussed in this section to understand the critical components needed to enable wideband operation in beamforming arrays. The wideband RF beamforming architecture employs feedback power amplifiers (PAs) and active PSs with T-coil wideband matching networks as shown in [38] and illustrated in Figure 3.32. The ultrawideband with high output power beamformer is implemented on a 4 × 1 array from 17–55 GHz. The RF input signal is first divided using a wideband 1 : 4 Wilkinson divider network and distributed to four wideband RF beamforming channels. Following this, each channel consists of:

- a single-ended to differential (S2D) active stage,
- 5-bit vector modulator-based PS,
- 4-bit variable gain amplifier (VGA), and
- a wideband two-stage PA.

Each of the four channels is designed to be fully differential to preserve stability, reduce coupling between elements, and connect to the wideband differential antenna. Each sub-component is explained in the following sub-sections.

3.4.3.1 Broadband Wilkinson Divider Network and Single-Ended to Differential Stage

The wideband 1:4 Wilkinson divider network with the active S2D stage is shown in Figure 3.33. To meet the required characteristic impedance in the power

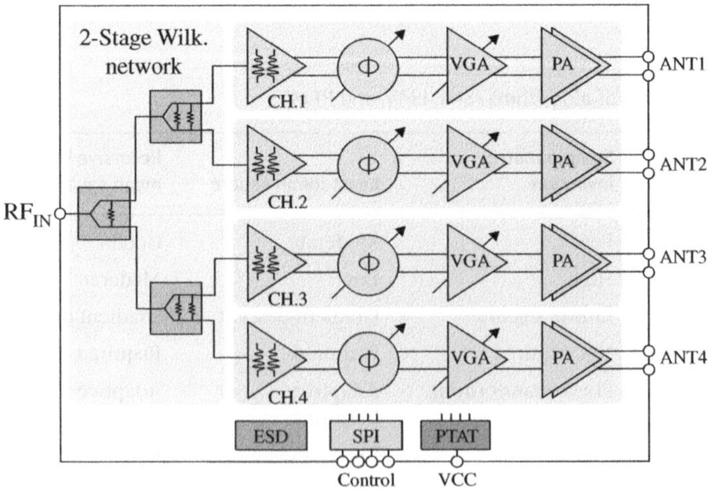

Figure 3.32 Block diagram of the wideband 17.1–52.4 GHz 4-channel Tx phased-array beamformer. Source: Reproduced with permission from Ref. [38]/IEEE.

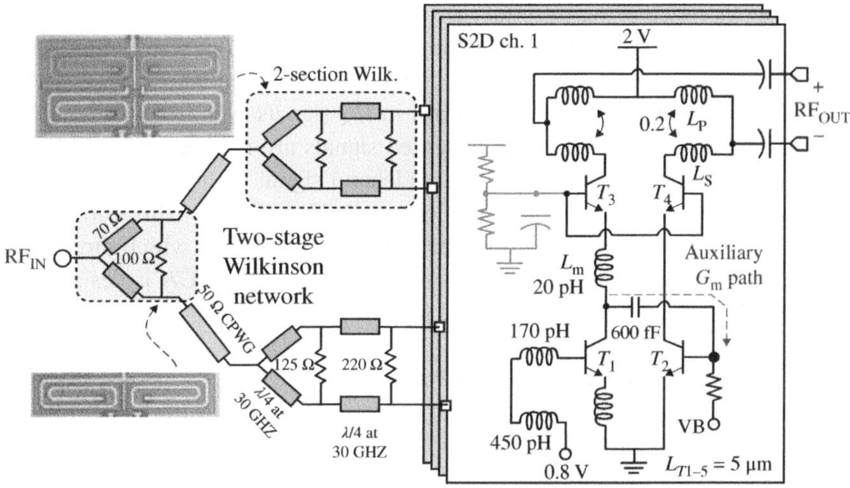

Figure 3.33 Wideband 1 : 4 Wilkinson divider network with the active single-ended to differential stage. Source: Reproduced with permission from Ref. [38]/IEEE.

divider, the $\lambda/4$ transmission lines are constructed in different chip layers for signal and ground, respectively. Resistors are then used to introduce the 100, 125, and 220 Ω resistances to the power divider, resulting in insertion losses of 2.7–3 dB from 15–57 GHz (from simulations). On the other hand, an active balun is placed between a source and a load when a differential and balanced RF functional block needs to be connected to a single-ended and grounded one to neutralize the gain and phase imbalance between the output nodes. It is made up of four transistors (T_1–T_4): a primary transconductance (Gm) at stage T_1 and a compensation path that uses an auxiliary Gm at stage T_2. Cascode transistors (T_3 and T_4) are used to increase the gain of the S2D stage and improve reverse isolation. To reduce phase errors at high frequencies, inductors can also be introduced in the main channel to compensate for the existing line-based delays. For input matching, an inductor-based L-matching network is introduced. Finally, to widen impedance matching to the next stage, a small T-coil network is applied.

3.4.3.2 Ultrawideband Active Phase Shifter and Variable Gain Amplifier

To enable operation over ultrawide frequency bands, two of the most important active components to enable this feature are the PS and VGA. Considering the scope of this chapter, they are described briefly as follows, and interested readers can find more details from [38]. The active PS used in this work enables 5-bit of phase control used. It consists of a quadrature all-pass filter (QAF) network and an analog adder with two current-steered VGAs. This QAF network produces the differential in-phase and quadrature (I/Q) signals. To decrease the influence of capacitance loading from the analog adder and synthesize reliable wideband I/Q

signals, series resistors are introduced. The matching between the QAF network and the analog adder can be further improved by employing series and degeneration inductors. Besides that, these components also act as a compensation network, eliminating phase and gain error in I/Q signals over a wide bandwidth. Next, the VGA cells' output (I/Q) weighted signals are merged and coupled to a T-coil peaking network to provide a wideband signal with the desired phase response. The wideband VGA is implemented after the PS. This component provides 16 dB gain control, with a 4-bit DAC control circuit, which directs the current from the cascode stage. When set to low gain, the VGA is designed to provide sufficient output power to drive the PA.

3.4.3.3 T-Coil with Resistive Feedback Power Amplifier

A wideband two-stage PA with optimal matching network is designed next, as illustrated in Figure 3.34 consisting of:

- A differential T-coil input matching network
- A driver amplifier (DA) with resistive feedback
- An interstage matching network
- A main PA with resistive feedback
- A differential T-coil output-matching network

First, a T-coil matching network in a small area and moderate resistive feedback is designed to provide wideband and low-loss load impedance transformation. Next, a DA and a tertiary-coil interstage matching network with capacitive coupling serve to optimize and convert the input impedance from the DA to the main PA. For wideband input matching, the DA additionally employs an R-feedback of 350 Ω and a T-coil matching network. Finally, an output-matching network based

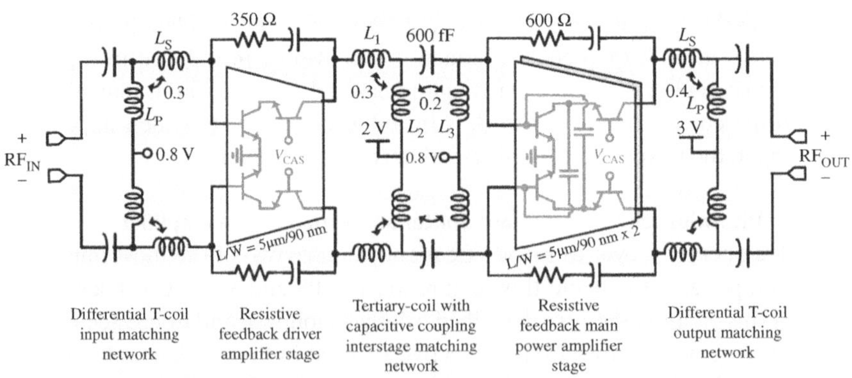

Figure 3.34 Wideband two stage PA with T-coil and resistive-feedback optimum load matching network. Source: Reproduced with permission from Ref. [38]/IEEE.

on the differential T-coil network is introduced. The PA has a simulated wideband gain of 24 dB with a 3-dB BW from 17 to 55 GHz, operating using 3 V, whereas the DA operates at 2 V for maximum efficiency.

The die micrograph of the wideband 4 × 1 transmit (Tx) beamformer designed in 90 nm SiGe BiCMOS SBC18S5 process is shown in Figure 3.35a. The single channel core area is 0.45 mm × 1.7 mm, and the chip size is 2.5 mm × 5 mm, with 380 μm-pitch bumps for wafer-level chip-scale packing limiting the chip size. Figure 3.35b illustrates the 4 × 1 phased-array beamformer's small-signal measured performance. Port 1 is designated as the chip common port, whereas Port 2 is designated as the antenna port. The peak gain was measured at 22 dB with a 3-dB bandwidth of 17.1–52.4 GHz. Figure 3.35c shows the large-signal measurements of the Tx 4 × 1 beamformer. The device obtained a wideband output power of 12.8–14.8 dBm P1dB in the entire 5G NR FR2 band range of 17–52 GHz. Peak PAE is 12.4% at 35 GHz and >10% between 25 and 38 GHz.

3.4.4 Architecture Example: Ultrawideband Programmable Digital Beamforming Receiver

An example of DBF capability operating in an ultrawideband mode is a programmable phased array (PPA) receiver from [39]. It is implemented in a 0.13-μm SiGe BiCMOS with the reconfigurable number of beams as presented in Figure 3.36, operating from 2 to 16 GHz with the following beamforming modes:

- eight antenna inputs with a single beamformer output.
- four antenna inputs with two simultaneous beamformer outputs.
- two antenna inputs with four simultaneous beamformer outputs.
- two antenna inputs with DBF outputs.

The output of the eight-antenna single-beam mode can be downconverted to an intermediate frequency before exiting the chip. This allows the chip to be used in DBF mode at the sub-array level (eight-element, four-element, or two-element sub-arrays) when a portion of the analog phase control channels are shunt off. Finally, additional mixers are introduced on-chip to enable DBF to function at the single-element level. The DBF modes require the use of an external LO signal. The chip is based on the all-RF architecture for the eight channels, and the signals are differential throughout the chip except at the RF input (RF_{IN})/output ports, which are single-ended (SE). To convert signals between the single-ended (input ports) and the differential circuits (output ports), high-linearity wideband active baluns (input ports) and wideband differential-to-single-ended (D2S) converters are employed. The entire chip is controlled via a serial-peripheral interface (SPI) and is powered by a single 2.5 V supply. The beamforming modes of the PPA chip can be summarized in Figure 3.37.

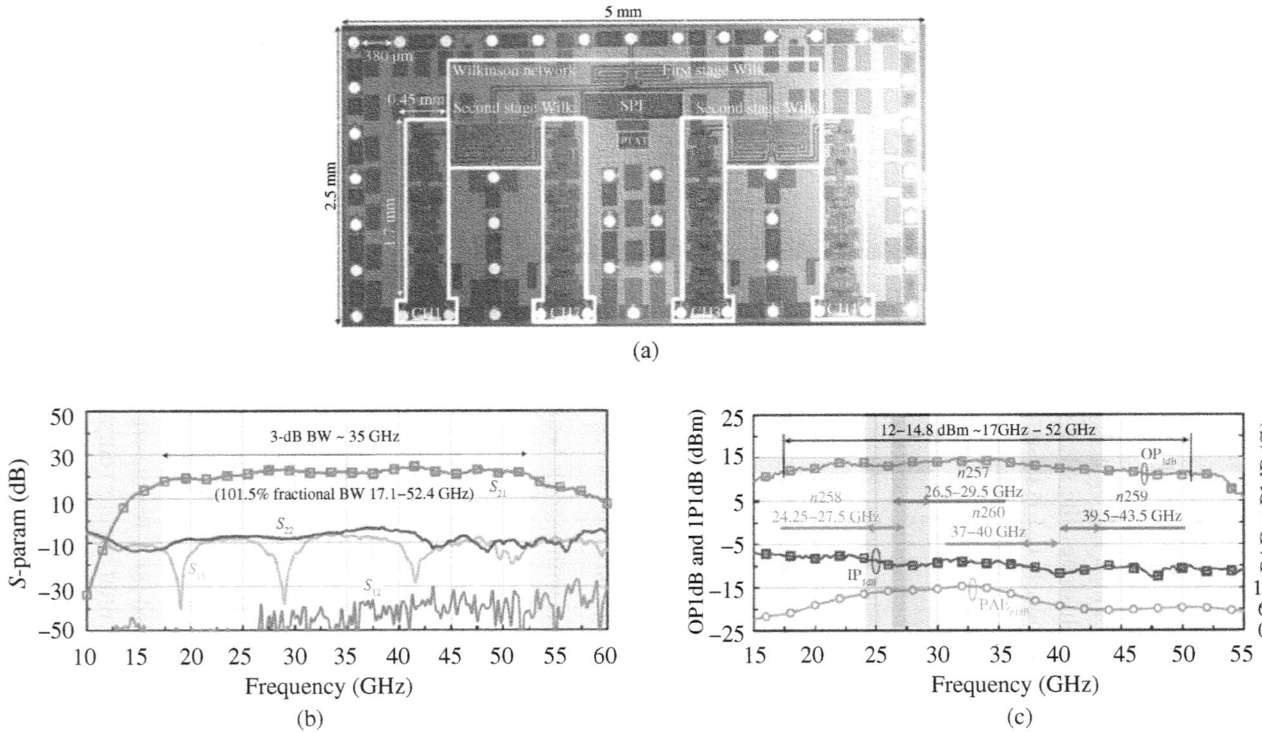

Figure 3.35 (a) 4-channel Tx wideband phased-array chip. (b) S-parameters of a small signal measurements of the 4 × 1 Tx beamformer. (c) Large-signal measurements: (a) OP1dB, IP1dB and PAE at P1dB. Source: Reproduced with permission from Ref. [38]/IEEE.

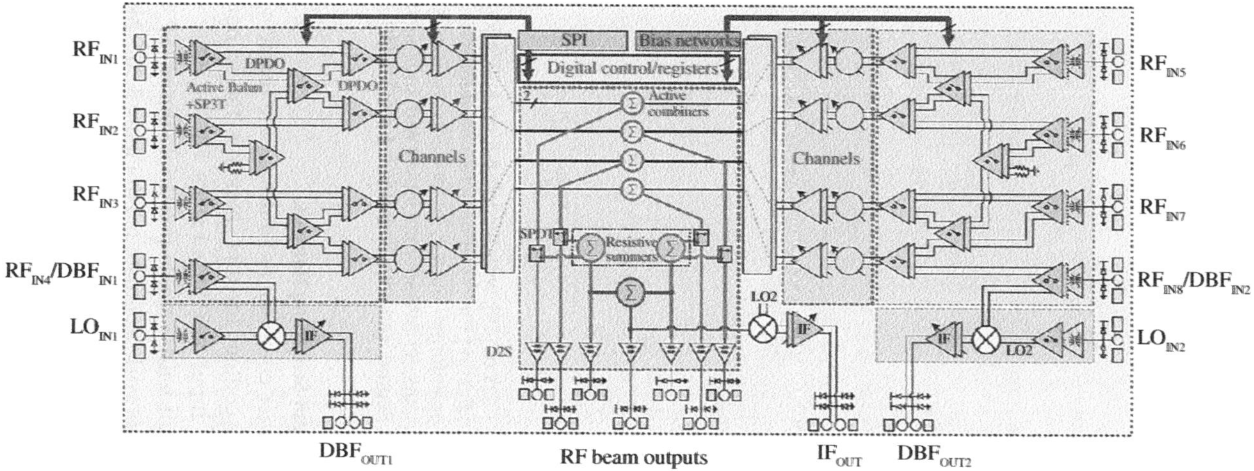

Figure 3.36 Chip architecture for the 2–16 GHz eight-element PPA. Source: Reproduced with permission from Ref. [39]/IEEE.

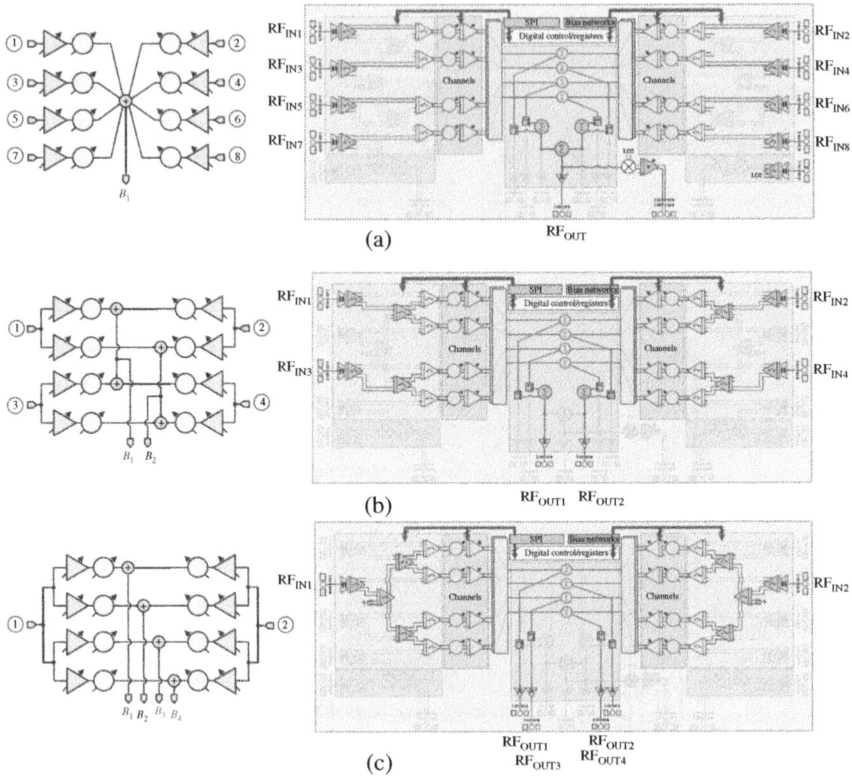

Figure 3.37 Beamforming modes of [39], with the left figures indicating the number of inputs, and the right figures indicating the resulting active beamforming modes: (a) Eight antenna inputs with a single-beamformer output, (b) four antenna inputs with two simultaneous beamformer outputs, (c) Two antenna inputs with four simultaneous beamformer outputs. Source: Reproduced with permission from Ref. [39]/IEEE.

3.4.4.1 Reconfigurable Input Switching Network

As [39] contains many other components to design the receiving digital beamformer, the most relevant content for this chapter is the reconfigurable active switching network on the input and output sides. On the input side, it consists of a wideband active balun with a single-pole-triple-throw (SP3T) followed by another active switching network, as illustrated in Figure 3.36. The input network routes the input RF signals to the required number of channels to result in the four different modes. As each mode uses a different set of switches (and different ports inside the same switch), therefore, introducing high isolation in the switch networks is important to minimize any leakage from the input RF ports to non-assigned channels. This is mainly due to the aim of keeping all eight channels (PS + VGA) always active. The first stage is a cascode amplifier with a grounded second input, realizing the balun function, and resistive feedback is used to provide a 50-Ω wideband input match. Input linearity of the amplifier is enabled by a one-bit control feedback network by lowering the gain. As compensation for the switch network loss, a 1.3-nH differential peaking inductance is used to increase the gain at high frequencies, resulting in a flat gain response at 2–16 GHz. The active switch topology can allow any outputs to be switched ON, whereas the loads of this section are matched to 70 Ω for a wideband output impedance.

3.4.4.2 Reconfigurable Output Combining Network

The output-combining network must operate under a variety of modes as shown in Figure 3.37. It contains active and passive combiners, together with routing switches. First, the left and right channels are added together in an active combiner and provide four different outputs. These signals are then combined together to result in one, two, or four independent outputs.

- Two-input/four-beam mode (Figure 3.37c): The four signals are directly routed using passive single-pole double-throw (SPDT) switches to active differential-to-SE converters (D2Ss).
- Four-input/two-beam mode (Figure 3.37b): The four signals are routed using SPDT switches to wideband resistive combiners and the two outputs are fed to two D2S blocks.
- Eight-input/one-beam mode (Figure 3.37a): The four signals are routed using SPDT switches to wideband resistive combiners, then to another active combiner, and then to the D2S block.

The wideband active combiner is implemented using two cascode amplifiers with merged outputs, and input and output resistive loading is used for a wideband response. The differential SPDTs are implemented using a single-series nMOS transistor. A simple series switch is used to produce a 10–15 dB isolation. Finally, a differential passive T-junction is used in the combiner for the four-input/two-beam and eight-input/one-beam modes. The chip results in an 8–12-dB gain at 2–16 GHz (depending on the PPA mode), a noise figure (NF) of 12 dB. The DBF function also results in a wideband response with a gain of 15–16 and an NF of 12–13 dB with high linearity.

3.5 Summary

This chapter discussed several important feeding techniques for conventional and MIMO antenna designs, both passive and active. Several passive feeding networks are suitable for MIMO antennas including the coaxial probe feed, aperture-coupling feed, microstrip line feed, and co-planar waveguide feed. Then, matching techniques and reconfigurable feed networks relevant to MIMO antennas were presented. PIN diodes are preferred in reconfigurable antenna designs due to their advantageous switching time and reconfiguration capabilities. Comparison of feeding techniques for analog beamforming indicated that the Butler Matrix is a widely used method to provide beamforming. While the matrix inversion algorithm can produce complex weights in the case of digital fixed beamforming, the RLMS approach provides good convergence for adaptive DBF over noisy situations. Finally, two examples of beamforming transmitters and receivers are presented to show the importance of such components in the overall system design of multi-antenna systems. Understanding of methods and components to enable beamforming is foreseen as crucial knowledge in paving the way toward the next-generation communication.

References

1. Pesel, R.G. (2015). Impedance matching networks for wireless applications. MS thesis. University of Waterloo.
2. Chen, Z.N., Liu, D., Nakano, H. et al. (2016). *Handbook of antenna technologies*. Springer Publishing Company, Incorporated.
3. Wing, A.H. and Eisenstein, J. (1944). Single- and double-stub impedance matching. *Journal of Applied Physics* 15 (8): 615–622. https://doi.org/10.1063/1.1707478.
4. Wang, Z., Zhao, L., Cai, Y., and Yin, Y. (2017). A simple printed MIMO antenna pair with high isolation performance. In: *2017 Sixth Asia-Pacific Conference on Antennas and Propagation (APCAP), Xi'an, China*, 1–3. IEEE.
5. Misra, D.K. (2012). *Radio-Frequency and Microwave Communication Circuits: Analysis and Design*. Wiley.
6. Suzuki, S., Yuan, Q., and Chen, Q. (2018). *Impedance matching approach of L-section circuit with ohmic loss in reactive components. Wireless Power Transfer* 5 (01): 64–74.
7. Wallace, J.W. and Jensen, M.A. (2004). Mutual coupling in MIMO wireless systems: a rigorous network theory analysis. *IEEE Transactions on Wireless Communications* 3 (4): 1317–1325. https://doi.org/10.1109/TWC.2004.830854.
8. Li, L., Venkatasubramanian, S., Lehtovuori, A., Icheln, C., Heino, M., and Haneda, K. (2015). T-shaped decoupling network for wideband isolation improvement between two strongly coupled antennas. *Loughborough Antennas & Propagation Conference (LAPC)*, 02-03 November 2015, Loughborough, UK, IEEE, pp. 1–4.

9 Pozar, D.M. and Kaufman, B. (1987). *Increasing the bandwidth of a microstrip antenna by proximity coupling. Electronics Letters* 23 (8): 368. https://doi.org/10.1049/el:19870270.

10 Mishra, R., Mishra, R.G., and Kuchhal, P. (2016). Analytical study on the effect of dimension and position of slot for the designing of ultra wide band (UWB) microstrip antenna. *Int. Conf. Adv. Comput. Commun. Informatics, ICACCI 2016*, September, pp. 488–493. http://doi.org/10.1109/ICACCI.2016.7732093.

11 Losee, F. (2005). *RF Systems, Components, and Circuits Handbook*. Artech.

12 Bombale, U.L. and Gupta, S. (2006). Design of ultra wideband microstrip antenna. *IEEE Antennas Propag. Soc. AP-S Int. Symp.*, March, pp. 3705–3708. http://doi.org/10.1109/APS.2006.1711426.

13 Rhazi, Y., Bri, S., and Touahani, R. (2013). Effect of microstrip antenna feeding in the K-band. *International Journal of Engineering and Technology (IJET)* 4 (6): 515–522.

14 Giauffret, L. and Laheurte, J.M. (1999). Parametric study of the coupling aperture in CPW-fed microstrip antennas. *IEE Proceedings-Microwaves, Antennas and Propagation* 146 (3): 169–174.

15 Garg, R., Bhartia, P., Bahl, I.J., and Ittipiboon, A. (2001). *Microstrip Antenna Design Handbook*. Artech House.

16 Zhao, X., Lee, Y., and Choi, J. (2013). Design of a compact MIMO antenna using coupled feed for LTE mobile applications. *International Journal of Antennas and Propagation* 2013: https://doi.org/10.1155/2013/837643.

17 Fujimoto, K. and Itoh, K. (2018). *Antennas for Small Mobile Terminals*. Artech House.

18 Elfergani, I.T.E., Abd-Alhameed, R.A., See, C.H., Sadeghpour, T., and Jones, S.M.R. (2011). Reconfigurable antenna design approach for mobile applications and a technique for harmonics suppression. *2011 Loughborough Antennas & Propagation Conference*, 14–15 November 2011, Loughborough, UK, IEEE, pp. 1–4.

19 Kholapure, A.S. and Karandikar, R.G. (2016) Emerging techniques for printed reconfigurable antenna: a review. *2016 Second International Conference on Research in Computational Intelligence and Communication Networks (ICRCICN)*, 23–25 September 2016, Kolkata, India, IEEE, pp. 57–61.

20 Nikolaou, S., Bairavasubramanian, R., Lugo, C. et al. (2006). Pattern and frequency reconfigurable annular slot antenna using PIN diodes. *IEEE Transactions on Antennas and Propagation* 54 (2): 439–448.

21 Karthika, K. and Kavitha, K. (2021). Reconfigurable antennas for advanced wireless communications: a review. *Wireless Personal Communications* 120 (4): 2711–2771.

22 Al-Husseini, M., El-Hajj, A., Tawk, Y., Kabalan, K.Y., and Christodoulou, C.G. (2010). A simple dual-port antenna system for cognitive radio applications.

2010 International Conference on High Performance Computing & Simulation, IEEE, pp. 549–552.

23 Lago, H., Zakaria, Z., Jamlos, M.F., and Soh, P.J. (2019). A wideband reconfigurable folded planar dipole using MEMS and hybrid polymeric substrates. *AEU-International Journal of Electronics and Communications* 99: 347–353.

24 Thummaluru, S.R., Kumar, R., and Chaudhary, R.K. (2019). Isolation and frequency reconfigurable compact MIMO antenna for wireless local area network applications. *IET Microwaves, Antennas & Propagation* 13 (4): 519–525.

25 Iannacci, J. (2017). RF-MEMS for high-performance and widely reconfigurable passive components—a review with focus on future telecommunications, Internet of Things (IoT) and 5G applications. *Journal of King Saud University, Science* 29 (4): 436–443.

26 Ebrahimi, E., Kelly, J.R., and Hall, P.S. (2011). Integrated wide-narrowband antenna for multi-standard radio. *IEEE Transactions on Antennas and Propagation* 59 (7): 2628–2635.

27 Khairnar, V.V., Kadam, B.V., Ramesha, C.K., and Gudino, L.J. (2018). A reconfigurable parasitic antenna with continuous beam scanning capability in H-plane. *AEU-International Journal of Electronics and Communications* 88: 78–86.

28 Soltani, S., Lotfi, P., and Murch, R.D. (2016). A port and frequency reconfigurable MIMO slot antenna for WLAN applications. *IEEE Transactions on Antennas and Propagation* 64 (4): 1209–1217.

29 Ahmed, I., Khammari, H., Shahid, A. et al. (2018). A survey on hybrid beamforming techniques in 5G: architecture and system model perspectives. *IEEE Communications Surveys & Tutorials* 20 (4): 3060–3097.

30 Zhang, S., Guo, C., Wang, T., and Zhang, W. (2018). ON–OFF analog beamforming for massive MIMO. *IEEE Transactions on Vehicular Technology* 67 (5): 4113–4123.

31 Krieger, G., Younis, M., Huber, S. et al. (2012). Digital beamforming and MIMO SAR: review and new concepts. *EUSAR 2012; 9th European Conference on Synthetic Aperture Radar*, 23–26 April 2012, Nuremberg, Germany, VDE, pp. 11–14.

32 Larsson, E.G., Edfors, O., Tufvesson, F., and Marzetta, T.L. (2014). Massive MIMO for next generation wireless systems. *IEEE Communications Magazine* 52 (2): 186–195. https://doi.org/10.1109/MCOM.2014.6736761.

33 Blass, J., "Multidirectional antenna - A new approach to stacked beams," 1958 IRE International Convention Record, New York, NY, USA, 1960, pp. 48–50, doi: 10.1109/IRECON.1960.1150892.

34 Ren, H., Zhang, H., and Arigong, B. (2020). Ultra-compact 3×3 Nolen matrix beamforming network. *IET Microwaves, Antennas & Propagation* 14: 143–148. https://doi.org/10.1049/iet-map.2019.0336.

35 Poyanco, J.M., Pizarro, F., and Rajo-Iglesias, E. (2022). Cost-effective wideband dielectric planar lens antenna for millimeter wave applications. *Sci Rep* 12: 4204 (2022). https://doi.org/10.1038/s41598-022-07911-z.

36 Quevedo-Teruel, O., Ebrahimpouri, M., and Ghasemifard, F. (2018). Lens antennas for 5G communications systems. In: *IEEE Communications Magazine*, 56 (7): 36–41, July 2018, doi: 10.1109/MCOM.2018.1700977.

37 Srar, J.A. and Chung, K.S. (2008). Performance of RLMS algorithm in adaptive array beam forming. *11th IEEE Singapore International Conference on Communication Systems*, 19–21 November 2008, Guangzhou, China, pp. 493–498. IEEE.

38 Alhamed, A.A. and Rebeiz, G.M. (2021). A global multi-standard/multi-band 17.1–52.4 GHz Tx phased array beamformer with 14.8 dBm OP1dB supporting 5G NR FR2 bands with multi-Gb/s 64-QAM for massive MIMO arrays. *2021 IEEE Radio Frequency Integrated Circuits Symposium (RFIC)*, 07–09 June 2021, Atlanta, GA, USA, IEEE, pp. 99–102.

39 Sayginer, M. and Rebeiz, G.M. (2016). An eight-element 2–16-GHz programmable phased array receiver with one,two, or four simultaneous beams in SiGe BiCMOS. *IEEE Transactions on Microwave Theory and Techniques* 64 (12): 4585–4597.

4

Passive MIMO Antenna Systems

MIMO antenna systems have been generally designed for operation in single, narrowband mode. Due to the need to improve the reliability and capacity of the current fifth generation (5G), there is growing research on such antenna systems operating in multiband and ultrawideband (UWB) modes. This chapter first aims to present the motivation MIMO antenna systems within the different generations of wireless communications. Then, the main principles of bandwidth widening and multiband techniques from conventional antenna elements will be discussed. Next, the recent advancements in the single, multiband, and wideband/UWB antennas to support MIMO operation in various wireless bands will be presented. These include design examples for passive MIMO antenna systems operating in the 5G New Radio Frequency Band 1 (FR1) and Frequency Band 2 (FR2). Besides that, the features of the different antenna types used in commercial products will also be underlined. Finally, two design examples, one each designed for the sub-6 GHz band and millimeter-wave band antennas be presented.

With the increasing demand for high data rate communication systems requirements within the next-generation wireless networks, there is a need for higher spectral efficiency (SE) and energy efficiency (EE), low latency, and a larger number of connection nodes [1]. This supports the transmission of a number of parallel data streams to a single user from the base station, which was then expanded to the multi-user MIMO technology. Multiple data streams can then be assigned to multiple users while maintaining the system's capacity and performance. In addition to lower frequency bands of below 6 GHz, 5G can operate in millimeter-wave frequency bands with higher data rates of 2–20 Gbps. As MIMO technology operates with broader bandwidths in comparison to previous 4G and long-term evolution-advanced (LTE-advanced) standards [2], continuous advancement in antenna design technologies requires a larger number of closely

MIMO Antenna Systems for 5G and Beyond, First Edition.
Xiaoming Chen, Ping Jack Soh, and Mohammad S. Sharawi.

spaced antennas with low correlation to ensure compactness for mobile and base station applications [3].

Single-band, multiband, and UWB antennas can be classified based on their frequency characteristics in support of MIMO communication systems. Single-band antenna often uses closely spaced antennas with improved pattern diversity to achieve higher data rates within a specific band. Multiband systems often use a combination of antennas operating in different bands to achieve higher data rates and improve coverage. Finally, UWB systems often use multiple antennas operating across the entire bandwidth to achieve high data rates and improve localization accuracy. Despite the clear advantages of using multiband and UWB antennas, the challenges that arise when designing such passive antennas go back to the fundamentals of multiband antennas, which include (but are not limited to):

- Size: For operation in a single band, suitable conventional single-band antennas can be adapted to the required wireless standard and form factor of the device (either compact mobile or larger indoor/outdoor units). Multiple units of such antennas can be closely spaced and fed, where the alleviation of coupling between them remains one of the last but important steps to be completed. The final size of such antennas is typically determined by the lowest operating frequency.
- Complexity: The need to integrate multiple bandwidth multiband operation in the basic radiator unit will result in more complex structures. Operation in additional band or wider band may require the addition of slots, parasitic elements, etc. Besides that, operation in multiple/wideband mode may require the integration of multiple radiators on a shared aperture, such as those typically used in base station MIMO arrays. This may further complicate the analysis and understanding of the electromagnetic radiation principles in the element, thus making optimization of such structures more tedious with additional physical parameters.
- Diversity: For single-band antennas, polarization diversity is typically used to enable spatial diversity. This is generally the main choice in mobile and compact devices due to space constraints to enable space diversity using such antennas. On the contrary, the use of additional bands via multiband and UWB antenna enables the transmission/reception of signals at different frequency bands (frequency diversity). This enables transmission of multiple independent data streams, whereas received signals can be combined to mitigate frequency selective fading to improve capacity and reliability. Table 4.1 compares several key characteristics of single-band, multiband, and UWB MIMO antennas.

To first be able to classify MIMO antennas into different single narrowband, multiband, and UWB operations, the classical bandwidth definition of antennas

Table 4.1 The key characteristics of narrowband, multiband, and UWB antennas.

Characteristic	Single band	Multiband	Ultrawideband
Bandwidth	Narrow	Moderate	Ultrawide
Antenna size	Small	Moderate/large	Moderate/large
Complexity	Simple	Moderate	Complex
Diversity	Polarization	Frequency, polarization	Frequency, polarization
Potential data rates	Moderate	High	Very high
Link reliability	Moderate	High	Very high

needs to be understood. It is the difference between the upper and lower cut-off frequencies (f_H and f_L, respectively):

$$BW - f_H - f_L \tag{4.1}$$

Considering the impedance-matching bandwidth of an antenna, the upper and lower operating frequency limits, f_H and f_L, respectively, are considered at the −10 dB (or −6 dB) crossing points of the magnitude of reflection coefficients (in dB), $|\Gamma|$. Besides that, it is also common to represent the difference in frequencies shown in Eq. (4.1) as the ratio $f_H : f_L$, which provides the number of times that f_H is higher than f_L (e.g. 2 : 1, 10 : 1, etc.). On the other hand, from quality factor (Q) analysis for a wideband antenna, Schantz derived a relation between them as [4].

$$Q = \frac{f_0}{BW} \tag{4.2}$$

where f_0 is the center frequency corresponding to the antenna-resonant frequency, estimated as

$$f_0 = (f_H + f_L)/2 \tag{4.3}$$

Finally, for an antenna to be considered a UWB antenna, its bandwidth should be more than 500 MHz as specified in [5].

4.1 Single-Band Implementations

The 5G new radio frequencies are divided into sub-6 GHz bands and millimeter wave bands. The Frequency Range 1 (FR1) band spans from 410 MHz to about 7.125 GHz, whereas the Frequency Range 2 (FR2) band includes millimeter-Wave (mm-Wave) frequencies from 24 to 71 GHz. While such antennas can be switched between sub-band channels, passive antennas are typically designed to be operated throughout the entire band required by standards. Moreover, the topic of active MIMO antennas will be further described in detail in the next chapter. By properly choosing the type of antenna element, the resulting radiation patterns can be designed to either operate with omnidirectional or directional radiation, depending on the needs of the application. Omnidirectional antennas are ideal for applications where the signal needs to be broadcast in all directions, while directional antennas are suited for applications where the signal needs to be focused toward a specific plane or direction. The following subsections briefly describe the classical single-band antennas, which can be designed in a dipole, monopole, loop, IFA, or PIFA.

4.1.1 MIMO Antenna for the FR1 Band

4.1.1.1 Dipole-Based Antennas

The dipole antenna is considered one of the most basic and yet important types of antenna. It consists of a conducting metal, such as copper, straight electric conductor that is interrupted at the center. An example of a compact dual polarized MIMO antenna based on magnetic dipoles is presented in [6]. Figure 4.1b–d illustrates the different views of the antenna, consisting of three main components: top radiator patch, T-shaped cross-coupling feed structure, and the feeding network. The top radiator is composed of four bow-tie dipole arms with exponential curvature, two crossed T-shaped strips, and four parasitic elements. It is designed on a Rogers R5880 substrate with a relative dielectric constant of 2.2, a loss tangent of 0.0004, and a thickness of 0.5 mm. The edges of the bow-tie arms follow the exponential function:

$$C = \pm\overline{\overline{R}}.a.e^{bx} + N \tag{4.4}$$

whereas a, b, and N are all constant coefficients that govern the shape of the curvature, $N=0$ and $x \in [0, x_{\max}]$. $\overline{\overline{R}}$ is the rotation matrix that can rotate the curve ±90°. The exponential function edges for the cross dipole contribute to the gradual size increment of the radiating slot, which widens bandwidth. Two cross T-shaped metallic strips are utilized for the coupled feed to the bowtie dipoles to enhance impedance matching and improve design freedom. Besides, the antenna is surrounded by a 3-D-printed wall and an outer perfect electric conducting sheet. The radiator is also supported by three Nylon pillars. A wideband ring hybrid coupler is introduced below the antenna ground to feed the dipoles with orthogonal phase differences. It is also designed on Rogers R4003 substrate (relative permittivity of 3.66 and a thickness of 0.25 mm). The two orthogonal bowtie dipoles are fed in phase to radiate at a −45° angle when Port 1 is fed. Likewise, when Port 2 is active,

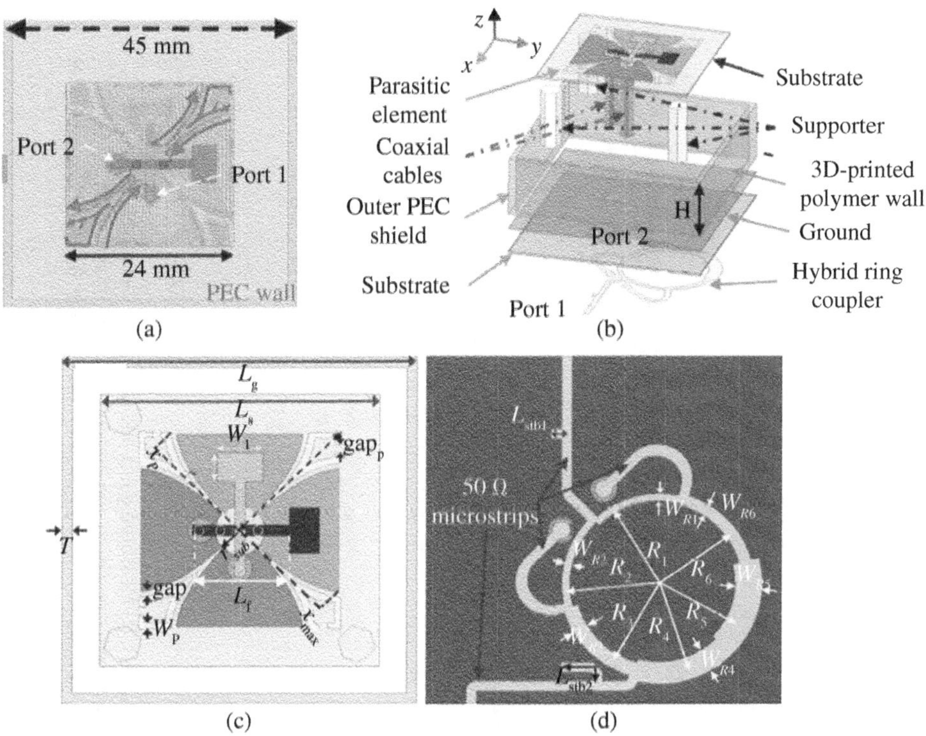

Figure 4.1 (a) Design with parasitic resonator. (b) Perspective view. (c) Top view. (d) Bottom view. Specifications: $a = 0.1$, $b = 245$, $L_g = 40$, $L_s = 37.5$, $L_f = 11.5$, $W_l = 4.5$, $W_t = 4$, $r_{sub} = 3$, $Fd_{loc} = 3.5$, $x_{max} = 13.85$, $H = 18$, $W_p = 1$, $gap_p = 1$, $gap = 0.25$, $x_p = 14.15$, $L_c = 18$, $L_{stb1} = 1.5$, $L_{stb2} = 3.25$, $R_1 = 10$, $R_2 = 10.25$, $R_3 = 10.15$, $R_4 = 10.1$, $R_5 = 10$, $R_6 = 10.15$, $W_{R1} = 0.75$, $W_{R2} = 0.7$, $W_{R3} = 1.25$, $W_{R4} = 1.75$, $W_{R5} = 2.25$, $W_{R6} = 1$, and $T = 1.25$. The units are millimeter except for b. Source: Reproduced with permission from Ref. [6]/IEEE.

the antenna emits a wave with a polarization angle of 45°. The coupling to the parasitic elements generates induced currents to operate as a magnetic dipole with a "matching load," as shown in Figure 4.1a.

The prototyped antenna is presented in Figure 4.2a, whereas Figure 4.2b illustrates the simulated and measured S-parameters. The antenna's performances vary slightly when different ports are excited. The bandwidth (with VSWR < 1.5) is 3.38–5.2 GHz (when Port 1 is fed) and 3.35–5.25 GHz (when Port 2 is fed). The reflection coefficients for both port excitations are slightly higher than −14 dB at the center frequency due to assembly non-idealities. Despite that, the operating band covers the needed FR2 5G bands. Figure 4.2c shows the measured and

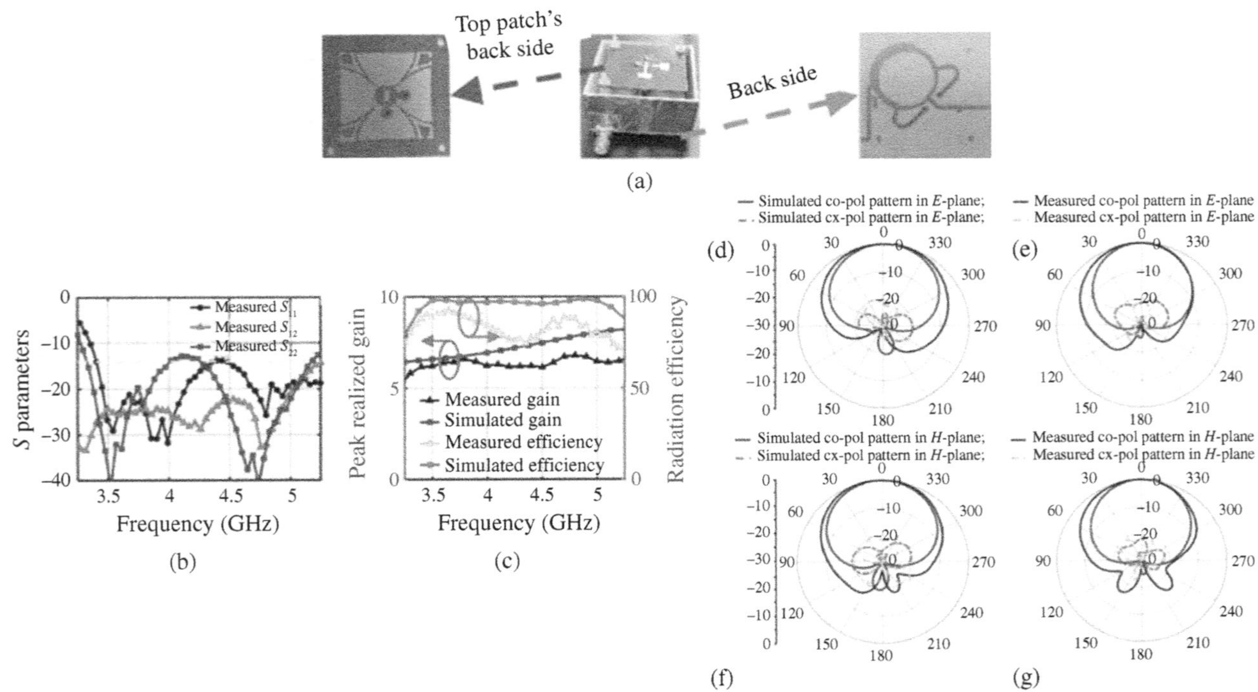

Figure 4.2 (a) Fabricated prototype antenna. (b) Measured *S*-parameters. (c) Measured gains and radiation efficiencies across the target band. Normalized co- and cross-polarization radiation patterns from simulation and measurement in *E*-plane: (d) at 3.5 GHz. (e) 4.9 GHz. the co- and cross-polarization radiation patterns from simulation and measurement in the *H*-plane: (f) at 3.5 GHz and (g) at 4.9 GHz. Source: Reproduced with permission from Ref. [6]/IEEE.

simulated peak gains and efficiencies when Port 1 is fed. The normalized co- and cross-polarization radiation patterns (simulated and measured) at 3.5 and 4.9 GHz are shown in Figure 4.2d–g. The results show that the XPD is greater than 10 dB for oblique angles within the range of [−60°, 60°] in both *E*- and *H*-planes at low and high frequencies. However, the fabricated prototype produced less directive radiation patterns.

4.1.1.2 Loop-Based Antennas

A loop antenna consists of a turn of an RF current-carrying coil to operate as an antenna. The coherent currents through the loop antenna generate a magnetic field perpendicular to the loop. Two predominant types of loop antennas are circular loops and rectangular loops. Besides that, other variants of loops such as delta and elliptical antennas have also been designed according to the same principle of operation [7].

An example of such an antenna structure is a four-port annular ring loop patch antenna producing four decoupled waves from 3.3 to 5.0 GHz for 5G MIMO access point applications, which is presented in [8] and shown in Figure 4.3. The ring loop patch is printed on a 0.4 mm-thick FR4 substrate with a relative permittivity of 4.4 and a loss tangent of 0.024. It is mounted 9.6 mm above a large circular ground plane with a diameter of 150 mm. This then produced a total antenna height of 10 mm (0.14λ at 4.15 GHz). The AR patch has a 16 mm width with an inner radius of 12 mm, and an outer radius of 28 mm (or a diameter of 56 mm, 0.77λ at 4.15 GHz). To introduce symmetry in the four-port structure with respect to the ground plane, a circular ground plane was selected. Besides that, the diameter of the ground plane is approximately two times the wavelength of the center frequency at 4.15 GHz. The larger ground plane diameter is also used to enable higher directivity and gain in its forward direction.

Ports 1–4 are equally spaced by 90° in the central axis with a radius of 20 mm of the annular ring patch, specifically at $\varphi = 0°$, 90°, 180°, and 270°. Each port is connected to a 50 Ω SMA connector located behind the circular ground plane via a 10 mm feed strip. To compensate for the inductance of the long feed strip, each port is encompassed by a ring gap, with a width of 3.1 mm (*g*) and an inner diameter of 3.8 mm, within the patch to improve impedance matching. Furthermore, a total of four gap-coupled shorting (GCS) strips is positioned at $\varphi = 45°$, 135°, 225°, and 315° to short the annular ring patch to the ground plane. The two adjacent ports are separated by a single GCS strip to improve decoupling. As aforementioned, the annular ring patch can be decomposed into four antennas of (Ant1–Ant4). Port 1 with two nearby GCS strips named S_1 and S_2 forms the first monopolar patch antenna (MPA) (Ant1), whereas Port 2 with the GCS strips S_2 and S_3, forms the second MPA (Ant2). Similarly, Port 3 with the GCS strips S_3 and S_4 forms the third MPA (Ant3) and port 4 with the GCS strips S_4 and S_1, is the fourth MPA

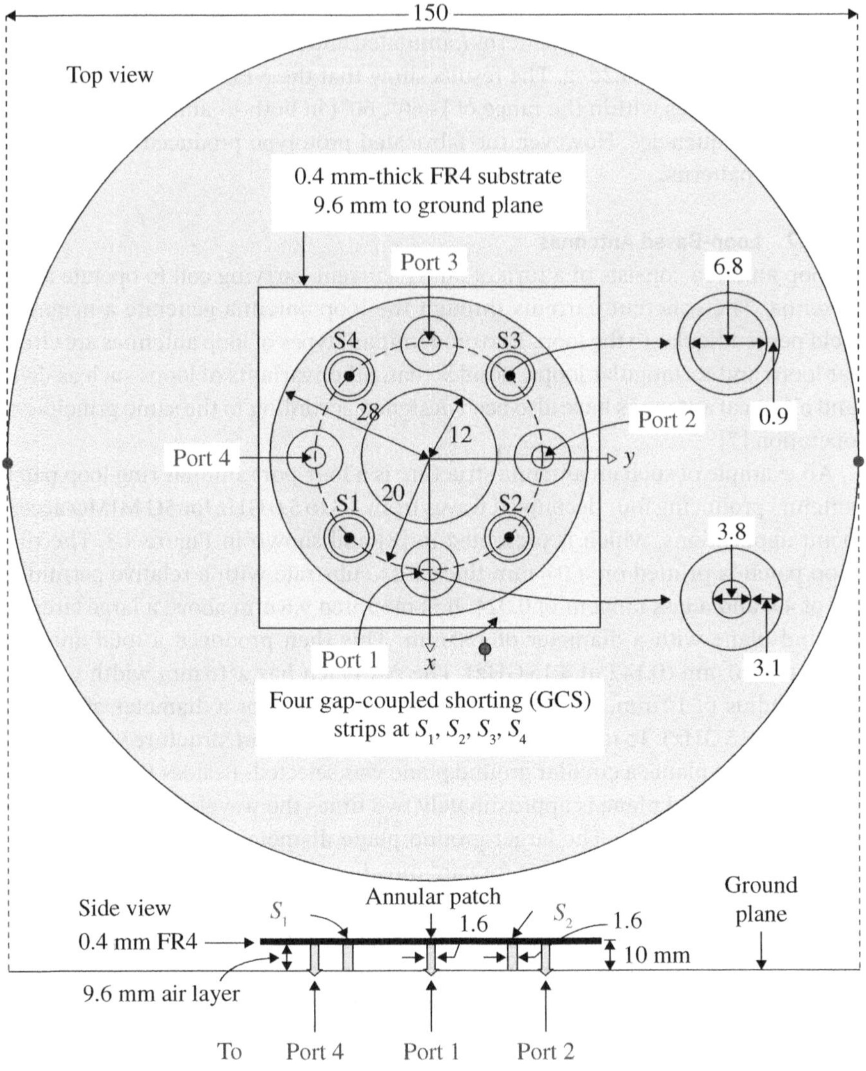

Figure 4.3 Geometry of a four-port wideband AR patch antenna for 5G MIMO in 3.3–5.0 GHz. Source: Reproduced with permission from Ref. [8]/IEEE.

(Ant4). To optimize isolation between two opposite antennas (Ant1 versus Ant3 or Ant2 versus Ant4), their distance (or the inner diameter of the ring) is chosen to be 24 mm, which is 0.25λ at 3.3 GHz, the lower-edge frequency of the operating band.

To evaluate the antenna performance, the simulated S-parameters of the four-port annular ring patch antenna with (Ant 1 only; Ports 2–4, S_3, and S_4 not

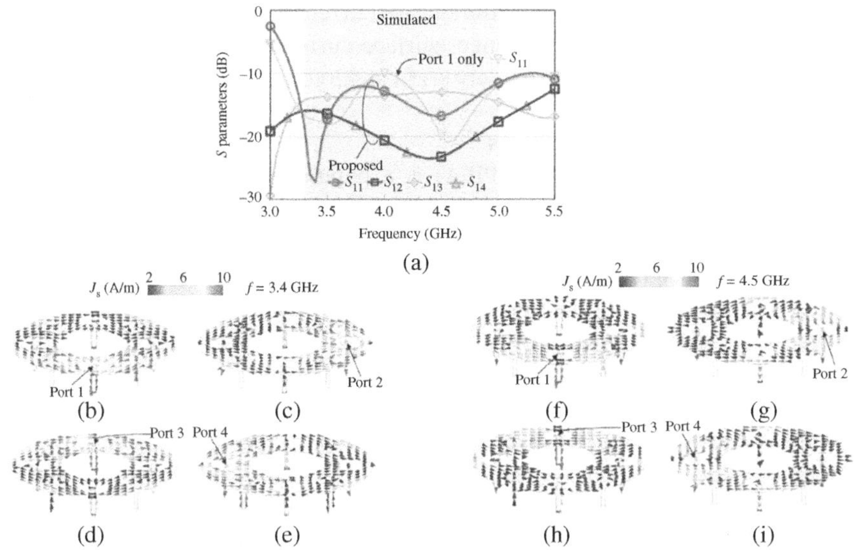

Figure 4.4 (a) Simulated S parameters of the four-port AR patch antenna and the case with Port 1 only (Ant1 only). Simulated surface current distributions at 3.4 GHz of the four-port AR patch antenna. (b) Port 1. (c) Port 2. (d) Port 3. (e) Port 4. Simulated surface current distributions at 4.5 GHz of the four-port AR patch antenna. (f) Port 1. (g) Port 2. (h) Port 3. (i) Port 4. Source: Reproduced with permission from Ref. [8]/IEEE.

included) are presented in Figure 4.4a. Only the results of Port 1 (S_{11}, S_{12}, S_{13}, and S_{14}) are shown due to the structural symmetry and for brevity. Two resonant modes around 3.4 and 4.5 GHz are produced in both cases, resulting in a wideband operation from 3.3–5.0 GHz. The findings demonstrate that the existence of Ports 2–4 has a negligible impact on the antenna performance of Port 1 in the four-port antenna. The simulated surface current distributions of ports 1–4 at 3.4 and 4.5 GHz for the four-port annular ring patch antenna are presented in Figure 4.4b–e and Figure 4.4f–i. Figure 4.4b–e indicates that the surface currents of each port are predominantly confined within the corresponding one-quarter arc segment of the ring patch. The transmission coefficients (S_{12}, S_{13}, S_{14}) from Port 1 to other ports are less than −14 dB at 3.4 GHz, where Port 1 exhibits good excitation with S_{11} better than approximately −27 dB. This is primarily attributed to the four GCS strips, which operate as bandpass resonant structures. The value of S_{13} for two diametrically located ports is greater than that of S_{12} and S_{14} for two adjacent ports. This difference can be attributed to the identical polarization of Ant1 and Ant3, and one GCS strip is positioned the center of two nearby ports to enhance the S_{12} and S_{14}. Furthermore, S_{13} can be further reduced by introducing additional shorting strips between the two diametrically opposed ports at the inner patch edge. In Figure 4.4f–i, the surface current distributions of Ports 1–4 at 4.5 GHz showed significantly stronger surface currents on the feed strip (sized at 10 mm in length and approximately 0.15λ at 4.5 GHz) for each port excitation.

4.1.1.3 Inverted-F-Based Antennas

The planar inverted-F antenna (PIFA) is a popular variant of the IFA when implemented on PCB substrates. It typically comprises a radiating patch and an electrically large ground plane, both of which are connected via a metallic pin or, in some cases, a metallic plane. In addition to that, the PIFA can use air as its dielectric instead of PCB materials, potentially increasing its efficiency relative to conventional IFAs. PIFAs can be constructed from either formed metal, on PCBs or on another non-conductive base with selective metallization techniques. PIFAs are widely used in mobile communication systems such as mobile phones and laptops due to their compact design. An example of MIMO antenna using PIFA is presented in [9]. The geometry of the quadruple-antenna module is illustrated in Figure 4.5, depicting the exploded, side, and plain views. The substrates employed are the FR-4 board ($\varepsilon_r = 4.4$, $\tan\delta = 0.02$). The circular antenna, with a diameter of 30.6 mm ($0.34\lambda_L$, $f_L = 3.3$ GHz), comprises three sections – the top antenna

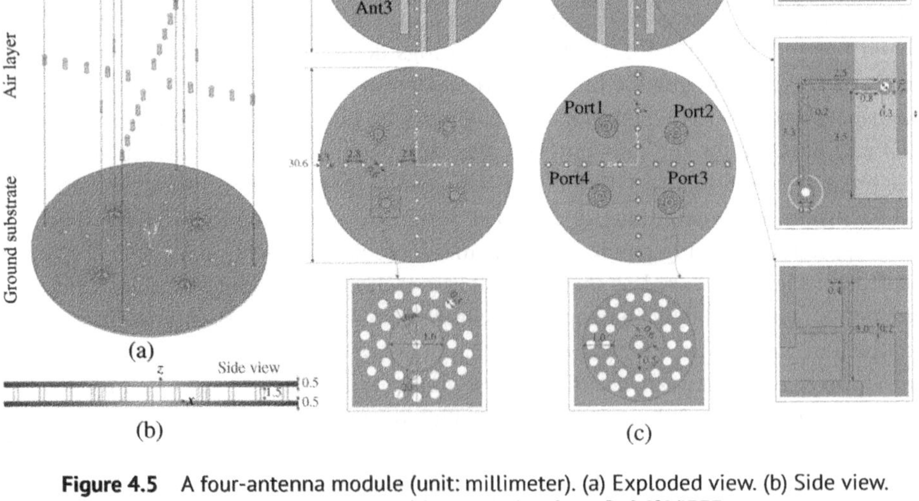

Figure 4.5 A four-antenna module (unit: millimeter). (a) Exploded view. (b) Side view. (c) Plain view. Source: Reproduced with permission from Ref. [9]/IEEE.

substrate, the central air layer, and the ground substrate, as shown in Figure 4.5a. It features a total effective height of 2 mm ($0.02\lambda_L$), including the 0.5 mm-thick antenna substrate and 1.5 mm-thick air layer. The top radiator consisting of four slotted PIFAs arranged in a sequentially rotated manner is depicted in Figure 4.5a, c.

The antenna element is designed to be a quarter metallic circle, with its inner edge integrated with five shorting vias, leaving the other two edges open. To enhance the isolation of neighboring elements, an L-shaped slot with a width of 1.4 mm and two length segments of 6.1 and 13.2 mm are added. In addition to that, a cross-line section with a length of 3 mm and a width of 0.2 mm is centered on the structure to connect all four longer segments. Figure 4.5c depicts the lower surface of the antenna substrate, where a microstrip line connects a metallized via (with a diameter of 0.3 mm) from the L-shaped line to feed the radiating aperture. The via is linked to a metallic circle with a diameter of 1 mm. A non-metallic ring with a width of 0.1 mm is used outside the metallic circle for impedance matching. Figure 4.5a shows a multitude of metallized vias with varying diameters. The larger vias, measuring 0.6 mm in diameter, are used to connect the radiating aperture and ground plane. On the other hand, the smaller 0.3 mm diameter vias connects the L-shaped line and feeding cable's inner conductor. In practice, these vias are all constructed using red copper wires. Figure 4.5c shows the ground substrate with a metallic circle of 0.6 mm diameter at the bottom surface, on which the feeding via is placed. There is a 1 mm wide metallic ring outside the circle, which is connected to the top ground plane through two metalized vias. Between the metallic circle and ring, there is a non-metallic ring of 0.5 mm width, and above it, a non-metallic circle with a diameter of 1.6 mm is etched at the top surface. During measurements, the feeding cable is utilized at the bottom surface, where the inner conductor is soldered to the metallic circle and the outer conductor to the outside metallic ring. Figure 4.5a displays the simulated S-parameters. Due to the presence of rotational symmetry, only parameters associated with Ant1 are presented. The S_{11} curve shows two resonances at 3.34 and 4.03 GHz. Additionally, the −6 dB reflection coefficient indicates operation from 3.25 to 4.21 GHz (960 MHz, 25.7%) for the 5G band N77 (3.3–4.2 GHz).

For neighboring elements featuring orthogonal polarization, the isolation remains nearly flat with at least 10.7 dB throughout N77 band. On the other hand, evaluation of diagonal elements with the same polarization indicates a decreasing trend in the isolation curve. The highest level of isolation, measuring 10.9 dB, is recorded at the lower resonance of 3.34 GHz. Figure 4.6b illustrates the simulated total efficiency and envelope correlation coefficient (ECC). The efficiency curve appears to be relatively constant at around −3 dB with some fluctuations, and the average efficiency for N77 stands at −3.2 dB (47.9%). The ECC values for neighboring elements are quite small, owing to their orthogonal polarization, and

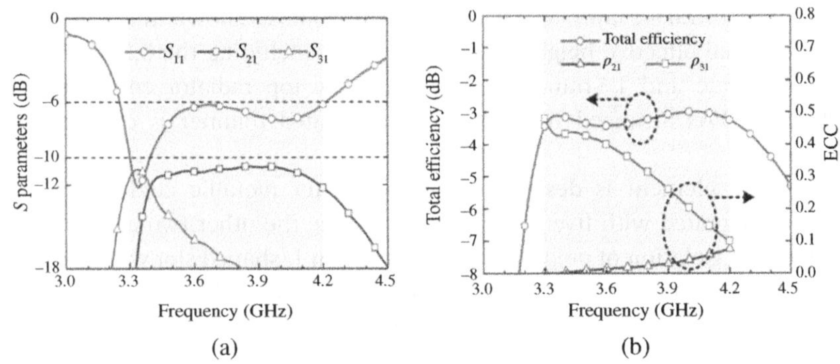

Figure 4.6 Simulated results. (a) *S* parameters. (b) Total efficiency and ECC. Source: Reproduced with permission from Ref. [9]/IEEE.

remain less than 0.08 across N77. The ECC curve for diagonal elements exhibits a similar decrease as the isolation curve and the ECC across N77 is at least 0.48.

4.1.2 MIMO Antennas in the FR2 Band

Besides the FR1 band, several antennas operating in the FR2 (millimeter-wave, mmWave) band have been developed. They include the widely popular microstrip-based antennas, a dielectric resonator (DR)-based antenna, and microstrip-based antennas. These examples will be described in the following subsections.

4.1.2.1 Dielectric Resonator-Based Antenna

In [10], a DR antenna with MIMO is developed for improved isolation in 5G FR2 applications. Figure 4.7 illustrates the MIMO DRA's geometry and dimensions, which are designed with enhanced isolation. The structure consists of two rectangular DRs with a relative permittivity of 9.8 mounted on a Rogers 5880 substrate with an ε_r of 2.2, tanδ of 0.0009, and thickness of 0.254 mm. Each DR is equipped with a microstrip-fed rectangular exciting slot for excitation purposes. Additionally, a metal strip with a length of L_p and width of W_p is printed on the upper surface of each DR to improve the isolation between the two antenna elements. Two DRs that are identical in size with the dimensions $a \times b \times d$ have been installed onto a metal ground plane that measures 20 mm × 20 mm. The TE^y_{311} mode is excited via a rectangular slot that is fed by a microstrip. The resonant frequency of a rectangular DR, for the TE^y_{pqr} mode, can be obtained through the wavenumbers k_x, k_y, and k_z inside the DR, whereby the subscripts p, q, and r represent the number of standing waves along the x-axis, y-axis, and z-axis, respectively. Approximated using the Marcatili's method [11], the DR

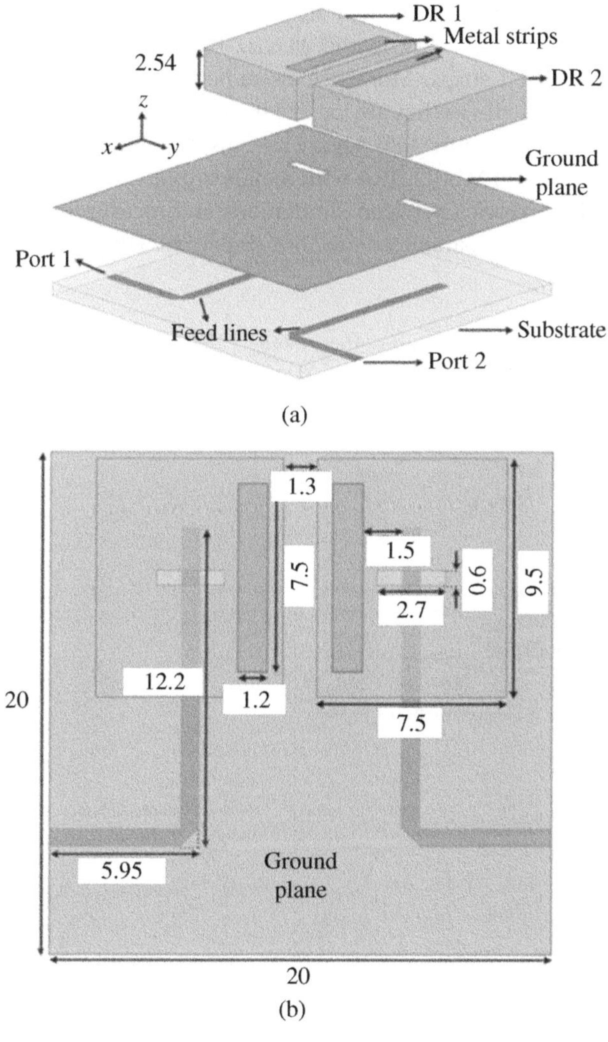

Figure 4.7 Geometry structure of the MIMO DRA. (a) Exploded 3-D view. (b) Top view. Source: Reproduced with permission from Ref. [10]/IEEE.

resonate at 28 GHz using the TE^{y}_{311} mode. A simple metallic strip is introduced onto the upper layer of each DR to enhance isolation between two separate antenna components.

A prototype of the MIMO DRA fabricated using a printed circuit board is measured, with its S-parameters illustrated in Figure 4.8a. The 10 dB measured reflection coefficient of the MIMO DRA with metal strips is from 27.19–28.48 GHz. Besides, measured isolation exceeds 24 dB in the 28 GHz band, indicating the effectiveness of the simple metal strips. Moreover, its radiation pattern is not significantly impacted by metal strips as they are located in an area where the DR's electric field is weak, as shown in Figure 4.8b. Figure 4.8c displays the simulated and measured radiation patterns of MIMO DRA with metal strips at 28 GHz, with its gain in Figure 4.8d. Discrepancies between simulations and measurements, however, are dominated by the comparable connectors' size to the DRA, besides effects from testing cables.

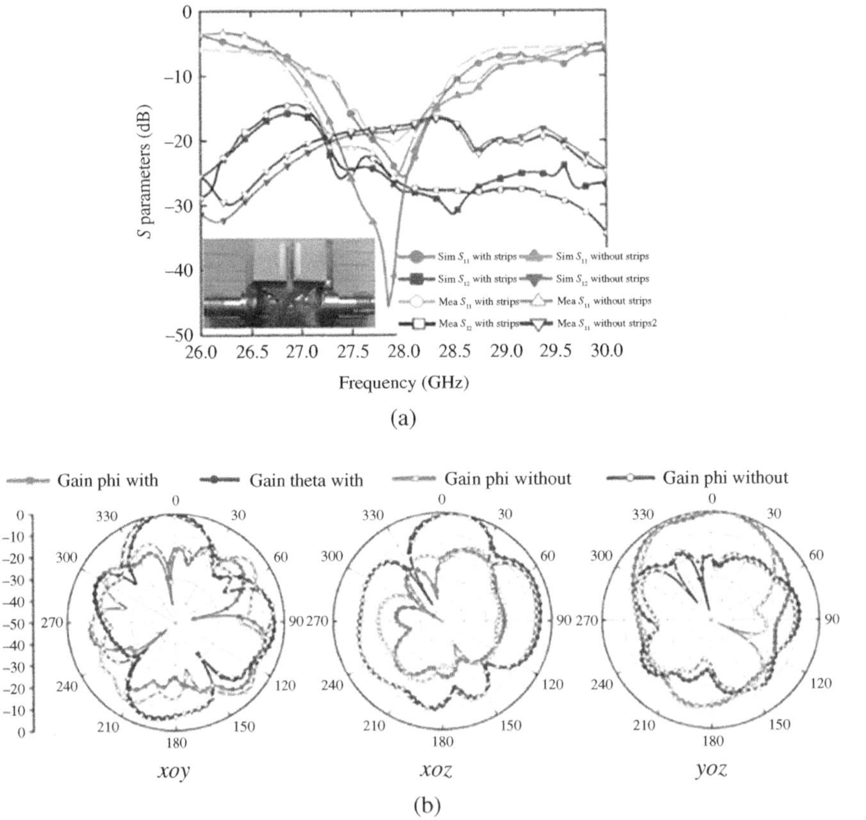

Figure 4.8 (a) Simulated and measured S-parameters of the MIMO DRA with and without metal strips. (b) Radiation patterns of the MIMO DRA with and without strips. (c) Measured and simulated radiation pattern of the MIMO DRA. (d) Realized gain and ECC of the MIMO DRA. Source: Reproduced with permission from Ref. [10]/IEEE.

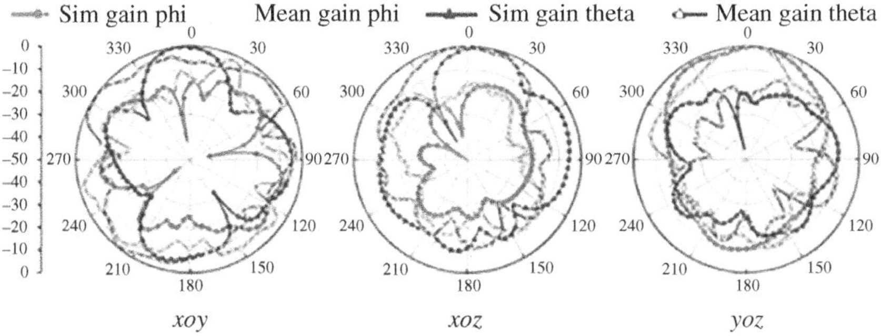

(c)

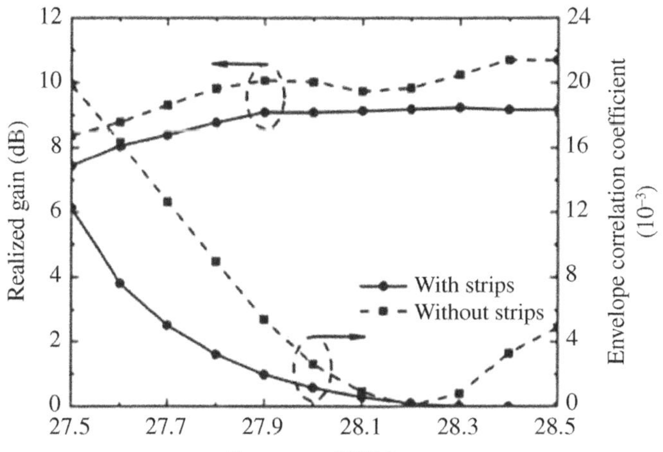

(d)

Figure 4.8 *(Continued)*

4.1.2.2 Microstrip-Based Antenna

Next, a dual-polarized broadband microstrip patch antenna designed for 5G FR2 band is presented in [12], focusing on low-cost implementation on a high-loss material (FR4 substrate). The T-resonator technique was used to determine its relative permittivity (ε_r) and loss tangent (tanδ), as shown in Figure 4.9. On the other hand, Figure 4.10a illustrates the design of a 1×4 dual-polarized antenna array. Each single antenna element consists of three substrates with identical thicknesses. The distance (L) between each antenna array element is set to 9.5 mm, with drill holes in each substrate perforated for alignment. Figure 4.10b

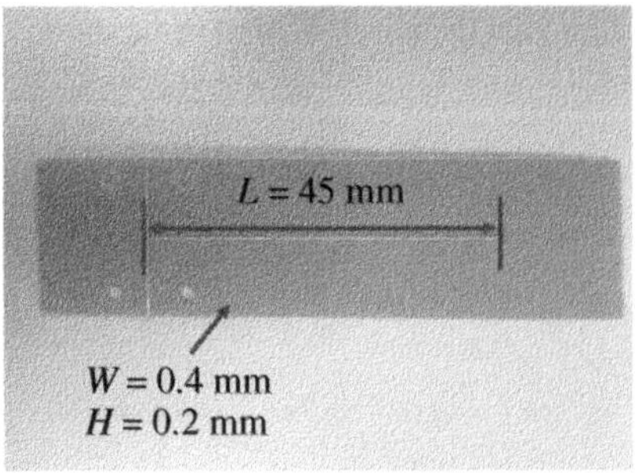

Figure 4.9 Fabricated microstrip T-resonator n FR4 substrate.
Source: Reproduced with permission from Ref. [12]/IEEE.

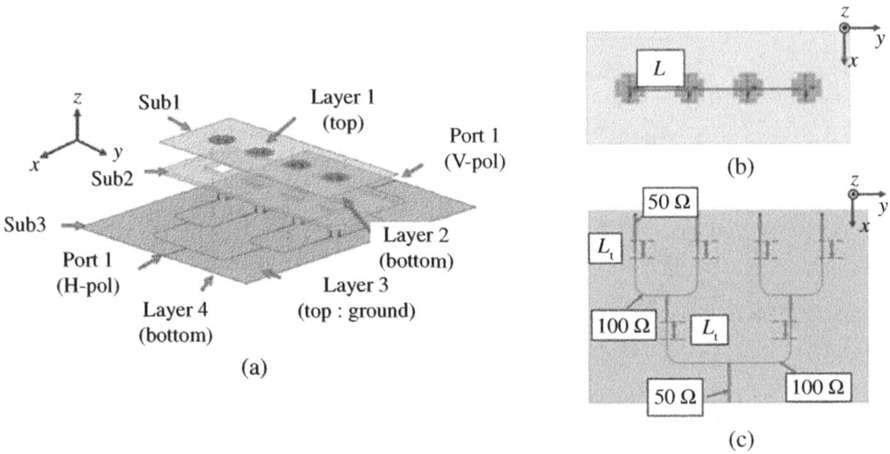

Figure 4.10 (a) 3D geometry of the dual-polarized 1 × 4 antenna array. (b) Top view of sub1 (layer1-top, layer2bottom) and (c) 1 × 4 feeding network. Source: Reproduced with permission from Ref. [12]/Sangkil_Kim.

shows the top view of sub1 (layer1-top, layer2-bottom) antenna. A feeding network for the antenna array is introduced under Sub3. Figure 4.10c displays the feeding network's half plane of the antenna array consisting of 50–100 Ω linear tapered microstrip line (with $L_t = 4.8$ mm ($\cong 3/4\lambda_g$)).

The fabricated antenna prototype is illustrated in Figure 4.11a, assembled using nylon bolts and nuts and the array denoted in the red box. The reflection coefficients ($|S_{11}|$ and $|S_{22}|$) and gain is presented in Figures 4.11b, c, respectively. The measured and simulated $|S_{11}|$ and $|S_{22}|$ values showed a high level of agreement. In addition, the measured H-/V-pol gain values are 10 ~ 11 dBi and the measured cross-pol suppression is about 20 dB. The normalized radiation patterns at 24, 26, and 28 GHz are displayed in Figure 4.11d–g, while the measured 3 dB beamwidth for azimuth (H-pol) and elevation (V-pol) were 12 and 60, respectively.

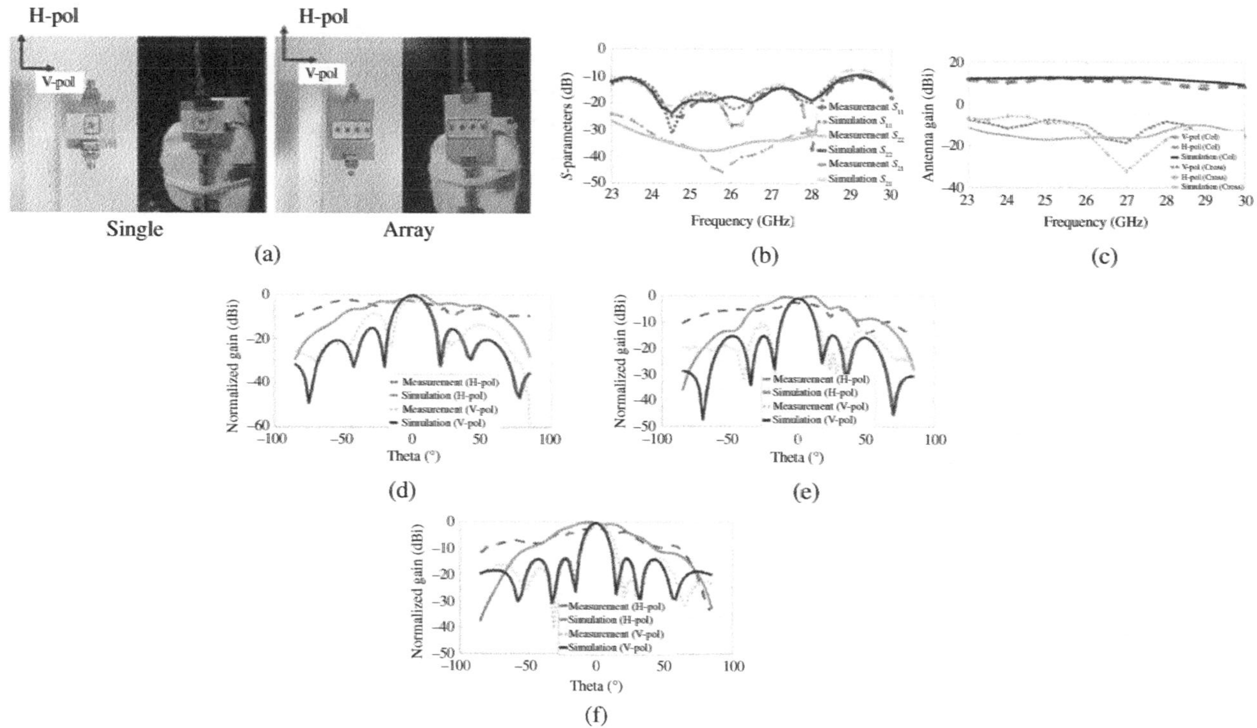

Figure 4.11 (a) Fabricated antennas and measurement environment. (b) Measured *S*-parameters of the 1×4 antenna array. (c) Measured antenna gains of the 1×4 antenna array. Measured radiation patterns of the 1×4 antenna array at (d) 24 GHz, (e) 26 GHz, and (f) 28 GHz. Source: Reproduced with permission from Ref. [12]/Sangkil_Kim.

4.2 Multi-Band MIMO Antennas

Mobile communication systems may include a multiband antenna capable of operating in distinct frequency bands, which are not immediately adjacent to each other. Examples of such antennas may include multiband operation of the existing GSM and LTE bands together with the 5G FR1 bands (around 3.5 GHz); or existing WLAN (2.4 GHz or 5 GHz) bands with the FR1 band; or even GSM/LTE bands and 5G FR2 (millimeter-wave bands, 28 or 38 GHz). Multiband techniques can be implemented on various types of antennas to achieve multiband operation, and several fundamental techniques for this purpose are elaborated in the subsequent sections.

4.2.1 Higher-Order Resonances

The use of higher-order resonances is a fundamental technique for achieving multiband operation. As illustrated in Figure 4.12a, the implementation of a monopole antenna with a length of $\lambda/4$ results in resonance at f_0, with an electric field minimum at the feed. However, a minimum electric field at the feed condition can also be achieved when the length of the same antenna corresponds to $3\lambda/4$, as depicted in Figure 4.12b. This enables the monopole antenna to resonate at $3f_0$, and other higher resonances can also be achieved at higher frequencies, such as $5f_0$. Different types of antennas, including dipoles, helices, patches, and slots, utilize higher-order resonances. As an example of this principle, a work in [13] presented a normal mode helical antenna mounted on a typical mobile phone with resonances at f_0 and $2.6f_0$.

4.2.2 Resonant Trap

Resonant traps can be either parallel inductor-capacitor circuits or short-circuited transmission line stubs that are designed to be anti-resonant (having essentially

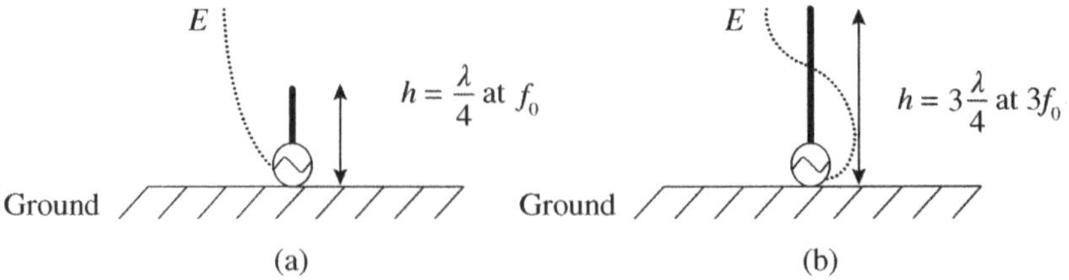

Figure 4.12 (a) A monopole antenna resonating at f_0 (b) Same antenna resonating at $3f_0$ (*E* is the electric field magnitude). Source: Reproduced with permission from Ref. [13]/John Wiley & Sons.

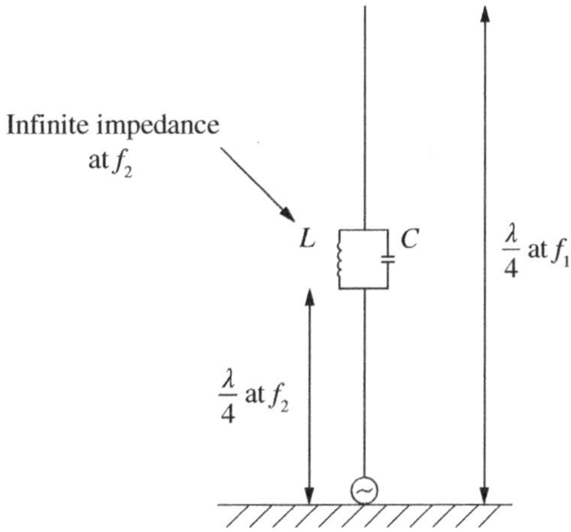

Figure 4.13 Antenna with a resonant trap. Source: Reproduced with permission from Ref. [13]/John Wiley & Sons.

an infinite input impedance) at a particular frequency. They enable the antenna to work at two different resonant frequencies. At higher frequencies, the trap is adjusted to be close to anti-resonance. At the lower frequency, the trap is well below resonance and is inductive, as shown in Figure 4.13 [13].

4.2.3 Multiple Resonant Structures

Another common method for obtaining a multiband antenna system is by using multiple resonant structures. This technique involves the use of two or more resonant structures that are in proximity or co-located in space with a single feed. The application of this technique is illustrated in Figure 4.14a, where the antennas are designed to operate at center frequencies of operation, f_1 and f_2 using a corporate feed. The technique of multiple resonant structures is frequently employed in mobile communication systems to achieve multiband mobile antennas. The work in [14] proposed dual-frequency antenna systems for handsets that are comprised of a combination of monopole and helical antennas, as shown in Figure 4.14b. This approach enabled multiple resonances at two distinct frequencies for dual-band operation at GSM 900 and 1800 bands.

4.2.4 Parasitic Resonators

An alternative approach to achieve multiband characteristics is to incorporate parasitic resonators into the antenna system. This technique includes extra parasitic elements to operate with the antenna feed, which permits operation at diverse frequencies. An example of antennas based on this technique is the Yagi-Uda antenna. Unlike the typical contact feed, multiple different elements

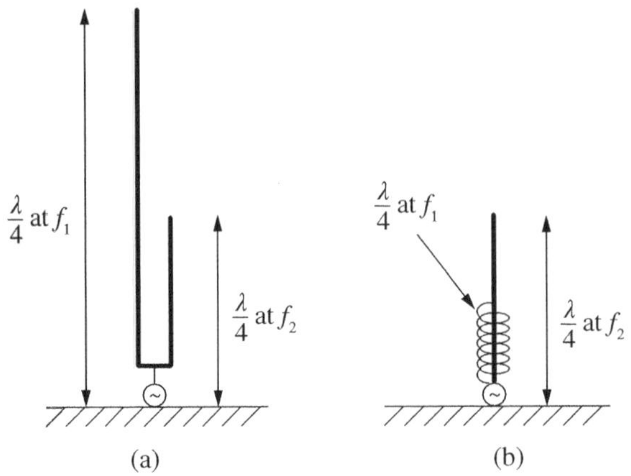

Figure 4.14 Two monopole antennas for dual-band operation (b) A helical antenna resonating at the frequency f_1 and a monopole antenna resonating at the frequency f_2 for dual-band operation. Source: Reproduced with permission from Ref. [13]/John Wiley & Sons.

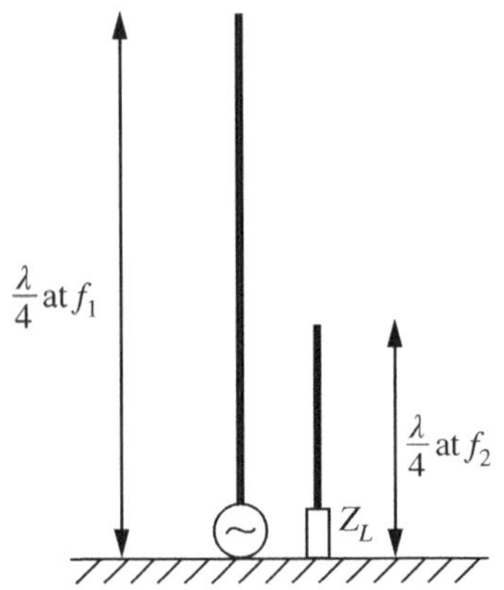

Figure 4.15 Parasitically coupled monopole antennas. Source: Reproduced with permission from Ref. [13]/John Wiley & Sons.

are parasitically coupled from the near field of the exciting element to resonate at other distinct frequencies. Figure 4.15 shows two monopoles, one of which is fed, whereas the other is parasitically coupled via the near field of the fed antenna. The parasitic antenna with a load can be used for tuning. This load would be typically reactive to maintain high antenna efficiency.

4.2.5 Example Implementations of Multiband MIMO Antennas

This section provides several examples of multiband MIMO antennas. [15] uses multiple resonant structures to design a compact multiband internal antenna, focusing on maintaining design simplicity. It combines a monopole and a loop antenna to generate resonant modes for multiband operation as shown in

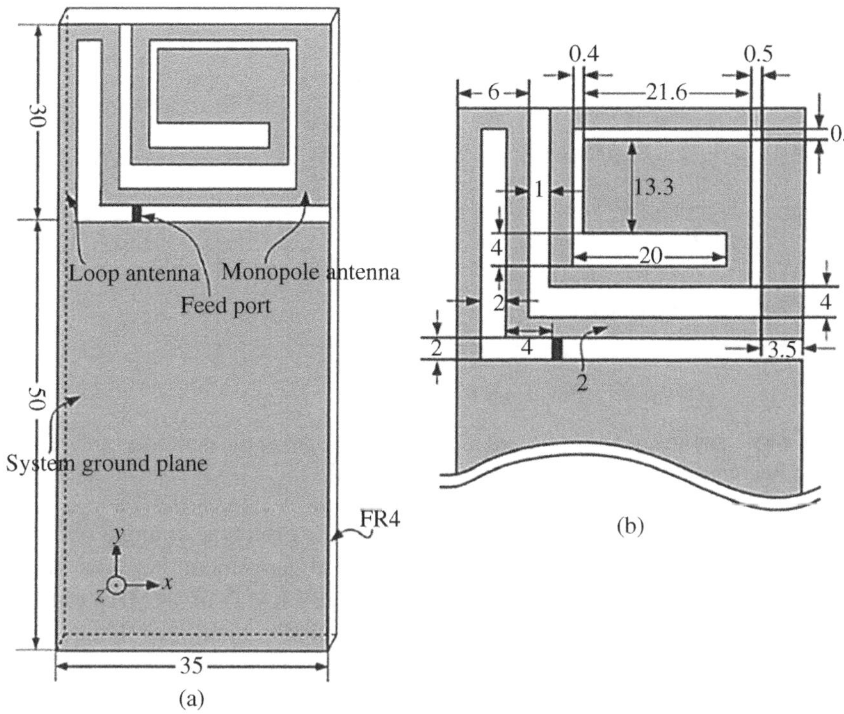

Figure 4.16 Geometry of the antenna. (a) Configuration of antenna. (b) The dimension of antenna (in mm). Source: Reproduced with permission from Ref. [15]/IEEE.

Figure 4.16a. The antenna can be printed on the circuit board of a mobile phone and connected to the system ground plane to form a loop-type structure. The loop structure widens impedance bandwidth without increasing the implementation complexity. The resonant element of the lower band is a spiral monopole, whereas both the monopole and loop antennas operate in the upper band. The coupling effect between the loop-type antenna and the spiral monopole antenna resulted in the shift of the third resonant mode or 1.5-wavelength mode of the monopole toward the second resonant mode to form a very wide operating band. The antenna pattern of total area 35 mm × 30 mm is printed on a 1.6-mm-thick FR4 substrate of size 80 mm × 35 mm. The detailed dimensions of the antenna pattern are shown in Figure 4.16b. The FR4 substrate has a relative permittivity of 4.4 and a loss tangent of 0.0245. The bottom of the spiral monopole forms a slot with the ground plane for exciting the mode at resonance of the loop antenna at about 6 GHz.

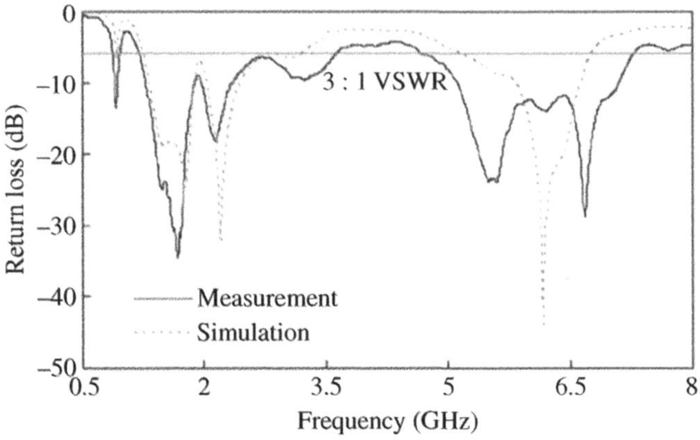

Figure 4.17 Simulated and measured reflection coefficient of the multiband antenna. Source: Reproduced with permission from Ref. [15]/IEEE.

Figure 4.17 shows the reflection coefficient of the prototype antenna obtained from simulations and measurements with good agreement between them. The lower band has impedance bandwidths (VSWR < 3) of 89 MHz (9.64%), spanning from 877 to 966 MHz, while the middle band has bandwidths of 2422 MHz (100.16%), covering the frequency range of 1210–3632 MHz. The higher band, on the other hand, has bandwidths of 2581 MHz (43.15%), covering frequencies ranging from 4688 to 7269 MHz. The measured impedance bandwidth is comprehensive and covers the entire frequency range for multiband operations of GSM/DCS/PCS/UMTS/WLAN/WiMAX.

Figure 4.18 illustrates the measured reflection coefficients for the monopole antenna, loop antenna, and antenna. The monopole antenna exhibited four resonant modes at approximately 1.0, 1.7, 2.4, and 5.2 GHz. The loop antenna produced a higher band at around 6500 MHz. The radiation patterns obtained displayed near-omnidirectional characteristics in the x–z plane, while monopole-like patterns were observed in the y–z plane. In the frequency range of 0.5–6 GHz, the antenna demonstrated a measured peak gain.

Next, the work in [16] proposed the design of a dual-band eight-antenna array for MIMO applications in 5G mobile terminals. The antenna operates in the LTE band 42 (3400–3600 MHz) and band 46 (5150–5925 MHz). The whole array is designed on a FR4 ($\varepsilon_r = 4.4$, $\tan\delta = 0.025$) substrate with a size of 140 mm × 70 mm × 1 mm. It comprises eight L-shaped slot antennas based on stepped impedance resonators (SIRs) as shown in Figure 4.19. The SIRs are used to achieve dual resonance by adjusting the impedance ratio and electrical length

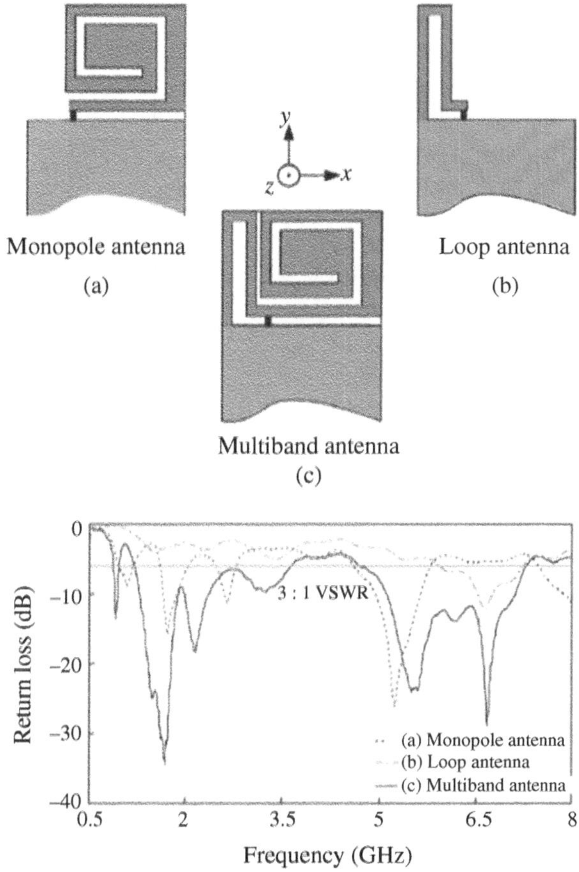

Figure 4.18 Measured return loss for (a) monopole antenna, (b) loop antenna, and (c) multiband antenna. Source: Reproduced with permission from Ref. [15]/IEEE.

ratio to enable impedance matching for each antenna element. Each slot antenna element based on a SIR is excited by a 50 Ω microstrip line etched on the top side of the substrate. For each microstrip feed line, with one end connected with a SMA connector, whereas the other end is shorted to the ground to produce an inductance for impedance matching. The feedline for this eight-element MIMO antenna array is depicted in Figure 4.20a. The microstrip feed lines are located on the substrate's top side that are connected to the ground plane through vias.

Figure 4.20b compares the simulated and experimental reflection coefficients for only Ant 1 and Ant 2, due to the array's symmetric structure. Despite small fabrication variation and the effect of the SMA connector, there are minimal differences between the measured and simulated results. The measured 10-dB

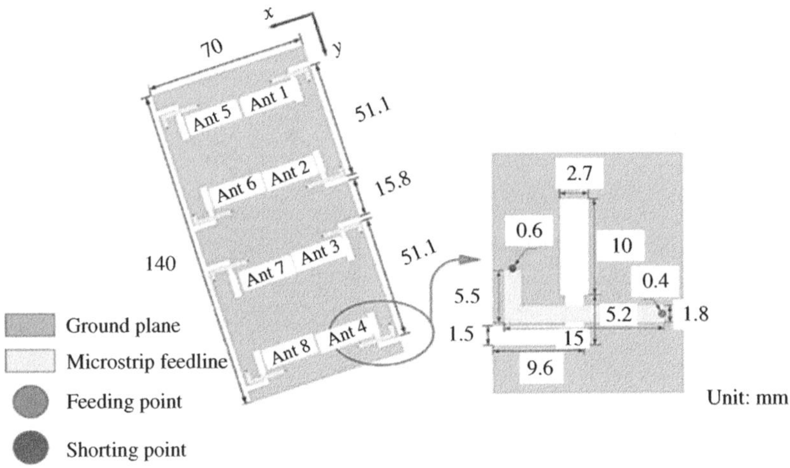

Figure 4.19 Geometrical configuration of the 8-port dual-band MIMO antenna array. Source: Reproduced with permission from Ref. [16]/IEEE.

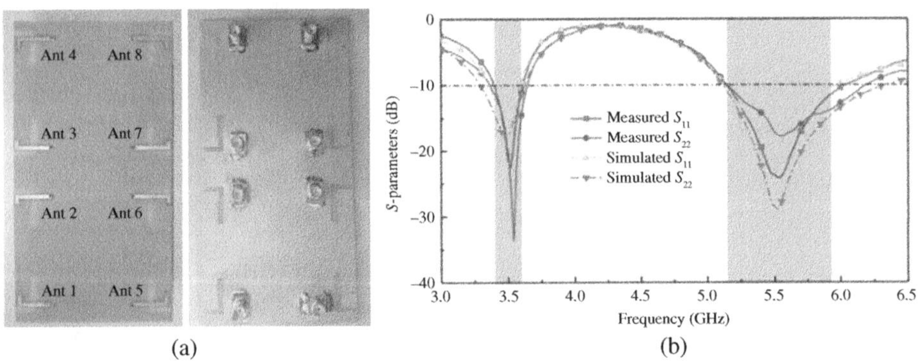

Figure 4.20 (a) Prototype of the manufactured dual-band eight-antenna MIMO array. (b) Simulated and experimental reflection coefficient of Ant1 and Ant2.
Source: Reproduced with permission from Ref. [16]/IEEE.

impedance range from 3390 to 3610 MHz (220 MHz) and from 5120 to 6040 MHz (920 MHz).

The *S*-parameters for the eight-element MIMO antenna array are presented in Figure 4.21. Reflection coefficient measurements for Ant *i* ($i = 3$; 4) are provided due to structural symmetry. The 10-dB impedance bandwidths for Ant 3 range from 7.4% (3380–3640 MHz) to 23.6% (5140–6440 MHz), while those for Ant 4 are 6.3% (3400–3620 MHz) to 17.4% (5120–6080 MHz). Measured isolations (S_{12} and S_{23}) are larger and are about 11.2 dB better in comparison to the simulations.

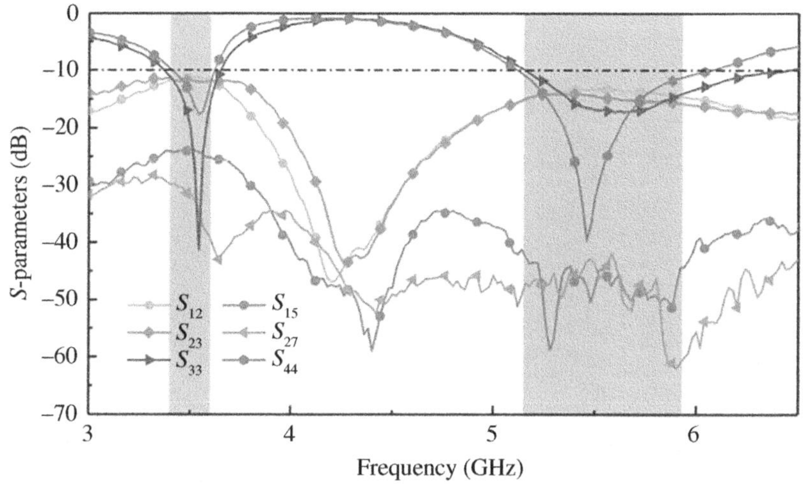

Figure 4.21 Experimental S-parameters of the MIMO antenna array. Source: Reproduced with permission from Ref. [16]/IEEE.

The measured isolation between neighboring antennas on the narrow side, i.e., S_{15}, is smaller and greater than 23 dB within the 3500-MHz band and 5500-MHz band, respectively. These results are consistent with the simulations.

A MIMO antenna system for multiband millimeter wave 5G and wideband 4G mobile communications is presented in [17]. It operates across three frequency bands, namely (28, 37, and 39 GHz) for 5G, besides the wider (1.8–2.6 GHz) for 4G. Its structure is presented in Figure 4.22a, and is fabricated on an RT/duroid-5880 substrate ($\varepsilon_r = 2.2$, $\tan\delta = 0.0009$, and thickness $t = 0.381$ mm) with the overall size of 70 × 77.8 mm² ($L_{sub} \times W_{sub}$). Four antenna elements (Ant1–Ant4) are each located on all four sides of the substrate, while microstrip feeding lines are placed on the top layer. Tapered slots shown in Figure 4.22c are etched in the ground plane located on the bottom layer of the substrate. Two different methods, F_1 and F_2, are utilized to excite a slot as an open-ended and the same slot as a tapered slot antenna, respectively, covering a frequency range of 1.8–2.6 GHz and 25–40 GHz. When the slot is fed from the edge (F_2), it works as a tapered slot antenna, providing high gain end-fire radiation patterns for 5G FR2 band. When fed from the center (F_1), the antenna works as an open-ended slot antenna with an omni-directional radiation at 2 GHz band to meet the requirements of 5G and 4G bands.

The simulated and measured reflection coefficients for the 5G band shown in Figure 4.23a indicated operation from 25 to 40 GHz. Figure 4.23b shows the isolation curves between F_1 and F_2, indicating a minimum of 25 dB measured isolation. Measured isolation between 5G ports (F_2, F_3, F_5, and F_7) is found to be over 50 dB.

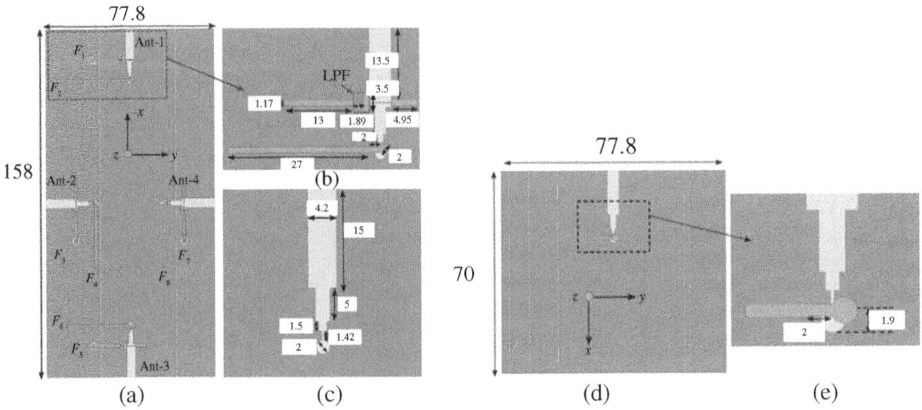

Figure 4.22 The geometry of the four-element MIMO antenna (a). (a) Top view (full design), (b) Ant-1 (enlarged view), and (c) tapered slot in the ground plane. (d) Top view and (e) enlarged view. Source: Reproduced with permission from Ref. [17]/IEEE.

On the other hand, the simulated and measured reflection coefficients for the 4G band are shown in Figure 4.23c. It indicated −6 dB impedance matching bandwidth from 1.8 to 2.6 GHz. Isolation between various 4G and 5G ports is presented in Figure 4.23d. Due to the perpendicular electric field of the tapered slot antenna to the slot orientation, the polarization of Ant-1 and Ant-3 aligns in the *y*-direction,

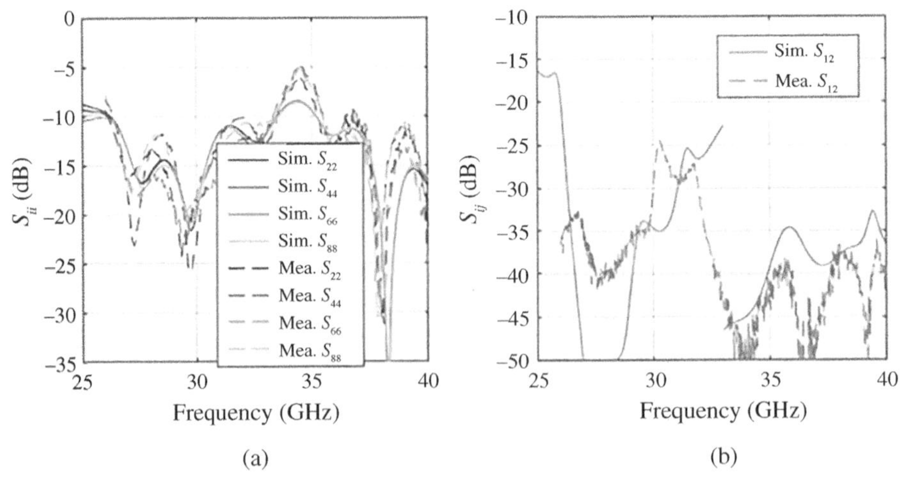

Figure 4.23 Simulated and measured *S*-parameters for 5G band. (a) S_{ii} and (b) S_{ij}. Simulated and measured *S*-parameters for 4G band. (c) S_{ii} and (d) S_{ij}. Source: Reproduced with permission from Ref. [17]/IEEE.

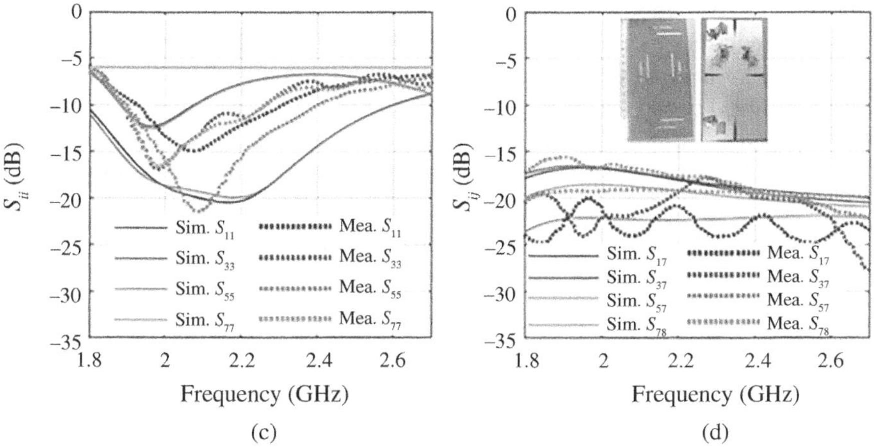

Figure 4.23 *(Continued)*

while the polarization of Ant-2 and Ant-4 aligns with the x-direction. Between F_5 and F_7, a minimum of 16 dB measured isolation is observed. This polarization diversity significantly enhances the isolation of different MIMO antenna elements. The maximum measured gain in the 5G band is 7.2 dBi at 39 GHz, whereas this is 3.6 dBi in the 4G band at 1.8 GHz. Measured efficiency is between 60–85% and 70–90% in the 5G and 4G bands, respectively. Both Ant-1 and Ant-2 exhibited directional radiation patterns in the xy-plane, with a directive beam toward the $\phi = 180°$ (Ant-1) and toward the $\phi = -90°$ (Ant-2) for the 4G band at 2.0 GHz. Conversely, in the xz and yz planes, both antennas display omni-directional radiation patterns.

Next, a single-layered MIMO antenna system is presented in [18] for operation in the WLAN (2.45, 5.2 GHz), 4G LTE (2.6 GHz) and 5G FR2 (24, 28 GHz) bands. It consists of a four-element dual-band monopole that operates in the 2.45 and 5.2 GHz bands. A ground plane with a decoupling structure is designed with a tapered slot as a high gain traveling wave antenna at 28 GHz. It is fabricated on Rogers RT/duroid 5880 with a relative permittivity of 2.2, a loss tangent of 0.0009 at 10 GHz, and a thickness (t) of 0.51 mm. Figure 4.24a illustrates its geometry, fabricated on a square-shaped substrate with a bottom ground plane. The monopole antenna (Ant-1) comprises two primary radiating arms with length of $L_1 = 26$ mm and $L_2 = 9$ mm fed by a 50 Ω transmission line. Dimensions L_1 and L_2 are optimized for operation in the WLAN and LTE frequency bands at (2.45, 2.6, 5.2) GHz. Upon optimizing the monopole design (Ant-1), a copy (Ant-2) is placed perpendicular to Ant-1 as shown in Figure 4.24b.

The overall MIMO antenna is presented in Figure 4.24c, depicting a symmetrical arrangement of all antenna elements on a square-shaped substrate

(104 mm × 104 mm × 0.51 mm). Its low frequency operation is slightly affected by the four adjacent monopole antenna elements for 4G and WLAN bands, besides the need to incorporate the four 5G tapered slot antennas in the ground plane. This is solved upon a slight re-optimization of the entire structure, with the final prototyped antenna shown in Figure 4.24d–e. The reflection coefficients for Ant-1 to Ant-4 depicted in Figure 4.24f indicated its operation in the WLAN and LTE bands, specifically 2.45, 2.6, and 5.2 GHz, with -10dB impedance bandwidth from 2.40 to 2.80 GHz and 5.1 to 5.6 GHz. Furthermore, the isolation between Ant-1 to Ant-4 is greater than 16 dB, as seen in Figure 4.24g, representing an improvement

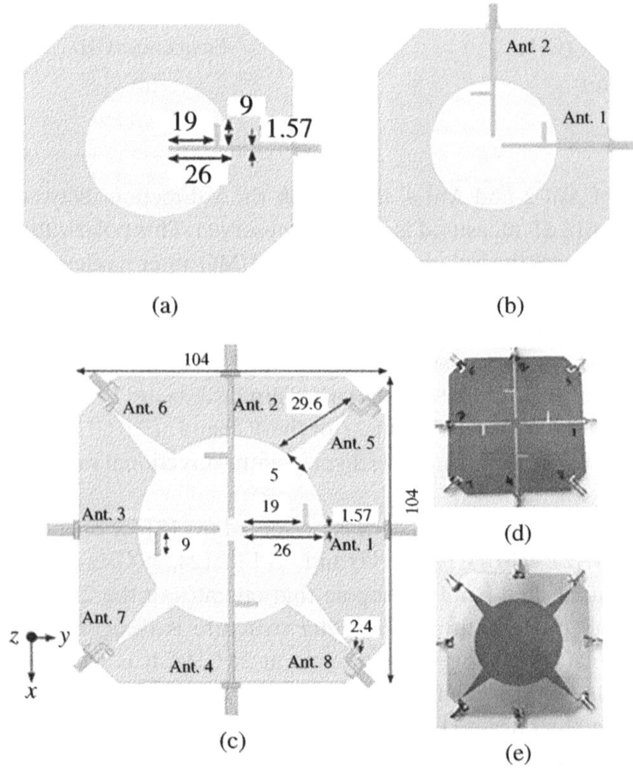

Figure 4.24 (a) Dual-band monopole antenna and (b) its simulated reflection coefficient. (c) Geometry of the MIMO antenna system and fabricated prototype. (d) Top view. (e) Bottom view. *S*-parameters for MIMO antenna system. (f) Reflection coefficient in low-frequency band. (g) Isolation in low-frequency band. (h) Reflection coefficient in high-frequency band. (i) Isolation in high-frequency band. Source: Reproduced with permission from Ref. [18]/IEEE.

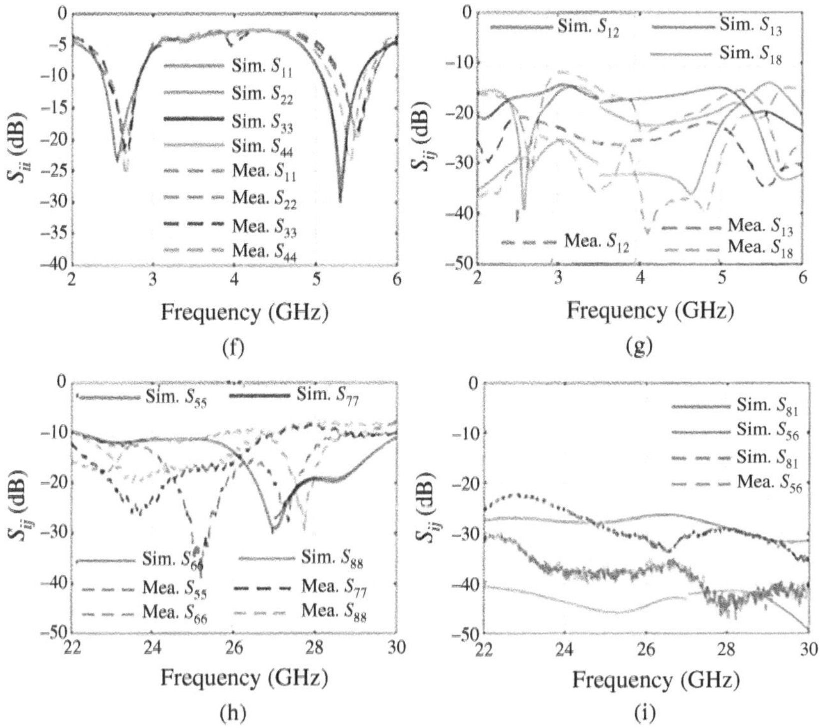

Figure 4.24 *(Continued)*

of at least 8 dB. The reflection coefficients for Ant-5 to Ant-8 are depicted in Figure 4.24h, whereas isolation are illustrated in Figure 4.24i for the FR2 band. Results indicate that Ant-5 to Ant-8 operated with broad bandwidths, ranging from 23 to 30 GHz, accommodating the 24 and 28 GHz bands for 5G applications.

4.3 Wideband and Ultra-Wide Band (UWB) MIMO Antennas

Several main potential wideband and UWB antenna types include the Vivaldi antenna, quasi-Yagi antenna, and planar monopole antenna. Despite their readily wideband characteristics, researchers have also included customized features for MIMO applications such as additional band notches and improving angle tilting of the beams in these antennas. The main types of UWB antennas and their design examples will be presented in the following subsections.

4.3.1 UWB MIMO Antenna for the Sub-6 GHz Band

4.3.1.1 Vivaldi-Based UWB Antennas

The Vivaldi antenna is a broadband antenna with high directivity and stable radiation patterns, designed as a planar structure typically on a dielectric substrate. The transmission line is surrounded by a feedline that terminates in a sector-shaped area, which encircles the circular space. This allows the radiating energy to be distributed in exponential patterns through the symmetrical slot lines prior to arriving at the resonant points [19]. There are typically three types of Vivaldi antennas, namely the tapered slot Vivaldi antenna (TSVA), antipodal Vivaldi antenna (AVA), and balanced antipodal Vivaldi antenna (BAVA). These parameters are discussed in more detail in [20–22]. Researchers have also modified Vivaldi antennas by introducing slots to improve compactness, gain, and radiation efficiency.

An example is a Vivaldi antenna with a wide bandwidth of 23.19 GHz (3.06–26.25 GHz) and a compact size of $45 \times 35 \times 0.79\ \text{mm}^3$ from [23]. This Vivaldi antenna features an exponential double-sided corrugated tapered slot. Its design process involves five steps on a Rogers RT 5880 substrate, with the first step being the introduction of a simple tapered Vivaldi radiator shown in Figure 4.25a. Two exponential tapered slots are symmetrically placed on top of the substrate, and a microstrip line feeding connects them. The dimensions of its radiating patch profile are designed using:

$$x = C_1 e^{Rz} + C_2, \tag{4.5}$$

where

$$C_1 = \frac{x_2 - x_1}{e^{Rz_2} - e^{Rz_1}}, \tag{4.6}$$

$$C_2 = \frac{x_2 e^{Rz_2} - x_1 e^{Rz_1}}{e^{Rz_2} - e^{Rz_1}} \tag{4.7}$$

The flared opening is determined by the coordinates of the endpoints (x_1, z_1) and (x_2, z_2), with a taper rate of 0.135, resulting in a wide opening of 25.34 mm at the edge of the flares. Figure 4.25b shows the introduction of four corrugated slots on the top two sides of the metallic plates, with two slots for each flare and a width of 2 mm. The lengths of the slots are 4 mm and 4.5 mm, respectively. In the third step, an additional six corrugated slots are created, and their dimensions optimized for improved performance relative to the previous steps, with the slots dimensioned at 5 and 5.5 mm. The configuration of the antenna is illustrated in Figure 4.25c, where two 5 mm-radius circular slots are included in the fourth step. Figure 4.25d depicts the corrugated circular slotted tapered Vivaldi antenna.

The introduction of the slots and corrugations increased bandwidth, gain, and efficiency. In addition, a via (with a height of 0.86 mm and radius of 0.6 mm) is

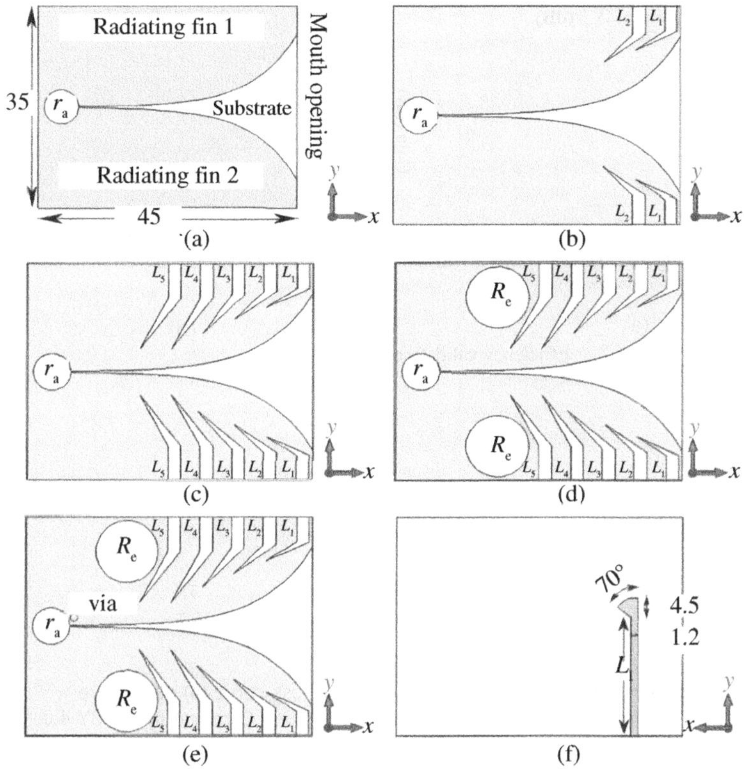

Figure 4.25 Evolution process of the Vivaldi antenna: (a) Step 1 (front view); (b) Step 2 (front view); (c) Step 3 (front view); (d) Step 4 (front view); (e) Step 5 (front view); (f) back view. Source: Reference [23]/Hindawi Publishing Corporation/CC BY 4.0.

used to connect the feed line and metal flare as shown in Figure 4.25e. This is to further increase antenna's bandwidth by modifying its surface current distribution. The Vivaldi's feed structure is shown in Figure 4.25f. Various techniques were employed to enhance the antenna's performance, including corrugated both-sided slots, circular slots, and via. After the parameter sweep process, a wideband compact tapered slot Vivaldi antenna was developed.

The modified flares increased the electrical length and improved antenna gain beyond 8 dBi. Note that substrate selection can also significantly affects bandwidth and efficiency. Figure 4.26a–c shows the reflection coefficient, gain, and efficiency of the Vivaldi antenna, which operates from 3.06 to 26.25 GHz, with a maximum gain of 10.2 dBi, and an average radiation efficiency exceeding 95%.

The next example of UWB MIMO Vivaldi antenna is enhanced with dual band-notched characteristics [24]. Featuring a compact size of $26 \times 26\ mm^2$, a

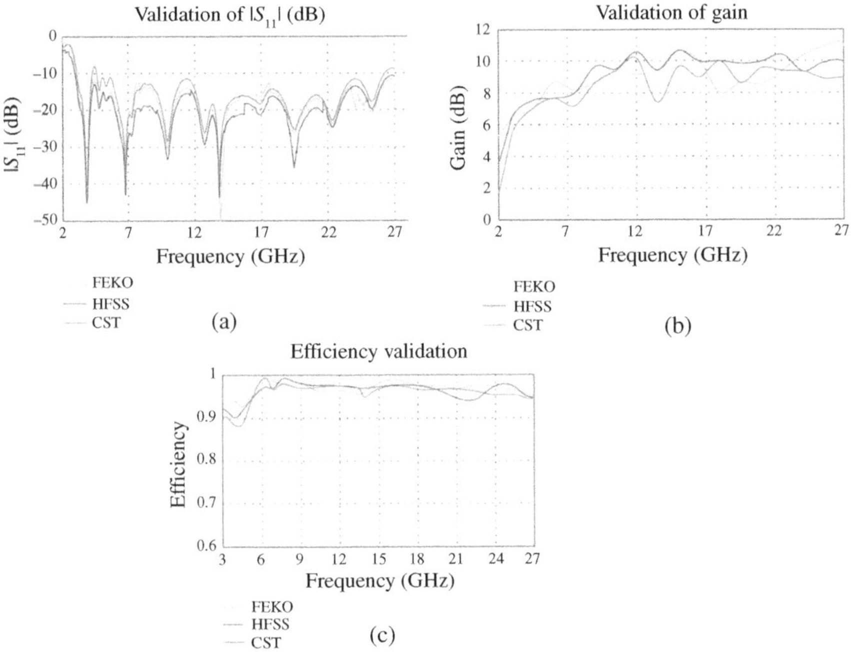

Figure 4.26 Validation of the Vivaldi antenna (a) return loss curve (b) gain curve (c) efficiency curve. Source: Reference [23]/Hindawi Publishing Corporation/CC BY 4.0.

T-shaped slot is introduced onto the ground plane to enhance inter-element isolation. The antenna is fed by two 50 Ω microstrip lines and consists of MIMO Vivaldi antenna and four SRRs. To enable dual-band notched characteristics, two split-ring resonators (SRRs) of varying sizes are incorporated next to the microstrip feed lines. The structure of the overall antenna fabricated on 0.762 mm-thick Taconic RF-35 (dielectric constant ε_r of it is 3.5 and the loss tangent is 0.0018) substrate is shown in Figure 4.27. The tapered curve function of the Vivaldi antenna is calculated similarly as conventional antennas, as follows:

$$y = C_1 e^{ax} + C_2 \tag{4.8}$$

Here, the a is 0.2. C_1 and C_2 can be determined by L_1 and d.

Figure 4.28 illustrates the four design steps of the dual band-notched MIMO Vivaldi antenna. First, as shown in Figure 4.28a, a dual-port Vivaldi antenna is designed and referred to as Antenna 1. This is then incorporated with a rectangular slot etched on the ground plane to reduce inter-element coupling, as shown in Figure 4.28b (Antenna 2). Then, a T-shaped slot is added on the ground plane to decrease surface current on the ground plane between two ports and further increase the port isolation (Antenna 3 in Figure 4.28c). Besides that, matching

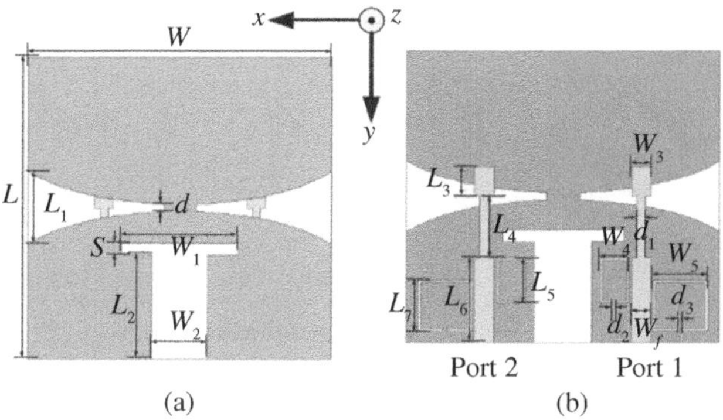

Figure 4.27 Configuration of the designed antenna. (a) Top view. (b) Bottom view. Source: Reproduced with permission from Ref. [24]/IEEE.

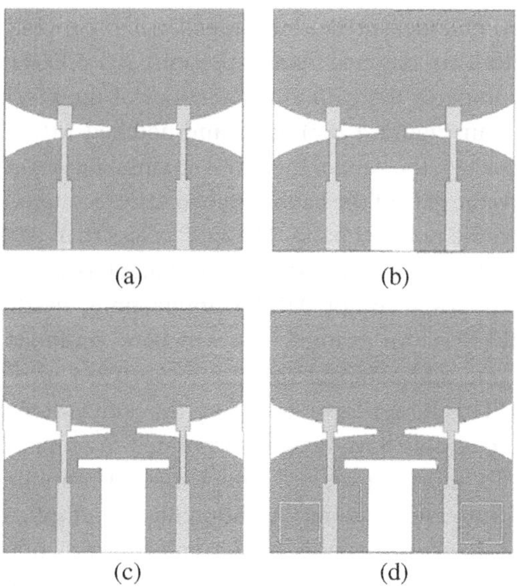

Figure 4.28 Design steps of the designed MIMO antenna. (a) Antenna 1. (b) Antenna 2. (c) Antenna 3. (d) Antenna 4. Source: Reproduced with permission from Ref. [24]/IEEE.

in the lower band also improved, improving impedance bandwidth. As the final step, two different SRRs are incorporated near each microstrip line to introduce band-notches in the WLAN and X-bands as shown in Figure 4.28d. The longer length of the SRR is 17.6 mm, whereas the shorter length of the SRR is 12.4 mm. The final designed antenna featured an impedance bandwidth of 2.9–11.7 GHz with two notched bands at 5.2–5.8 GHz and 7.8–8.5 GHz.

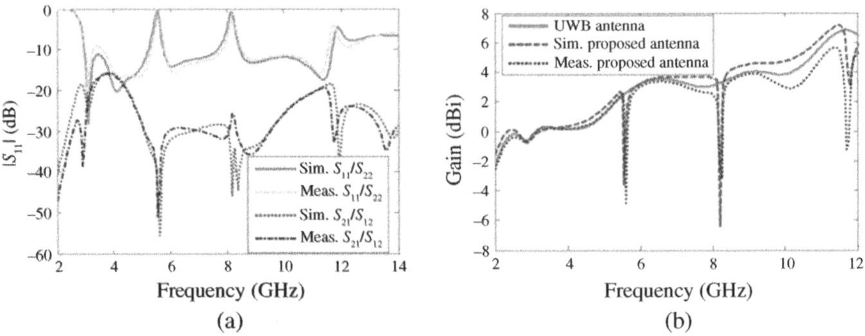

Figure 4.29 (a) Measured *S*-parameters. (b) Measured gain. Source: Reproduced with permission from Ref. [24]/IEEE.

The measured S_{11}/S_{22} and S_{21}/S_{12} are depicted in Figure 4.29a indicating good agreements with simulation results. The measured 10-dB impedance bandwidth ranges from 2.9 to 11.6 GHz, with two notched bands present: 5.3–5.8 GHz and 7.85–8.55 GHz. Figure 4.29b compares the gain characteristics of the UWB without the SRR rings (Antenna 3 in Figure 4.28c) and band-notched MIMO antennas with port 1 excited. As expected, the gain of the UWB antenna increased consistently over frequency, whereas the gain of the band-notched MIMO antenna experienced a sharp decrease within the notched band. The ECC is less than 0.02 within the entire impedance bandwidth, indicating its diversity characteristics.

Vivaldi antennas have also been designed for MIMO applications in the millimeter-wave band for 5G application. An example is a wideband antipodal Vivaldi antenna (AVA) with high gain and compact size ($12 \times 5.5 \times 0.254$ mm^3) in [25], as depicted in Figure 4.30. This design operates using the conventional AVA geometry above 20 GHz, whereas slots have been incorporated to enhance performance at lower frequencies. To improve gain and reduce mutual coupling, a metasurface (MS) is introduced into the antenna. Additionally, an elliptical dielectric lens has been applied to focus the energy in the end-fire direction, without altering the impedance performance. The two Vivaldi radiator and ground arms are integrated on both the top and bottom sides of a Rogers 4350B ($\varepsilon_r = 3.48$) dielectric substrate with a thickness of 0.254 mm.

The length and width (W) of the basic antenna geometry are 10 and 5.5 mm, respectively, calculated based on [11], and indicated as Evol1 in Figure 4.30a. To enhance impedance bandwidth and radiation characteristics in the second step (Evol2), three slots are introduced as corrugation edges on the outline of the antipodal tapered slot antenna. In the third step (Evol3), a pair of four-branch MS structures shown in Figure 4.30b is introduced on the antenna to minimize mutual coupling and enhance gain in the lower frequencies. To further enhance

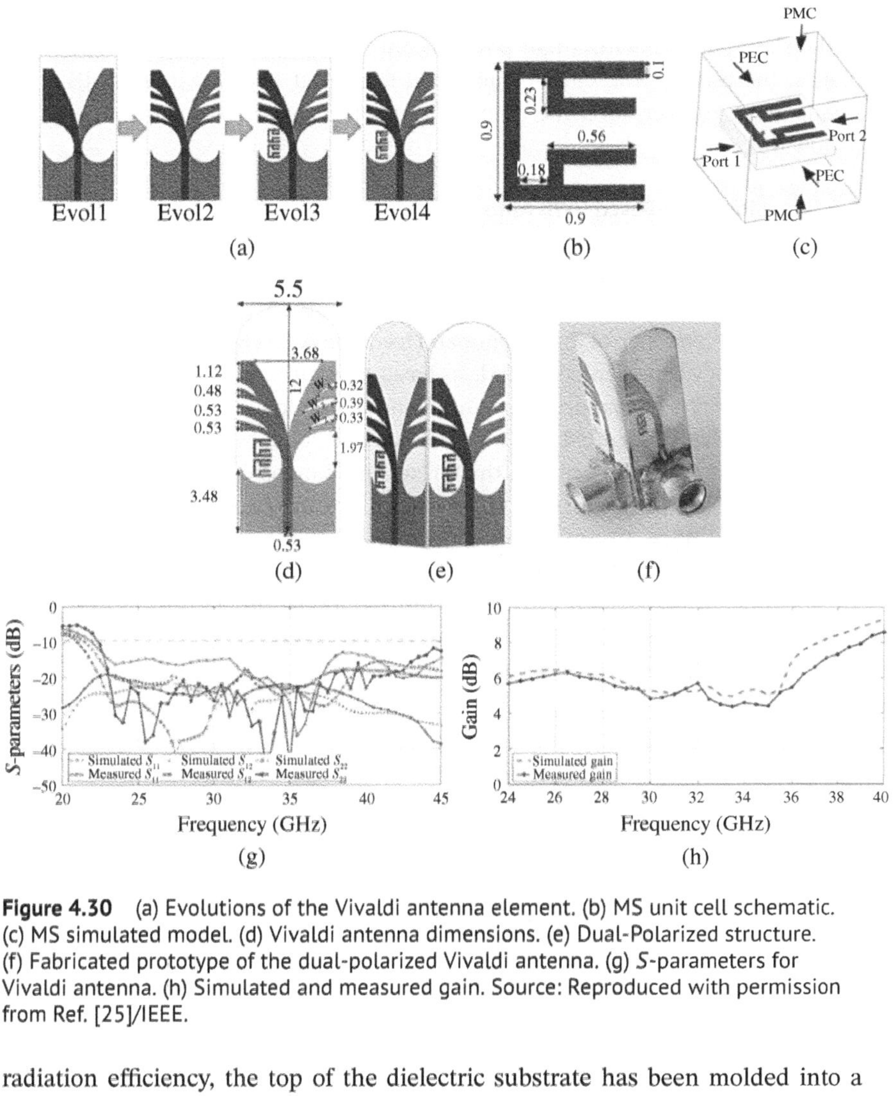

Figure 4.30 (a) Evolutions of the Vivaldi antenna element. (b) MS unit cell schematic. (c) MS simulated model. (d) Vivaldi antenna dimensions. (e) Dual-Polarized structure. (f) Fabricated prototype of the dual-polarized Vivaldi antenna. (g) *S*-parameters for Vivaldi antenna. (h) Simulated and measured gain. Source: Reproduced with permission from Ref. [25]/IEEE.

radiation efficiency, the top of the dielectric substrate has been molded into a half-elliptical dielectric lens, as demonstrated in the fourth evolution. The Vivaldi antenna element and its dual-polarized structure are presented in Figure 4.30d, e.

A fabricated antenna prototype is shown in Figure 4.30f fed using SMPM connectors. The *S*-parameters, both simulated and measured, are shown in Figure 4.30g to operate from 20 to 45 GHz. Measurement results indicated reflection coefficients exceeding −14 dB and a mutual coupling of over −20 dB, which are suitable for 5G FR2 band from about 24–43.5 GHz. Simulated and measured gain which is almost identical for both polarizations is shown in Figure 4.30h.

The antenna exhibits end-fire radiation targeted toward the symmetrical axis of the slot aperture, with a half-power beam width of at least 65° for both *E*/*H* planes. In addition, the cross-polarization measured is better than 15 dB at all frequencies. Furthermore, the antenna's gain is at least 5 dBi across the entire bandwidth.

4.3.1.2 Monopole-Based UWB Antennas

Another popular UWB antenna type is the planar monopole antenna. It typically consists of a radiator (of any shape), a feed line (microstrip, coplanar waveguide, etc.) on the same side of a substrate, and a partial ground plane, located on either the same or reverse side of the radiator and feedline. A UWB MIMO antenna based on this topology is presented in [26]. Its basic structure consists of a metallic disc of radius R and it has a partial ground plane with a length L_m and a width W_m placed as shown in Figure 4.31. It is designed on FR4 substrate having a relative permittivity of 4.3 and 0.8 mm of thickness. The dimensions obtained after optimization are: length of substrate (L_s) = 43 mm, width of substrate (W_s) = 40 mm, L_m = 12.2 mm, W_m = 40 mm, length of feed line (L) = 12.5 mm, width of feed line (W) = 1.5 mm, and R = 12.4 mm. As a MIMO antenna, two different configurations are integrated with two identical circular disc monopole antenna elements to study two antenna systems.

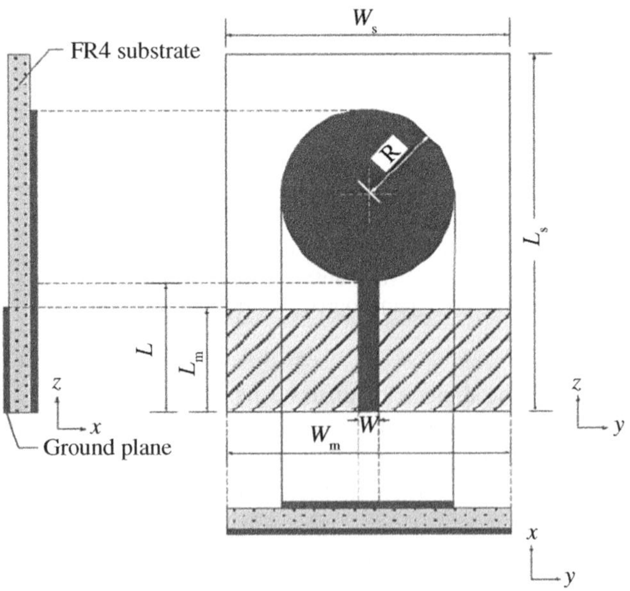

Figure 4.31 Circular disc monopole antenna layout. Source: Reproduced with permission from Ref. [26]/Yvan Duroc.

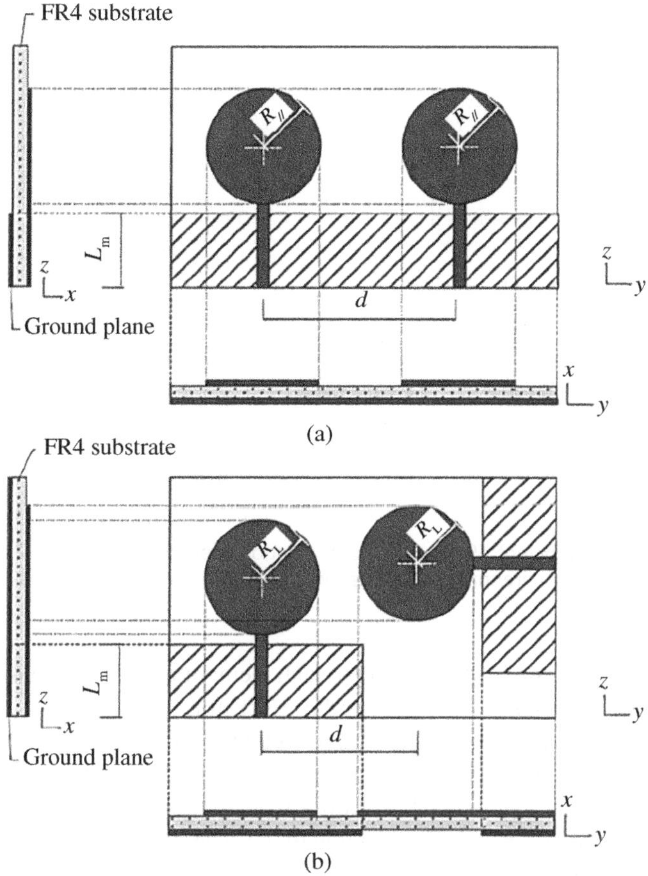

Figure 4.32 The developed antenna systems: (a) System 1 (parallel configuration); (b) System 2 orthogonal configuration. Source: Reproduced with permission from Ref. [26]/Yvan Duroc.

The parallel and perpendicular arrangements of the elements are denoted as System 1 and 2, respectively, as shown in Figure 4.32. In both systems, two antenna elements are separated by a distance *d* and are placed on a substrate dimensioned at $43\times\ 80\,\text{mm}^2$ (length × width). In System 1, the antennas are placed symmetrically, separated by a distance *d* and over a common ground plane dimensioned at $12.2\times80\,\text{mm}^2$, as shown in Figure 4.32a. The distance d and the radius of the disc, $R_{//}$ are optimized to be $d=43.5\,\text{mm}$ and $R_{//}=13.7\,\text{mm}$ for minimum reflection coefficient and mutual coupling. On the other hand, in System 2, the antenna elements are placed orthogonally without sharing a common ground plane, as shown in Figure 4.32b. Besides d, the radius of disc

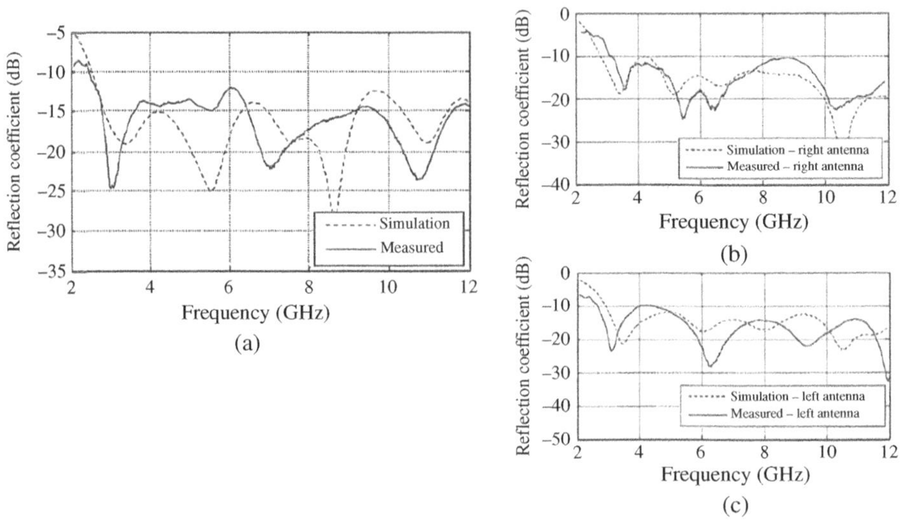

Figure 4.33 (a) Reflection coefficient of System 1; (b) Reflection coefficient of System 2 of the right antenna; and (c) Left antenna. Source: Reproduced with permission from Ref. [26]/Yvan Duroc.

($R^{\perp}$) and ground plane dimensions ($L_m \times W_m$) for each element are optimized to be $L_m = 12.2$ mm; $W_m = 34$ mm; $R^{\perp} = 10.7$ mm; and $d = 37.8$ mm.

Figure 4.33a shows the simulated and measured reflection coefficients, S_{11} in dB for System 1, whereas for System 2, both reflection coefficients for the right and left elements are presented in Figure 4.33b, c. Notice that for System 2, there are some differences in the curves for each element due to the asymmetrical arrangement. However, after integration of the two elements, impedance matching is still acceptable in both antenna systems.

The radiation patterns for each element in both systems are observed at three frequencies (3.1, 5.5, and 10 GHz). Plots for System 1 and 2 are illustrated in Figures 4.34a and 4.34b, respectively. The radiation patterns are nearly omni-directional at lower frequencies in the H-plane. In System 1, the radiation patterns for each element are symmetric in terms of magnitude, but non-symmetric with respect to position, e.g. exactly 180° change at 10 GHz. This is caused by the proximity of the second antenna element.

The simulated and measured correlations for both systems are illustrated in Figure 4.35 and are generally lower than the diversity threshold value of 0.5 [27]. Due to the measured S_{11} and S_{12} for System 1, measured correlation is lower also than the simulations within 6.0–7.8 GHz, as shown in Figure 4.35a. Although it seems to be a wide difference, its effects are not significant as the correlation values obtained are very low. On the other hand, correlation levels for System 2 shown in

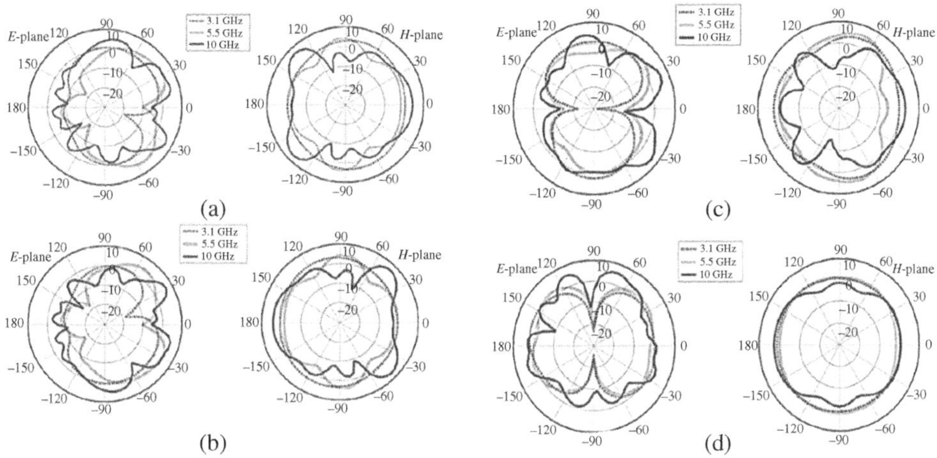

Figure 4.34 Radiation patterns of System 1: (a) Left antenna (b) Right antenna. (b) Radiation patterns of System 2: (a) Left antenna (b) Right antenna. Source: Reproduced with permission from Ref. [26]/IEEE.

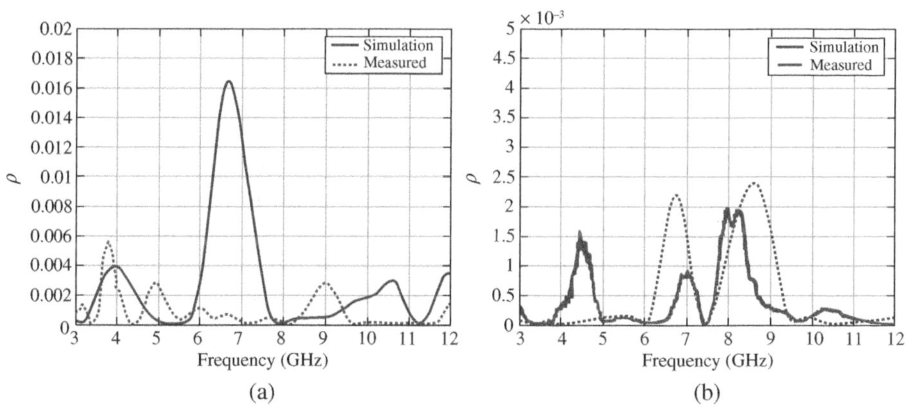

Figure 4.35 (a) Average correlation: (a) System 1 and (b) System 2. Source: Reproduced with permission from Ref. [26]/IEEE.

Figure 4.35b are low due to its low mutual coupling resulting from its orthogonal element orientation. System 2 also offers better size compactness and diversity performance, whereas System 1 eases integration with circuits due to the location of the excitation ports on the same side.

Next, a planar monopole with more complex features is proposed in [28]. Two cone-shaped radiating elements are introduced inside the T-shaped protruded region of the ground plane for operation from 3.1–5.8 GHz, as shown

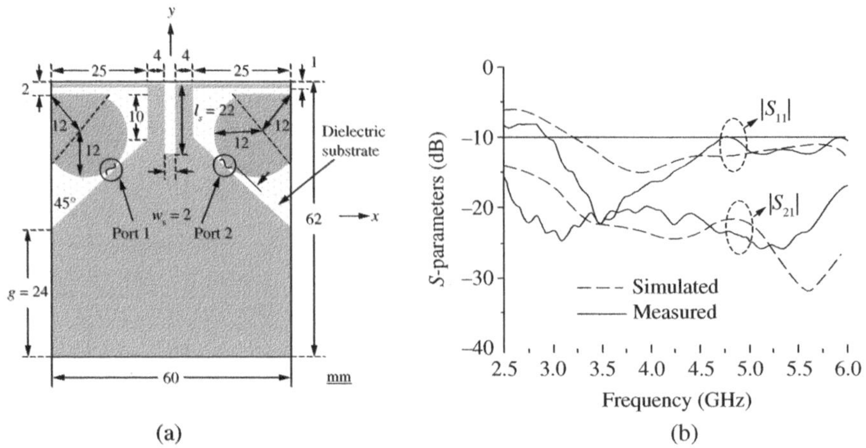

Figure 4.36 (a) Antenna geometry. (b) Simulated and measured S_{11} and S_{21}. Source: Reproduced with permission from Ref. [28]/IEEE.

in Figure 4.36a The height of the cone is 24 mm and the radius of the rounded arc is 12 mm. The antenna is printed on a 1.52 mm-thick Rogers 4003 substrate ($\varepsilon_r = 3.38$) with an overall dimension of 60 mm × 62 mm. The gap between the radiator and the system ground plane is 1 mm. Each radiator is excited by a 50 Ω co-axial probe of radius 0.6 mm at the midpoint along the circumference of the arc. The T-shaped ground plane protrudes from the main ground plane and is separated into two sections by a central slot measuring $w_s \times l_s = 2 \times 22$ mm. The length of the vertical section of the half T-shape is 4 × 12 mm and the top horizontal strip is 25 × 2 mm. It helps enhance the isolation between the two feeding ports and serves as a reflecting plate for the two antennas to provide spatial diversity. Figure 4.36b shows the antenna's simulated and measured reflection coefficient, S_{11} and isolation S_{21} between the two ports with an impedance bandwidth of 3.1–5.8 GHz ($S_{11} < -10$ dB). By adjusting the length l_s of the central slot, the isolation S_{21} between the two feeding ports can be optimized across the impedance bandwidth. Figure 4.37a, b shows the measured radiation pattern of the antenna at 3.5 GHz in three principal planes (x–y plane, $\varphi = 45°/225°$ plane, and $\varphi = 135°/315°$) when excited at Port 1 and Port 2, respectively. Measurements indicated stable radiation pattern with acceptable gain and cross-polar discrimination.

4.3.1.3 IFA-Based MIMO Antennas

Another UWB antenna type suitable for MIMO systems is the IFA-based antenna. An example of this is a UWB 4 × 4 MIMO antenna constructed by two PIFA-pairs for 5G smartphones proposed in [29]. Two PIFAs with inverted-T

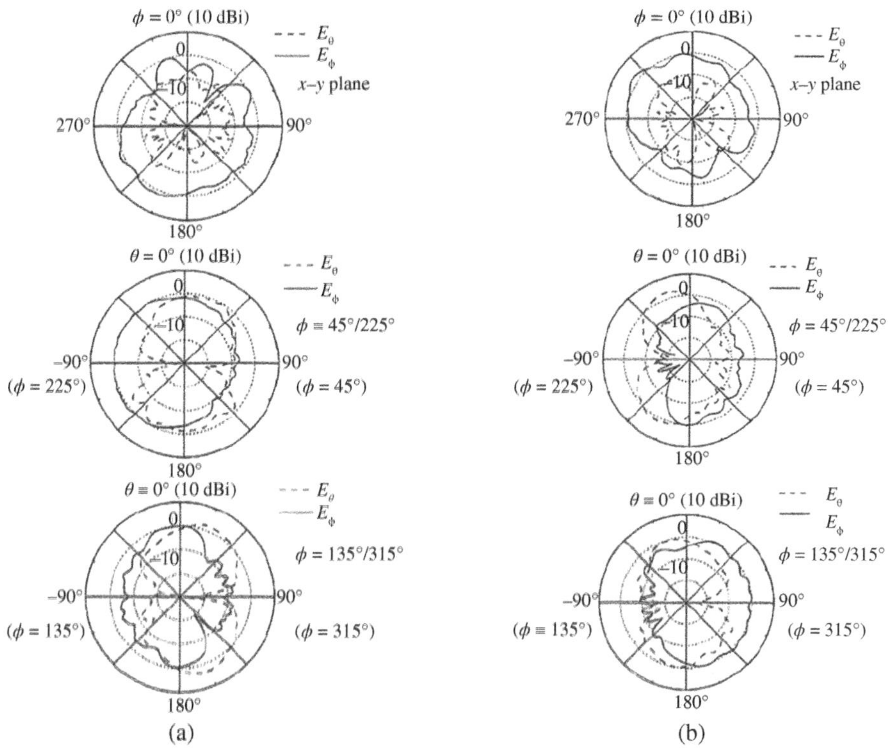

Figure 4.37 Radiation patterns at 3.5 GHz (a) with Port 1 excited; (b) with Port 2 excited. Source: Reproduced with permission from Ref. [28]/IEEE.

shaped open slot have been combined to form a PIFA-pair that operates over a wide bandwidth of 78% from 3.3 to 7.5 GHz for 5G NR frequency bands such as n77/n78/n79 and LTE band 46. Despite being co-located closely, the two PIFAs in the PIFA-pair maintained an isolation of more than 10 dB without requiring extra decoupling structures. The shorting stub in the PIFA-pair ensured high inter-element isolation, allowing current to flow from the resonant PIFA in the PIFA-pair to the ground rather than to the non-resonant PIFA. Finally, a compact 4 × 4 MIMO antenna consisting of two PIFA-pairs with an overall dimension of 150 × 75 × 7 mm^3 on a 0.8 mm-thick FR4 substrate (permittivity of 4.4 and a loss tangent of 0.02) is designed, as seen in Figure 4.38. The MIMO antenna has been printed on the inner side surface of smartphone with no clearance on the ground. To widen the bandwidth, the integration of an inverted-T shaped open slot is introduced. This integration serves to incorporate additional resonant modes (loop mode and slot mode).

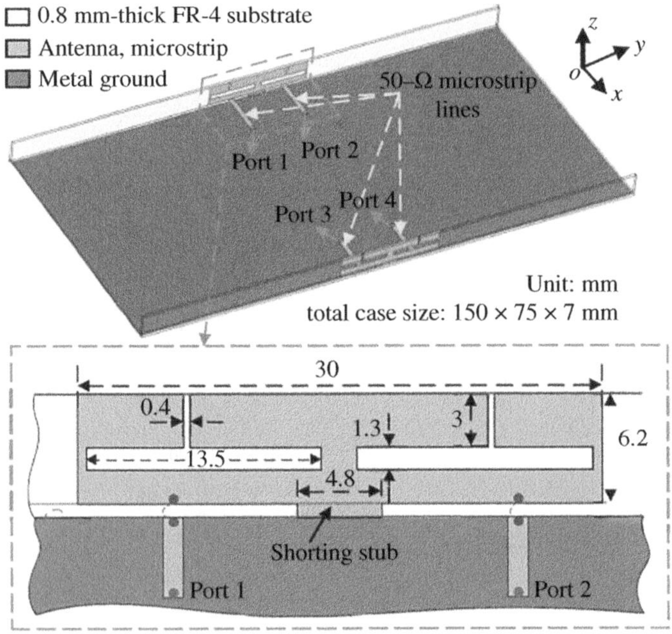

Figure 4.38 Geometry and detailed dimensions of the MIMO antenna. Source: Reproduced with permission from Ref. [29]/IEEE.

Figure 4.39a–c depicts the fabricated prototype, wherein four SMA connectors are situated beneath the ground to feed four antenna elements. Figure 4.39d–e illustrates the reflection coefficients and transmission coefficients. It is shown that a 6 dB bandwidth of 78% is achieved. On the other hand, isolation between the two PIFAs in a PIFA-pair is at least 10 dB, whereas that of different PIFA-pairs is higher than 22 dB. The total efficiencies measured across the desired bands in Figure 4.39f range from approximately 40–78%. Figure 4.39g shows the simulated and measured ECCs between PIFA-elements 1 and 2, 1 and 3, and 1 and 4, which are all observed to be below 0.06.

Another wideband eight-element MIMO handset antenna operating from 3.3 to 7.1 GHz is presented in [30]. The antenna is designed on an FR4 substrate (with a relative permittivity of 4.4 and loss tangent of 0.025) with an overall size of 150 mm × 75 mm × 0.8 m, as shown in Figure 4.40a. A metal ground is printed on the rear surface, while four vertical side frames (height = 7 mm) are electrically connected to the edges of the ground plane. The inner sides of the frames are copper plated to excite the four metal rims of the smartphone. A total of eight

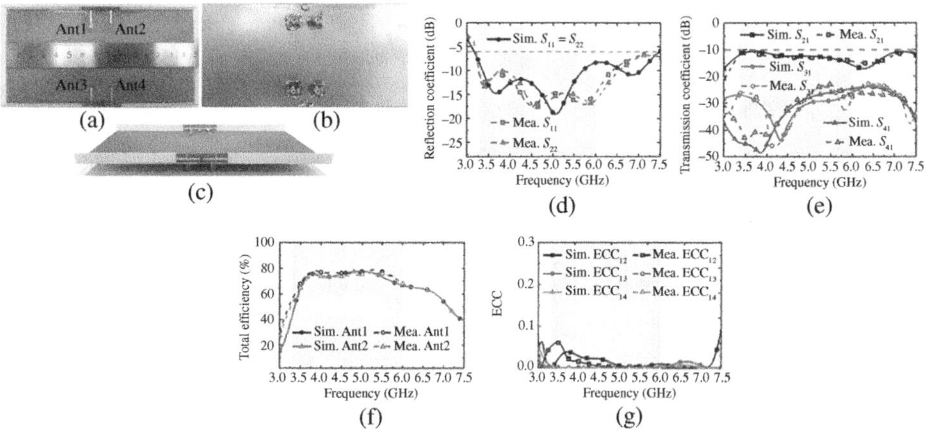

Figure 4.39 Photographs of the fabricated MIMO antenna. (a) Front view. (b) Back view. (c) Side view. Simulated and measured (a) reflection coefficients and (b) transmission coefficients of MIMO antenna. Simulated and measured (a) antenna efficiencies and (b) ECCs between some adjacent PIFA-elements. Source: Reproduced with permission from Ref. [29]/IEEE.

centrosymmetric antenna elements (Ant 1–Ant 8) are located along the two long edges, with a metal ground located 1 mm above the metal rims, as depicted in Figure 4.40b to allow sufficient space for the placement of the display module. The detailed dimensions of the antenna element are presented in Figure 4.40c, indicating a narrow rectangular slot clearance (12.4 mm × 1.5 mm) etched on the metal ground. Besides that, a rectangular gap is etched onto the long side-edge metal rim to form an *L*-shaped slot aperture.

Figure 4.41a, b presents the fabricated MIMO antenna, whereas Figure 4.41c shows the reflection coefficients for Ants 1–4. The transmission coefficients between adjacent antenna elements (S_{21}, S_{32}, and S_{43}) illustrated in Figure 4.41d are better than 11 dB. The isolation between Ant 1 and Ant 3 (S_{31}) is observed to be higher than 17 dB due to the larger separation distance. Besides that, isolations of other antenna elements opposite each other (S_{51}, S_{62}, and S_{52}) are at least 25 dB. The total efficiency ranges from 49% to 76%, whereas its realized gains are 2–4 dBi (Ant1) and 0.9–6 dBi (Ant2), as presented in Figure 4.41e.

4.3.1.4 Dielectric Resonator-Based UWB Antennas

DR antennas (DRAs) have evolved as an important class of antennas due to their various features and advantages such as rectangular, hemispherical, cylindrical, cylindrical-ring, conical, and hybrid shapes combining multiple basic geometries.

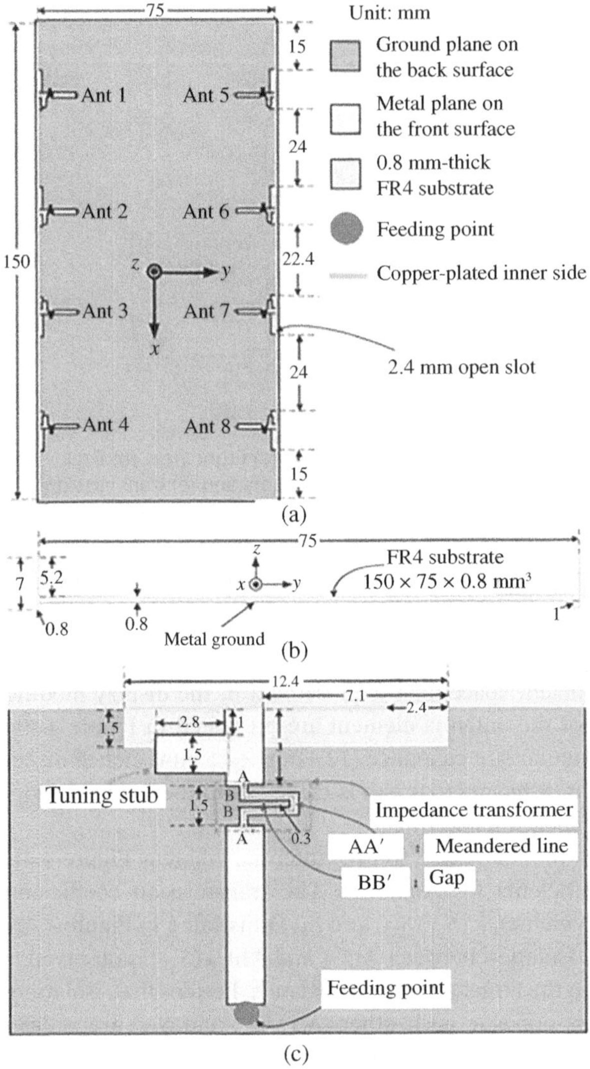

Figure 4.40 (a) Structure of the proposed antenna; (b) Side view of the antenna, (c) Top view of the antenna element. Source: Reproduced with permission from Ref. [30]/IEEE/Public Domain.

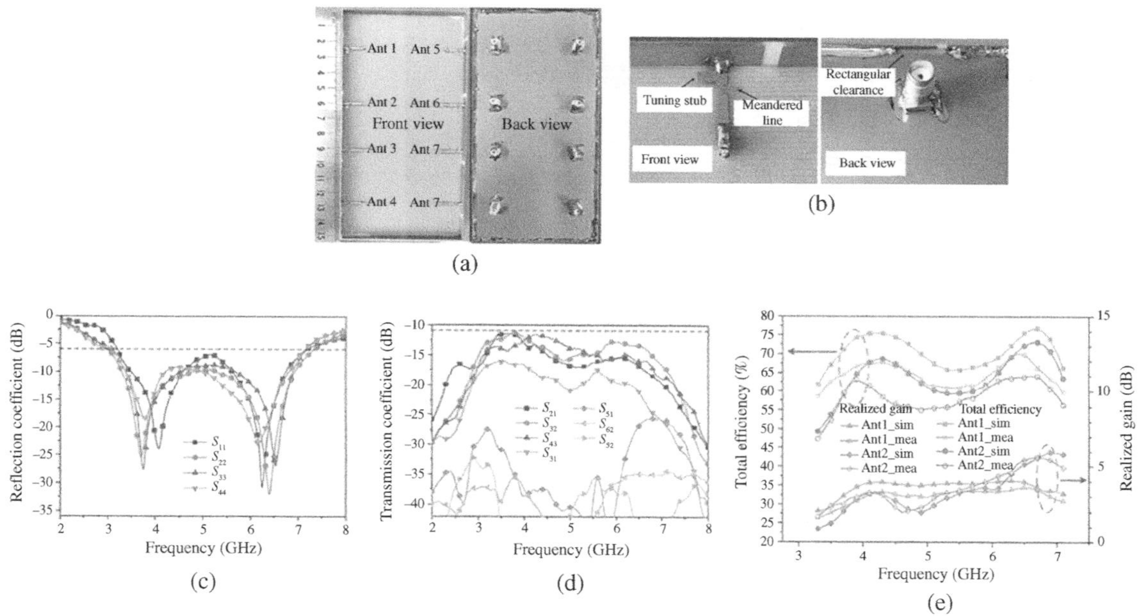

Figure 4.41 Photos of the fabricated MIMO antenna: (a) front and back view of the antenna; (b) front and back view of the antenna element. Measured (c) Reflection coefficients; and (d) Transmission coefficients of the fabricated antenna. (e) Total efficiency and realized gain of Ant 1 and Ant 2. Source: Reproduced with permission from Ref. [30]/IEEE/Public Domain.

Proper excitation of dielectric material with a suitable geometry and shape can contribute to multiple modes, enabling its wideband operation. Apart from that, DRAs offer very high efficiency relative to microstrip-based printed antennas for MIMO applications. Exciting various modes in a single DR is one of the most popular techniques for diversity, as DRs can effectively work as a cluster of multiple antennas. Orthogonal modes excited in a single DR (known as degenerated modes) facilitate the introduction of pattern and polarization diversity with a low correlation coefficient. Besides degenerated modes, DR may also generate multiple modes from structurally symmetric DR, multiple DR elements, use of hybrid shapes, and array-based design.

An example of the DRA MIMO antenna is presented in [31]. The stair-shaped dual rectangular DR antenna (RDRA) is fed with a coplanar waveguide (CPW). Its structure is shown in Figure 4.42a, with its detailed dimensions (in mm) as follows: $W = 48$, $W_1 = 3$, $W_2 = 8$, $W_3 = 3.6$, $W_4 = 1.6$, $W_5 = 1$, $W_{dr} = 3$, $h_{dr} = 10$, $h_{substrate} = 0.8, L = 48, D_s = 30, L_{dr1} = 10, L_{dr2} = 12, h_1 = 3, h_2 = 2.6_{\varepsilon dr^2}, \varepsilon_{dr^4} = \varepsilon_r = 9.8$, $\varepsilon_{dr^1} = \varepsilon_{dr^3} = 20$. The design features two pairs of dual stairs shaped RDRA units

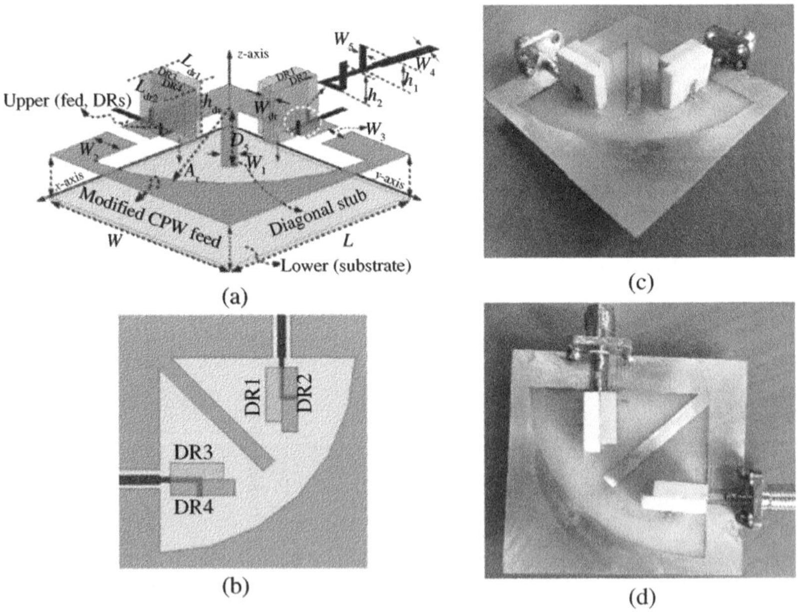

Figure 4.42 Schematic of antenna (a) 3-D view, (b) top view; prototype structure of the antenna; (c) 3-D view; and (d) top view. Source: Reproduced with permission from Ref. [31]/IEEE/Public Domain.

with different permittivity values ($\varepsilon_r = 9.8$ and $\varepsilon_r = 20$) arranged orthogonally, as depicted in Figure 4.42c, d. A symmetrical overextended diagonal stub (D_S), oriented at a 45°, and the modified arc shaped ground plane (A_r) on the CPW feed improve isolation between the dielectric radiators by altering the current path.

The antenna is designed in three main steps. First, a single RDRA is designed and fed by a CPW line, resulting in a wideband operation at higher frequencies, as shown in Figure 4.43a, b. Next, a second RDRA with the same permittivity and feeding is located orthogonally to the first. The impedance bandwidth is then observed to become a triple band, with increased isolation of approximately 11 dB. The RDRAs are then modified into their stair-shaped form by removing a small portion along the DRA length side. This created operation in two bands, while maintaining the same level of isolation. Finally, different permittivities of the RDRA are used, resulting in the merger of the two operating bands and widening of its impedance bandwidth. The permittivity of DR1 and DR3 are $\varepsilon_r = 20$, whereas DR2 and DR4 are $\varepsilon_r = 9.8$.

Figure 4.43c, d highlights the design steps and resulting S-parameter of the stair-shaped RDRA with different permittivities and different feed structures. The initial RDRAs are fed using a straight CPW feed enabled by a wideband operation at lower frequencies. This feed type is subsequently modified into an L-shaped feed for the RDRAs, resulting in a decrease in impedance bandwidth. The isolation between radiators remained nearly the same across the operating frequency range for both feed types, as shown in Figure 4.43d. Finally, the integration of a diagonal stub (D_S) and an arc-shaped ground plane (A_r) in Step 3 then improved isolation between the RDRA radiators.

The final antenna fabricated and evaluated indicated reasonable agreement with simulations, as depicted in Figure 4.44a, b. Measurements showed that the UWB MIMO DRA operated from 1.6 to 12.2 GHz with 153.6% impedance bandwidth, as shown in Figure 4.44a. The antenna also featured isolation of less than 25 dB, as shown in Figure 4.44b, owing to the modified arc-shaped ground plane (A_r) and diagonal stub (D_S). Throughout the operating band, a gain of 4.72 dBi, a low ECC of less than 0.003, and a diversity gain of approximately 10 dB are observed, with a measured variation of channel capacity loss (CCL) of under 0.7 bps/Hz. In addition, the radiation and total efficiency are about 89% and 88%, respectively.

4.3.1.5 Dipole-Based Antennas

Dipole-based antennas are another type of MIMO antenna which can be used as a wideband antenna. An example of a magnetoelectric (ME) dipole antenna fed by a fork-shaped printed ridge gap waveguide (PRGW) line is shown in Figure 4.45.

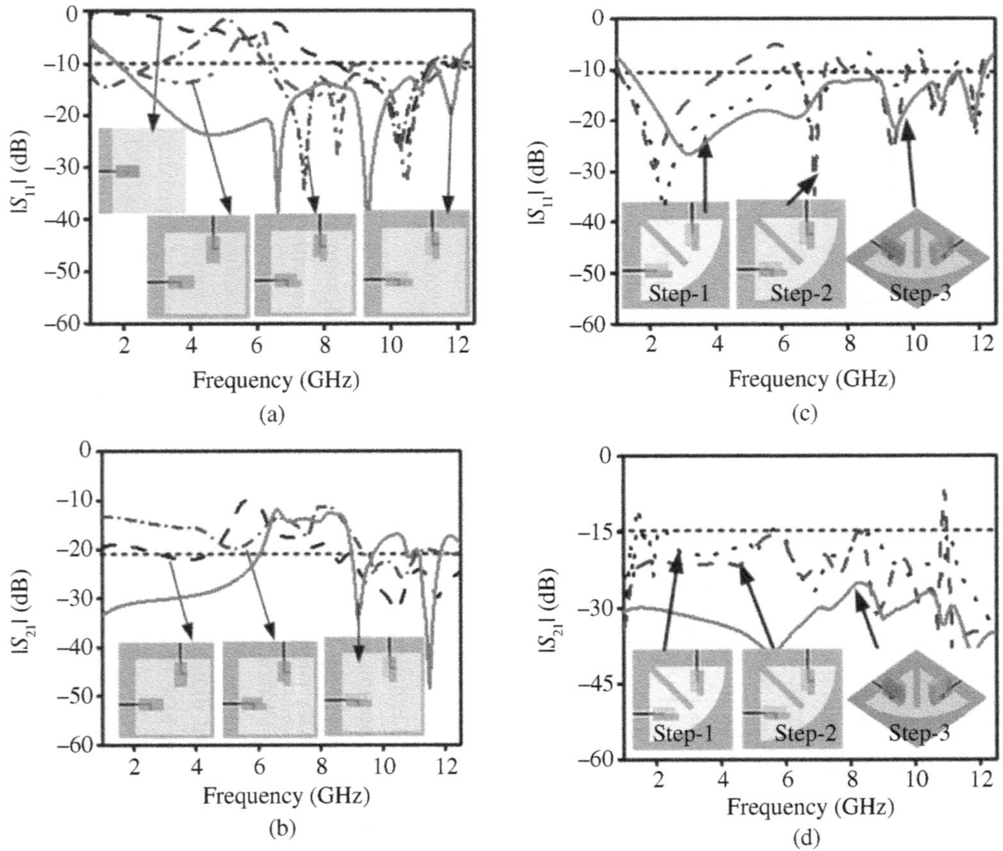

Figure 4.43 (a, b) Simulated results of S_{11} and S_{21}; simulation results for different feed steps of (c) S_{11} and (d) S_{21} for different arc ground plane (A_r) and with the diagonal stub (D_s). Source: Reproduced with permission from Ref. [31]/IEEE/Public Domain.

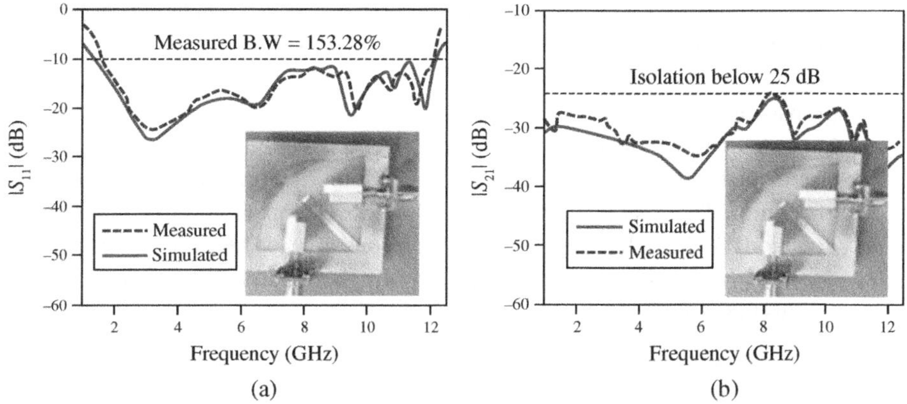

Figure 4.44 The simulated and measured S-parameters of the antenna (a) S_{11} and (b) S_{21}. Source: Reproduced with permission from Ref. [31]/IEEE.

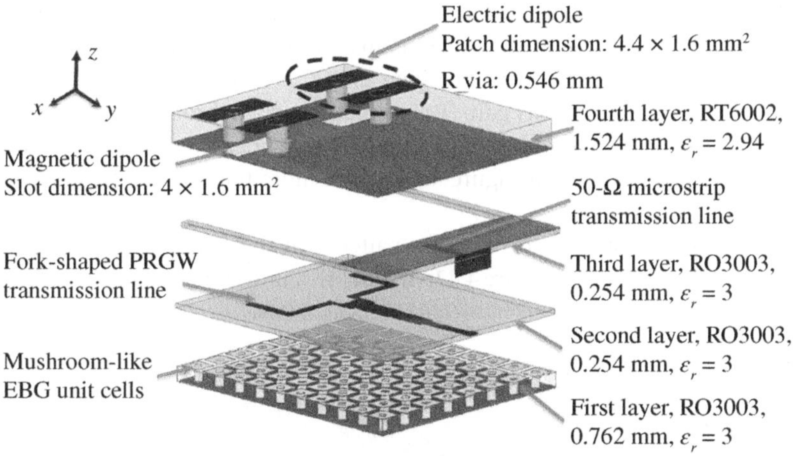

Figure 4.45 3-D view of the ME-dipole antenna fed by PRGW line (from [32], © 2020 IEEE, reprinted with permission).

The antenna is made up of four substrate layers stacked on top of each other, and the specifics of each layer are described below: overall consists of four layers:

- Bottom-most layer (Layer 1): an array of mushroom EBG unit cells offers a band-stop characteristic. The EBG cells block surface wave propagation toward the undesired directions.

- Layer 2: comprises a fork-shaped microstrip line which connects to the antenna input via a 50 Ω transmission line situated on the bottom of the third layer.
- Layer 3: spacer with constant air-gap dimensions for quasi-TEM mode propagation.
- Layer 4: a parallel plate ground plane with two rectangular slots of $4 \times 1.6\ \text{mm}^2$ size etched out, acting as magnetic dipole sources with a separation distance of 3.2 mm. These slots allow uniform EM wave propagation in the broadside direction. Besides that, four patch dimensions at $4.4 \times 1.6\ \text{mm}^2$ are printed on the top side of this layer, each connected with the ground plane via a vertical via (radius 0.546 mm).

The final prototype consists of a four-port ME-dipole antenna integrated with the mushroom EBG SRR meta-lens. This meta-lens consists of three off-centered SRR layers, each with 12×12 unit cells, as shown in Figure 4.46a, b. The SRR layers are placed above the four-port ME-dipole antenna. Due to its symmetry, only Port 1 and Port 3 were evaluated. The simulated and measured S-parameters at Port 1 and Port 3 depicted in Figure 4.47a show operation from 24 to 36 GHz. Moreover, mutual coupling between all ports remains less than −25 dB throughout this frequency range. Diversity performance such as ECC is less than 0.2, the TARC is below −30 dB, and the directivity gain (DG) is more than 9.8 throughout the entire bandwidth. It can be observed that the measured gain varies from 11.3 to 13.4 dBi over 26.5–36 GHz, as shown in Figure 4.47b, with more than 85% radiation efficiency.

The measured and simulated radiation patterns in the E-plane illustrated in Figure 4.47c–h for Port 1 and Figure 4.47i–n for Port 3, indicated the beam steering

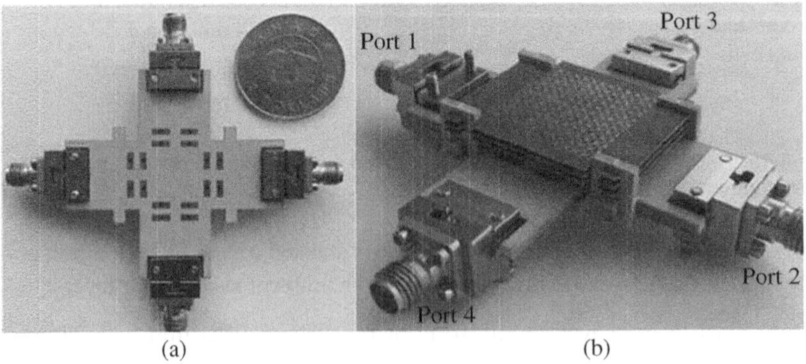

Figure 4.46 Four-port ME-dipole antenna loaded with SRR meta-lens. (a) Top view. (b) 3-D view with the SRR meta lens in place. Source: Reproduced with permission from Ref. [32]/IEEE.

capabilities of the antenna within 26.5–40 GHz. Results indicate that feeding Port 1 while terminating the other ports steers the E-plane main beam toward approximately −40°, as shown in Figure 4.47c–h. Similarly, Figure 4.47i–n with Port 3 fed and remaining ports terminated resulted in the beam directed toward approximately +40° in the E-plane. The antenna can be potentially integrated with an SP4T switch at the ports to electronically steer beams from −40° to +40° in each quadrant.

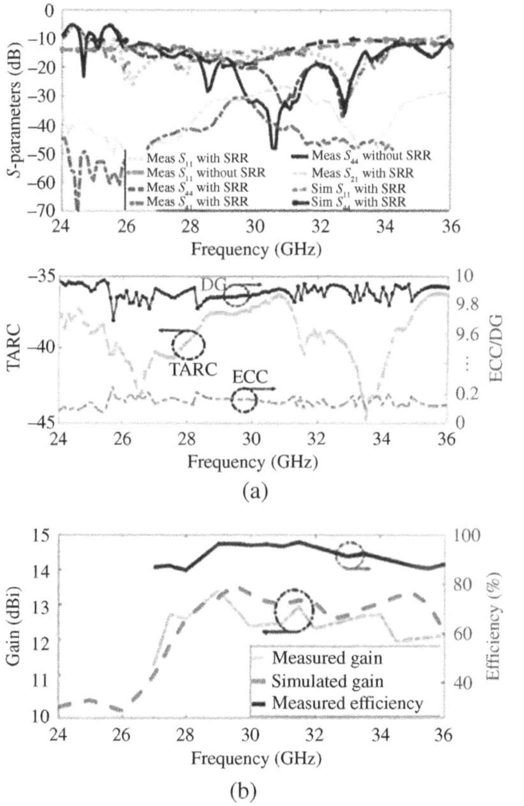

Figure 4.47 Simulated and measured (a) S-parameters; and (b) ECC, DG, and TARC parameters of the four-port ME-dipole antenna. Normalized simulated and measured radiation patterns of the four-port ME-dipole antenna loaded with SRR meta-lens when Port 1 is excited at (c) 27 GHz; (d) 28 GHz; (e) 30 GHz; (f) 32 GHz; (g) 34 GHz; and (h) 36 GHz. Normalized simulated and measured radiation patterns of the four-port ME-dipole antenna loaded with SRR meta-lens when Port 3 is excited at (i) 27 GHz; (j) 28 GHz; (k) 30 GHz; (l) 32 GHz; (m) 34 GHz; and (n) 36 GHz. Source: Adapted from Ref. [33].

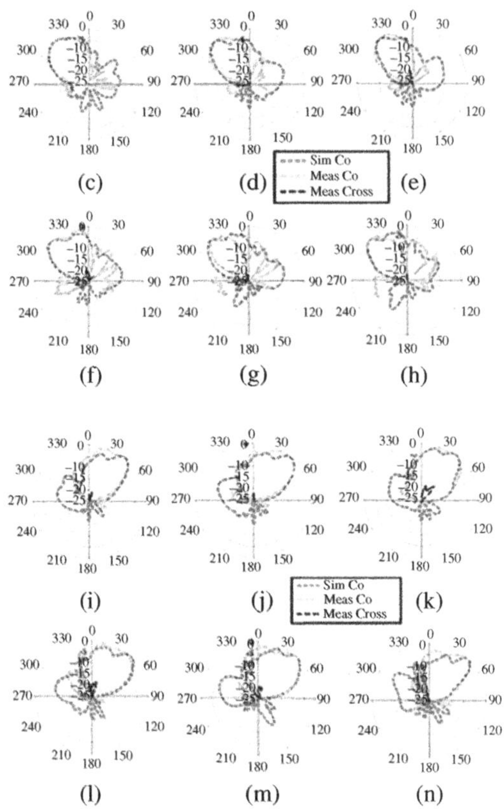

Figure 4.47 (*Continued*)

4.4 Practical Implementations

Various single-band, multiband, and wideband MIMO antennas can be used to improve the reliability of the communication system. Several examples of practical implementations for these antennas are presented in the following subsections.

4.4.1 Single-Band 5 GHz Linearly Polarized High Gain Patch Module Antenna

An example of a commercial single-band antenna is the 2 x 2 MIMO patch antenna (model OLX19M-097130-A) operating in the 5 GHz band. It serves as a module antenna, boasting a high peak gain of 13.8 dBi and 13.6 dBi in the 5150 to 5825 MHz range. It is specifically designed for Wi-Fi applications for high data throughput. The antenna has been designed to be linearly polarized, which enables a stable and efficient system signal strength operation. An embedded antenna has a peak

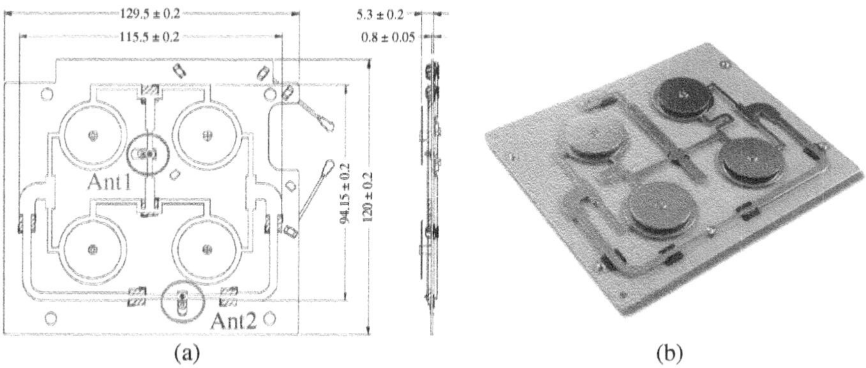

Figure 4.48 (a) Mechanical dimensions. (b) 2 x 2 linearly polarized high gain patch module antenna. Source: Reference [33]/with permission of LYNwave.

Table 4.2 Specifications of the 2 × 2 single-band 5 GHz module antenna.

Antenna	1	2
Frequency (MHz)	5150 ~ 5825	5150 ~ 5825
Peak Gain (dBi)	13.8	13.6
VSWR	2.0 : 1	2.0 : 1
HPBW (Azimuth)	30°	28°
HPBW (Elevation)	28°	34°
Impedance (Ω)	50	
Polarization	Linear	
Dimensions (mm)	129.5 × 120 × 5.3	
Power (W)	5	
Operating temp (°C)	−40°C ~+−85 °C	

gain of less than 2dBi in free space due to the high peak gain of 13.8/13.6 dBi of this patch antenna, which provides superior performance. Figure 4.48a shows the mechanical dimensions and Figure 4.48b shows 2 x 2 5 GHz linearly polarized high gain patch module antenna diagram. Table 4.2 shows the specification of the single-band 5 GHz high gain patch antenna [33].

4.4.2 Multiband Antenna

A commercial multiband antenna is presented next. The L-com HG65807XPPR-NF is a low passive intermodulation (PIM) rated, high-performance wall mount

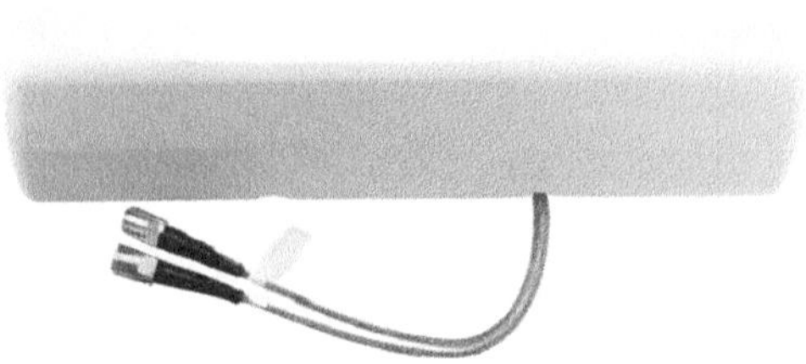

Figure 4.49 Multiband MIMO antenna. Source: Reference [34]/with permission of Infinite Electronics International, Inc.

Table 4.3 Antenna specifications by band.

Description	Band 1	Band 2	Band 3	Band 4	Band 5
Range (GHz)	0.617–0.698	0.698–0.96	1.71–2.7	3.3–4.2	4.9–6
Gain (dBi)	5.5	6	7	6	6.5
Horizontal HPBW	80°	70°	68°	55°	50°
Vertical HPBW	70°	70°	68°	50°	50°
Port-port isolation (dB)	20	20	20	28	30

MIMO panel antenna specifically designed for building wireless networks such as distributed antenna systems (DAS). The antenna has 5.5–7 dBi gain and can be used to distribute cellular and Wi-Fi signals throughout a building or area. It operates in four main bands between 617 and 6000 MHz for 5G, LTE, CDMA, LPWAN/IoT/M2M applications including LoRA, LTE-M, and NB-IoT. It features directional 60° patterns with ±45° slant polarization, with low PIM rated (of <−150 dBc). Figure 4.49 shows the antenna diagram, whereas Tables 4.3 and 4.4 show the specifications of this multiband antenna.

4.4.3 Wideband/UWB Antenna

A commercial wideband antenna for 5G NR low band in-building networks is STOC66G3AM2-P43 from Sanny Telecom [35]. This omnidirectional ceiling-mount antenna operates from 617 to 6000 MHz to cover current networks such as 3G/LTE/4G and 5G NR (low and mid bands) based on different worldwide standards. Multiband antenna arrays are concentrated into a single radome, with a low PIM value below 153 dBc to cater to low PIM requirements for LTE/4G/5G bands and DAS applications. Figure 4.50 shows the dual-polarized UWB

Table 4.4 Electrical, mechanical, and environmental specifications of the antenna.

Description	Minimum	Typical	Maximum
Electrical specifications			
Frequency range (MHz)	617		6000
Input VSWR			2 : 1
Gain (dBi)		5.5	7
Input power (W)			50
Passive intermodulation		−153	
Mechanical and environmental specifications			
Size		309.88 mm × 420.12 mm × 100.08 mm	
Polarization		±45° slant	
Temperature		−50 to +60°C	
Humidity		<95	
Weight		997.9 g	
Number of ports		2	

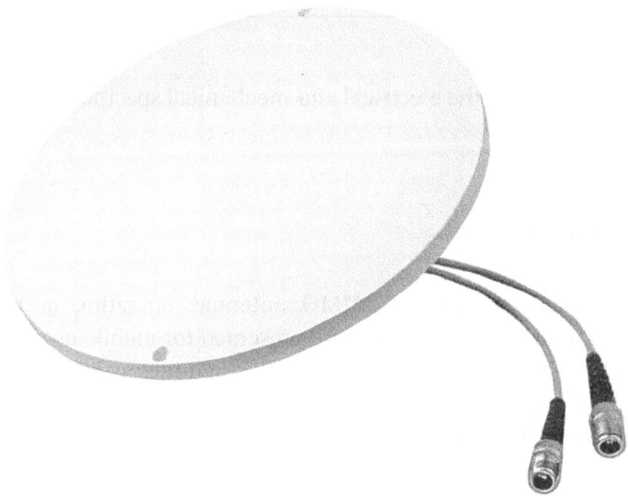

Figure 4.50 STOC66G3AM2-P43 UWB antenna. Source: Reference [35]/with permission of Sanny Telecom.

Table 4.5 Specifications of the STOC66G3AM2-P43 UWB antenna.

Frequency range (MHz)	617–806	806–960	1350–1550	3300–3800	4900–6000
Gain (dBi)	1.5 ± 0.5	2.2 ± 0.5	3.5 ± 0.5	6.0 ± 1.0	6.0 ± 1.0
Polarization	Linear horizonal × 2				
VSWR	≤1.8	≤1.8	≤1.8	≤1.8	≤1.8
Isolation (dB)	≥16	≥18	≥18	≥30	≥30
PIM	≤−153 dBc				
Maximum power			50 W		
Impedance			50 Ω		
Dimensions			Φ266 × 18 mm		
Antenna weight			0.5 kg		
Operational humidity			<95%		
Operational temperature			−40 to +65 °C		
Mounting type			Ceiling mount via holes		
Radome material			ABS, white, RAL9003		
Connector type			2 × *N*-female/4.3–10-female		
Lightning protection			DC ground		
Port			Port 1 and Port 2		

antenna, whereas Table 4.5 shows the electrical and mechanical specifications of the antenna.

4.5 Design Examples

In this section, two design examples of MIMO antennas operating in the sub-6 GHz band and the millimeter-wave band are presented for mobile user terminals. Readers may also find their corresponding CST design files as references. To complement practical applications, performance of these antennas is also studied in proximity to users' hand and finger models.

4.5.1 Antenna for the Sub-6 GHz Band

The performance of 5G MIMO antennas with differing numbers of antenna elements integrated into a mobile terminal in both free space and in the proximity of a user's hand is presented in [36]. The primary objective is to evaluate the physical and performance limits of up to 18 identical MIMO antennas integrated within a single mobile terminal chassis (i.e. 2, 4, 6, 8, 10, 12, 14, 18 elements) and to identify effective methods for their optimization. All MIMO antennas are designed to operate in the 5G frequency band between 5 and 6 GHz, which includes LTE band 46 from 5150 to 5925 MHz. The PIFA topology was selected for investigation due to its popularity in mobile devices. The antenna elementss are constructed using a copper sheet of 0.291 mm thickness and placed on a Rogers RT/Duroid 5880 substrate with dimensions of $150 \times 80 \times 1.575$ mm^3. The length of the chassis is between 2.50λ and 3.00λ, whereas its width is from 1.33λ to 1.60λ (based on the frequency of 5–6 GHz and the substrate permittivity of 2.2 and a loss tangent of 0.0009). The resonant frequency of the antenna element is tuned by chamfering the corners. A 50 Ω L-shaped microstrip feed line is used to feed the antenna. The chassis accommodates varying numbers of antenna elements, with their distribution for each M from 2 to 18. The antenna elements are positioned symmetrically around the central vertical axis on the outer frame of the chassis.

To perform this study, interested readers may enable/disable different antenna elements, depending on the number of desired elements in the chassis (M). Important parameters to be observed besides reflection and coupling coefficients are the efficiency, radiation patterns, and MIMO parameters, depending on the selection of different antenna elements. Another possible topic of interest is the performance of the antenna with a similar set of M_s, with and without the user's hand, to observe the effect of signal blockage. Interested readers are also encouraged to compare their findings from the published paper in [36].

4.5.2 Antenna for the Millimeter Wave Band

The next design example is a planar 4-port antenna designed for mobile terminal on a Rogers (RO3003) substrate centered at 28 GHz with a bandwidth of 2.5 GHz, as shown in Figure 4.51 [37]. Four apertures are etched on the copper material where the antenna elements are placed. To tune frequency, two edges of each antenna element were chamfered by about 0.55 mm. The edge-to-edge distance of each antenna element is $\lambda/2$. A transmission line of 50 Ω is placed on the substrate's reverse side. To connect the radiating elements with the transmission line, a metallic via with a radius of 0.14 mm is utilized. A finger model was placed on top of antenna at four different positions (P_1, P_2, P_3, P_4).

To perform this study, interested readers may enable/disable different antenna elements or relocate them, depending on their desired number elements in the chassis (M). Important parameters to be observed besides reflection and coupling coefficients are the efficiency, radiation patterns, and MIMO parameters, depending on the selection of different antenna elements. Another possible topic of interest is the performance of the antenna with a similar set of M_s, with

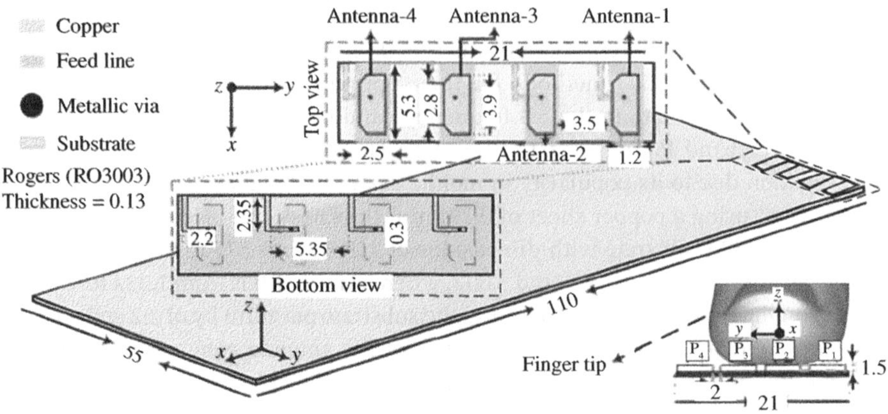

Figure 4.51 Antenna dimensions on PCB and the index finger placement. All dimensions are in millimeters. Source: Reproduced with permission from Ref. [37]/IEEE.

and without the user's hand, to observe the effect of signal blockage. Interested readers are also encouraged to compare their findings from the published paper in [37].

4.6 Summary

This chapter described passive MIMO antennas, which include single-band, multiband, and UWB MIMO antennas, which can be used to improve the reliability and capacity of wireless communication systems. The chapter starts with a brief explanation of the need and challenges for these different classes of antennas in the context of MIMO communication. Then, a definition of the bandwidth parameter for antennas which is widely used to classify them is explained. Next, a brief description of the classical antenna types is presented, followed by several examples of single-band MIMO antennas, typically less than 5% of fractional bandwidth. This is followed by the multiband antennas, typically designed to operate at multiple frequencies to operate across multiple bands. Several methods to enable multiband antenna operation are first presented prior to several antenna examples designed to operate in both the 5G FR1 and FR2 bands along with other existing wireless communication bands. Finally, wideband and UWB antennas, which are designed to operate at a very wide range of frequencies, are presented. Several practical examples of commercial MIMO antennas (single-band, multiband, and wideband) are also presented. The final section of this chapter presented two MIMO antenna design examples, one in the sub-6 GHz band, and another in the millimeter-wave band.

References

1 Chih-Lin, I., Rowell, C., Han, S. et al. (2014). Toward green and soft: a 5G perspective. *IEEE Communications Magazine* 52 (2): 66–73. https://doi.org/10.1109/MCOM.2014.6736745.

2 Khan, R., Al-Hadi, A.A., Soh, P.J. et al. (2018). User influence on mobile terminal antennas: A review of challenges and potential solution for 5G antennas. *IEEE Access* 6: 77695–77715. https://doi.org/10.1109/ACCESS.2018.2883788.

3 Ishteyaq, I. and Muzaffar, K. (2022). Multiple input multiple output (MIMO) and fifth generation (5G): an indispensable technology for sub-6 GHz and millimeter wave future generation mobile terminal applications. *International Journal of Microwave and Wireless Technologies* 14 (7): 932–948. https://doi.org/10.1017/S1759078721001100.

4 Schantz, H.G., Beeler, W., and Dickson, D. (2005). UWB planar bulbous dipole antennas. *IEEE Antennas and Propagation Society International Symposium*, 03–08 July 2005, vol. 2, Washington, DC, USA, pp. 475–478. http://doi.org/10.1109/APS.2005.1551848.

5 Kshetrimayum, R.S. (2009). An introduction to UWB communication systems. *IEEE Potentials* 28 (2): 9–13. https://doi.org/10.1109/MPOT.2009.931847.

6 Qin, Y., Zhang, L., Mao, C., and Zhu, H. (2023). A compact wideband antenna with suppressed mutual coupling for 5G MIMO applications. *IEEE Antennas and Wireless Propagation Letters* 22 (4): 938–942. https://doi.org/10.1109/LAWP.2022.3229020.

7 Galvan-Tejada, G.M., Peyrot-Solis, M.A., and Jardón-Aguilar, H. (2017). *Ultra-Wideband Antennas: Design, Methodologies, and Performance*. Taylor & Francis Group, LLC.

8 Wong, K.L., Chen, J.Z., and Li, W.Y. (2021). Four-port wideband annular-ring patch antenna generating four decoupled waves for 5G multi-input–multi-output access points. *IEEE Transactions on Antennas and Propagation* 69 (5): 2946–2951. https://doi.org/10.1109/TAP.2020.3025237.

9 Chang, L. and Wang, H. (2021). Miniaturized wideband four-antenna module based on dual-mode PIFA for 5G 4×4 MIMO applications. *IEEE Transactions on Antennas and Propagation* 69 (9): 5297–5304. https://doi.org/10.1109/TAP.2021.3069490.

10 Zhang, Y., Deng, J.Y., Li, M.J. et al. (2019). A MIMO dielectric resonator antenna with improved isolation for 5G mm-wave applications. *IEEE Antennas and Wireless Propagation Letters* 18 (4): 747–751. https://doi.org/10.1109/LAWP.2019.2901961.

11 Mongia, R.K. (1992). Theoretical and experimental resonant frequencies of rectangular dielectric resonators. *IEE Proceedings-Microwaves, Antennas and Propagation* 139 (1): 98–104. https://doi.org/10.1049/ip-h-2.1992.0016.

12 Kim, G. and Kim, S. (2021). Design and analysis of dual polarized broadband microstrip patch antenna for 5G mmWave antenna module on FR4 substrate. *IEEE Access* 9: 64306–64316. https://doi.org/10.1109/ACCESS.2021.3075495.

13 Huang, Y. and Boyle, K. (2008). *Antennas: From Theory to Practice*. Wiley.

14 Haapala, P., Vainikainen, P., and Eratuuli, P. (1996). Dual frequency helical antennas for handsets. *Proceedings of Vehicular Technology Conference – VTC,* 28 April–01 May 1996, vol. 1, Atlanta, GA, USA, pp. 336–338. http://doi.org/10.1109/VETEC.1996.503464.

15 Hsieh, H.W., Lee, Y.C., Tiong, K.K. and Sun, J.S. (2009). "Design of a multi-band antenna for mobile handset operations," in *IEEE Antennas and Wireless Propagation Letters*, vol. 8, pp. 200-203, 2009, doi: https://doi.org/10.1109/LAWP.2008.2011655.

16 Li, J.J., Zhang, X., Wang, Z., et al. (2019). Dual-band eight-antenna array design for mimo applications in 5G mobile terminals. *IEEE Access* 7: 71636–71644. https://doi.org/10.1109/ACCESS.2019.2908969.

17 Al Abbas, E., Ikram, M., Mobashsher, A.T., and Abbosh, A. (2019). MIMO antenna system for multi-band millimeter-wave 5G and wideband 4G mobile communications. *IEEE Access* 7: 181916–181923. https://doi.org/10.1109/ACCESS.2019.2958897.

18 Ikram, M., Nguyen-Trong, N., and Abbosh, A. (2019). Multiband MIMO microwave and millimeter antenna system employing dual-function tapered slot structure. *IEEE Transactions on Antennas and Propagation* 67 (8): 5705–5710. https://doi.org/10.1109/TAP.2019.2922547.

19 Balanis, C.A. (2016). *Antenna Theory: Analysis and Design*. Wiley.

20 Janaswamy, R. and Schaubert, D. (1987). Analysis of the tapered slot antenna. *IEEE Transactions on Antennas and Propagation* 35 (9): 1058–1065. https://doi.org/10.1109/TAP.1987.1144218.

21 Schaubert, D., Kollberg, E., Korzeniowski, T. et al. (1985). Endfire tapered slot antennas on dielectric substrates. *IEEE Transactions on Antennas and Propagation* 33 (12): 1392–1400. https://doi.org/10.1109/TAP.1985.1143542.

22 Yngvesson, K.S., Korzeniowski, T.L., Kim, Y.S. et al. (1989). The tapered slot antenna-a new integrated element for millimeter-wave applications. *IEEE Transactions on Microwave Theory and Techniques* 37 (2): 365–374. https://doi.org/10.1109/22.20062.

23 Paul, L.C. and Islam, M.M. (2021). A Super wideband directional compact vivaldi antenna for lower 5G and satellite applications. *International Journal of Antennas and Propagation* https://doi.org/10.1155/2021/8933103.

24 Li, Z., Yin, C., and Zhu, X. (2019). Compact UWB MIMO vivaldi antenna with dual band-notched characteristics. *IEEE Access* 7: 38696–38701. https://doi.org/10.1109/ACCESS.2019.2906338.

25 Azari, A., Skrivervik, A., Aliakbarian, H., and Sadeghzadeh, R.A. (2023). A super wideband dual-polarized Vivaldi antenna for 5G mmWave applications. *IEEE Access* 11: 80761–80768. https://doi.org/10.1109/ACCESS.2023.3300040.

26 Najam, A.I., Duroc, Y., and Tedjini, S. (2010). Design and analysis of MIMO antennas for UWB communications. *Proceedings of the Fourth European Conference on Antennas and Propagation*, 12–16 April 2010, Barcelona, Spain, pp. 1–5.

27 Ko, S.C.K. and Murch, R.D. (2001). Compact integrated diversity antenna for wireless communications. *IEEE Transactions on Antennas and Propagation* 49 (6): 954–960. https://doi.org/10.1109/8.931154.

28 Liu, L., Zhao, H., See, T.S.P., and Chen, Z.N. (2006). A printed ultra-wideband diversity antenna. *IEEE International Conference on Ultra-Wideband*, 24–27 September 2006, Waltham, MA, USA, pp. 351–356. http://doi.org/10.1109/ICU.2006.281575.

29 Yuan, X.T., Chen, Z., Gu, T., and Yuan, T. (2021). A wideband PIFA-pair-based MIMO antenna for 5G smartphones. *IEEE Antennas and Wireless Propagation Letters* 20 (3): 371–375. https://doi.org/10.1109/LAWP.2021.3050337.

30 Cai, Q., Li, Y., Zhang, X., and Shen, W. (2019). Wideband MIMO antenna array covering 3.3–7.1 GHz for 5G metal-rimmed smartphone applications. *IEEE Access* 7: 142070–142084. https://doi.org/10.1109/ACCESS.2019.2944681.

31 Sani, M.M. and Chowdhury, R. (2020). An ultra-wideband rectangular dielectric resonator antenna with MIMO configuration. *IEEE Access* 8: https://doi.org/10.1109/ACCESS.2020.3012793.

32 Kakhki, M.B., Dadgarpour, A., Antoniades, M.A. et al. (2021). Magneto-electric dipole antennas loaded with meta-lens for 5G MIMO pattern diversity applications. *IEEE Transactions on Antennas and Propagation* 70 (8): 7112–7117. https://doi.org/10.1109/TAP.2021.3111652.

33 LYNwave. 2 × 2 5GHz linearly polarized high gain patch module antenna, OLX19M-097130-A datasheet. https://lynwave.com/product/2x2-5ghz-linearly-polarized-high-gain-patch-module-antenna/ (accessed on 29 July 2024).

34 L-com. 617–960 MHz/1710–2700 MHz/3300–4200 MHz/4900–6000 MHz Low PIM Rated X-pol 5.5 dBi Panel DAS Antenna, Wall Mount, 2xType N Female Connector datasheet. https://www.l-com.com/Images/Downloadables/Datasheets/ds_HG65807XPPR-NF.pdf (accessed on 29 July 2024).

35 Sanny Telecom Indoor low PIM 617 MHz to 6000 MHz dual polarized omni ceiling antenna. https://www.sannytelecom.com/wp-content/uploads/2024/01/STOC66G3AM2-P43.pdf (accessed on 29 July 2024).

36 Elshirkasi, A.M., Al-Hadi, A.A., Soh, P.J., et al. (2020). Performance Study of a MIMO Mobile Terminal with Upto 18 Elements Operating in the Sub-6 GHz 5G Band with User Hand. *IEEE Access* 8: 28164–28177. https://doi.org/10.1109/ACCESS.2020.2971561.

37 Khan, R., Al-Hadi, A.A., and Soh, P.J. (2017). User effect on the efficiency of 4-port mobile terminal antenna at 28 GHz. *International Symposium on Antennas and Propagation (ISAP)*, 30 October–2 Novemberl 2017, Phuket, Thailand, pp. 1–2. http://doi.org/10.1109/ISANP.2017.8228840.

5

Active Reconfigurable MIMO Antenna Systems

Reconfigurable antennas are well known in the literature and practical implementations where the antenna characteristics are controlled and reconfigured based on the application at hand. For example, the frequency band covered can be changed and swept within a larger band to allow a certain device to operate in various geographic areas or within different operators that use different dedicated bands. The control can be also of the antenna radiation characteristics via the beam pattern change to direct the energy toward specific directions. The topic of reconfigurable antennas has been well documented in literature and some of the fundamental references on the topic are [1–3].

In this chapter, we will discuss several reconfigurable MIMO antenna systems using active elements. We will focus on active electrical elements rather than mechanical- or optical-based ones due to their ease of integration, low cost, wide usage, and stability of operation. In addition, reconfigurability can be made via the change of the material properties of the substrate but this mechanism will not be the focus in this chapter. The most practical reconfiguration elements are electrical-based ones, which usually are various kinds of electrical switches or variable capacitors (i.e. PIN diodes, varactor diodes, RF switches, etc.). This chapter will start by discussing frequency-reconfigurable MIMO antenna systems and some of the basic controlling circuits for switching-based and variable capacitor-based ones, then will move forward toward pattern and polarization reconfigurabilities. In each section, several examples from literature are shown to aid the reader in his/her future designs and understanding of the operating principles of such antenna systems.

5.1 Frequency-Reconfigurable MIMO Antennas

One way to change the resonance frequency of antenna is to load it with a reactive element. This loading will shift the resonance frequency (f_c) according to the

MIMO Antenna Systems for 5G and Beyond, First Edition.
Xiaoming Chen, Ping Jack Soh, and Mohammad S. Sharawi.

respective change in the capacitive (C_{eq}) or inductive (L_{eq}) portions according to the well-known relationship in Eq. (5.1). Depending on the antenna type and its equivalent circuit model (series resonant circuit or a parallel resonant one), the additional reactive element will either increase or decrease the equivalent value or its respective element, thus decreasing or increasing the resonance frequency. Varactor diodes are widely used in frequency-reconfigurable antennas where C_{eq} is changed via the change in the reverse bias voltage of the diode. Thus, multiple values of capacitances can be obtained via the change of the reverse bias voltage across the terminals of the varactor. For example, an SMV1233 varactor can provide a capacitance variation between 5.08 and 0.84 pF when the applied reverse bias voltage (V_R) changes from 0 to 15 V.

$$f_c = \frac{1}{2\pi\sqrt{L_{eq}C_{eq}}} \tag{5.1}$$

Another way to change the resonance frequency would be to connect or disconnect portions of the radiating structure to change the electrical current path or length. The use of switches or diodes is widely used for this kind of frequency control. PIN diodes (P-type, Intrinsic, N-type semiconductor junction) are widely used and simple to control by on-off switching elements used in frequency-reconfigurable antennas. While varactor diodes change their capacitances via the reverse bias voltage applied across their terminals, PIN diodes turn on when forward-biased. An example of a PIN diode is the NXP-BAP-50 with a forward bias voltage of 0.95 V and a forward resistance of 25 Ω.

Figure 5.1 shows several examples of biasing circuits using varactor and PIN diodes with two types of antenna elements, one printed inverted-F antenna, and another slot-based one. The generic circuit configurations are shown in Figures 5.1a, b. The use of a current limiting resistor (in the range of 1–2 KΩ, but depending on diode type and ratings) as well as the introduction of a radio frequency (RF) choke for blocking the high-frequency signals back to the biasing source are demonstrated. The forward and reverse bias voltages are denoted by (V_F) and (V_R) respectively. The location of the connecting point of the diode end to the antenna should be optimized via modeling and simulations. This point of contact is important as it might affect the matching of the antenna.

Some actual examples of biasing circuits along with their respective diode elements are shown in Figures 5.1c, d. The example in Figure 5.1c shows a PIFA antenna with two diodes connected to provide dual separated band operation, and also fine frequency tuning in the lower band [4]. The dual bands are separated via the use of a PIN diode. When the diode is off, a large portion of the PIFA is disconnected from its initial short section, thus the antenna will resonate at a higher band, i.e. 2.4 GHz. When the PIN diode was on, it connected to the other portion of the antenna, thus increasing its electrical length and reducing its resonating

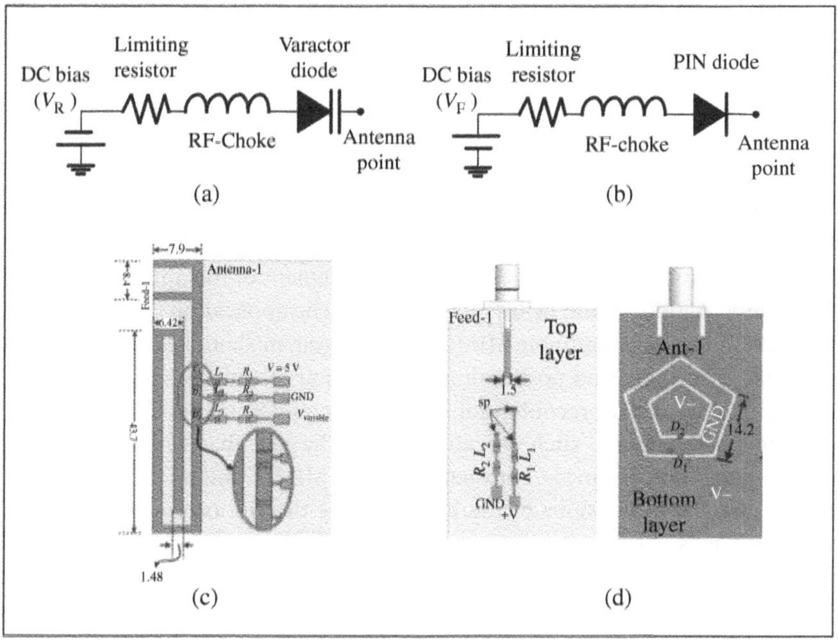

Figure 5.1 Biasing configurations for varactor and PIN diodes: (a) reverse bias circuit for a varactor diode, (b) forward bias circuit for a PIN diode, (c) an example of PIN and varactor diode use within a PIFA antenna. Source: Reference [4]/with permission of IEEE. (d) A slot antenna example with a varactor diode. Source: Reference [5]/with permission of IEEE.

bands. The lower bands of 570–680 MHz were covered via the variation of the reverse bias voltage of the varactor diode between 0 and 6 V. The second example shown in Figure 5.1d shows a dual-pentagonal slot antenna loaded with a varactor diode for frequency reconfigurability [5]. The presence of the two slots allowed two resonances. The presence of the varactor allows the frequency to operate from 1.32 to 1.49 GHz and 1.76 to 5 GHz via the use of reverse bias voltages between 0 and 10 V. This allows the capacitance values to change between 0.9 and 5.08 pF. The biasing was done using the configuration shown, where three shorting pins (SPs) were introduced between the top layer (where the elements are placed) and the ground (GND) plane layer at the bottom (where the slots are etched). Two varactor diodes were used for the tuning of the two slots (one per slot). The metal islands of reverse bias voltage, and GND are shown in white color marking the three SP locations. This is a clever way to ensure that proper biasing is achieved via the consideration of a GND island between the two slots.

In the modeling and analysis of such frequency-reconfigurable antennas, the biasing circuit and the model for the diode used should be included in the full-wave simulations or equivalent circuit simulations to achieve proper results that will resemble the fabricated prototype. All the diodes have equivalent circuits provided by the vendor that resembles their behavior within its range of operation, and the user should always include those values, which are usually functions of the biasing conditions. Such models do have internal parasitics and resistances that are important for the proper estimation of the resonances and losses.

Once the single element of the frequency-reconfigurable antenna system is tested and verified to give the desired frequency sweeps given the proper biasing values of the varactor and/or PIN diodes, the placement of the antenna within a MIMO configuration should be investigated next. The application and device size within which the reconfigurable MIMO antenna system should be implemented dictates the placement and positioning of the various elements. One of the very challenging designs with reconfigurable MIMO antennas are devices with small form factor and volume, such as cell phones. The fact that cell phones have standard sizes, the distances between adjacent MIMO antenna elements are almost set by the dimensions of the backplane or the rim of the device. Thus, novel techniques should be incorporated to satisfy the port isolation and field correlation requirements.

As mentioned in Chapter 2, to take advantage of MIMO technology, the antennas within the MIMO antenna system should have high port isolation values (i.e. low $|S_{xy}|$ values) to provide good radiation efficiencies, and should have low field correlation values (CC, or ECC) to provide the required spatial separation between the incoming data streams. Although the simple technique of placing the antennas far away from one another is considered the simplest and most straightforward technique in trying to satisfy these two requirements, at lower frequency bands, this will not be the case, as the shared back-plane (i.e. GND) will introduce higher port coupling levels as it might be part of the radiating structure. More descriptions of methods for improving port isolation of MIMO antenna systems can be found in Chapter 8.

The first example of a reconfigurable coupled loop MIMO antenna in the metal rim for mobile terminals is the one shown in Figure 5.2 following [6]. The design covers the low band (LB) 699–960 MHz and the high band (HB) 1710–2690 MHz of the long-term evolution (LTE) 4G standard. The top view of the dual two-element MIMO antennas is shown in Figure 5.2a. Note that elements 1 and 3 are identical, and the same goes for elements 2 and 4. The detailed geometry of elements 1 and 3 is shown in Figure 5.2b. The layout of elements 2 and 4 are similar but with a shorter metal rim portion and a matching network at the HB to increase the operating bandwidth. The longer-length elements, i.e. elements 1 and 3, cover the LB. The reconfiguration is achieved via the use of an RF switch that will provide a longer current path for LB coverage. Note that the length of the metal rim used to cover the LB is shorter than a quarter wavelength at 900 MHz and with 2 mm

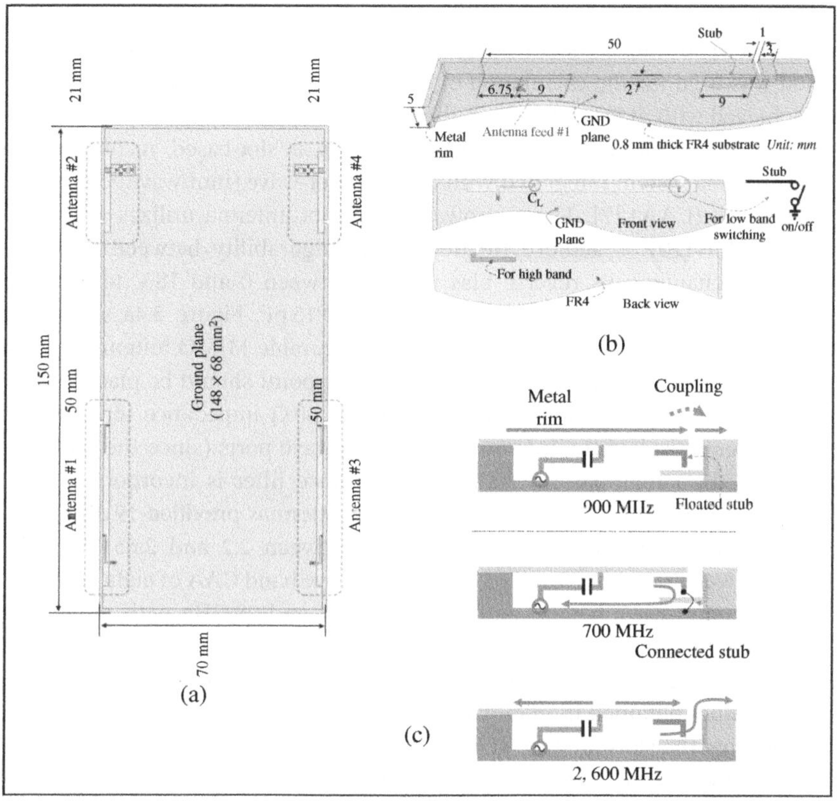

Figure 5.2 A reconfigurable in-rim MIMO antenna systems, (a) top view of the four-elements, elements 1 and 3 are identical (covering LB and HB), elements 2 and 4 are identical (covering HB), (b) detailed geometry of elements 1 and 3 covering the LB, (c) operating mechanism. Source: Reproduced with permission from Ref. [6]/IEEE.

of GND clearance only. This makes this configuration very attractive for practical handheld devices. The operating mechanism is illustrated in Figure 5.2c where the switch is OFF to cover the band around 900 MHz. When the switch is ON, the electric path is increased via the connection of the floated stub and the GND, thus enlarging the loop perimeter and reducing its resonance to cover the 700-MHz bands. For the higher bands, elements 2 and 4 are used along with the higher order mode of elements 1 and 3 as shown in the third sub-figure of Figure 5.2c.

The reconfigurable MIMO antenna system shown in Figure 5.2 had total system efficiencies between 40 and 65% for the LB and 50 and 85% for the HB. The ECC values varied between 0.1 and 0.45 for the LB and were less than 0.05 for the HB. This was attributed to the directional patterns as well as spatial separation at the HB compared to the close proximity between the elements (in wavelengths) at the LB. Figure 5.3 shows the radiation patterns from each antenna element with respect to the coordinate system of the design in Figure 5.2. The complex radiation method was used for the calculation of the ECC. Very good port isolation levels were achieved with at least 10.5 dB in the LB.

The second example that we will present is a slot-based reconfigurable MIMO antenna system integrated with a millimeter-wave (mm-wave)-connected antenna array (CAA) [7]. The microwave band slot antenna utilizes a varactor diode (MA46H120) to achieve frequency reconfigurability between 2.05 and 2.7 GHz via changing its reverse bias voltage between 0 and 18 V to load the slot with capacitance values between 1.34 and 0.15 pF. Figure 5.4a shows the geometry of the proposed integrated and reconfigurable MIMO antenna system with two antenna elements. Note that the feeding point should be placed closer to the edge of a slot antenna to achieve proper 50 Ω impedance matching to the feed line. To isolate the microwave and mm-wave ports (since the radiation slot is shared), a high-pass interdigital capacitance filter is incorporated with five fingers. The microwave reconfigurable slot antennas provided 19 dB of port isolation and gain values covering 3–4.5 dBi between 2.2 and 2.65 GHz. The efficiencies were between 55 and 65%. The mm-wave band CAA of eight elements covered two mm-wave bands centered around 24 and 28 GHz with more than 1 GHz of bandwidth in each band. Isolation levels with all other ports were more than 20 dB. The gain values covered 11–12 dBi with efficiencies of at least 70%. The ECC values obtained in microwave bands were less than 0.1, and this is due to the fact that the slots are placed on two perpendicular edges, thus making use of polarization mismatch to lower the ECC in addition to spatial separation. This example also shows the importance of shared-aperture designs in future mobile terminals that can reduce the size of microwave and mm-wave antenna systems. The reader is also encouraged to check recent shared aperture antennas in [8].

There are other examples of frequency-reconfigurable MIMO antennas in literature where different types of antenna elements were used and loaded with electrically controlled diodes or switches to serve in different applications. For example, in [9] a multi-band frequency-reconfigurable MIMO antenna based on PIFA elements was proposed for laptop and tablet applications. PIN diodes were used for band switching. It covered the LB (704–787 MHz) and HB (2500–2690 MHz) frequencies of the LTE standard (bands 17/13 and 20/7, respectively). Another volumetric PIFA-based reconfigurable MIMO for sub-6 GHz mobile terminals was proposed in [10]. PIN diodes were used for frequency reconfigurability, and three bands were covered, namely 0.94, 2.55, and 4.5 GHz. The antennas were placed at the top side of a standard cell phone backplane, and thus a cut in the GND was utilized to reduce mutual coupling.

A patch-based very low-profile reconfigurable MIMO antenna with four elements that is integrated into the back cover of standard mobile terminals was

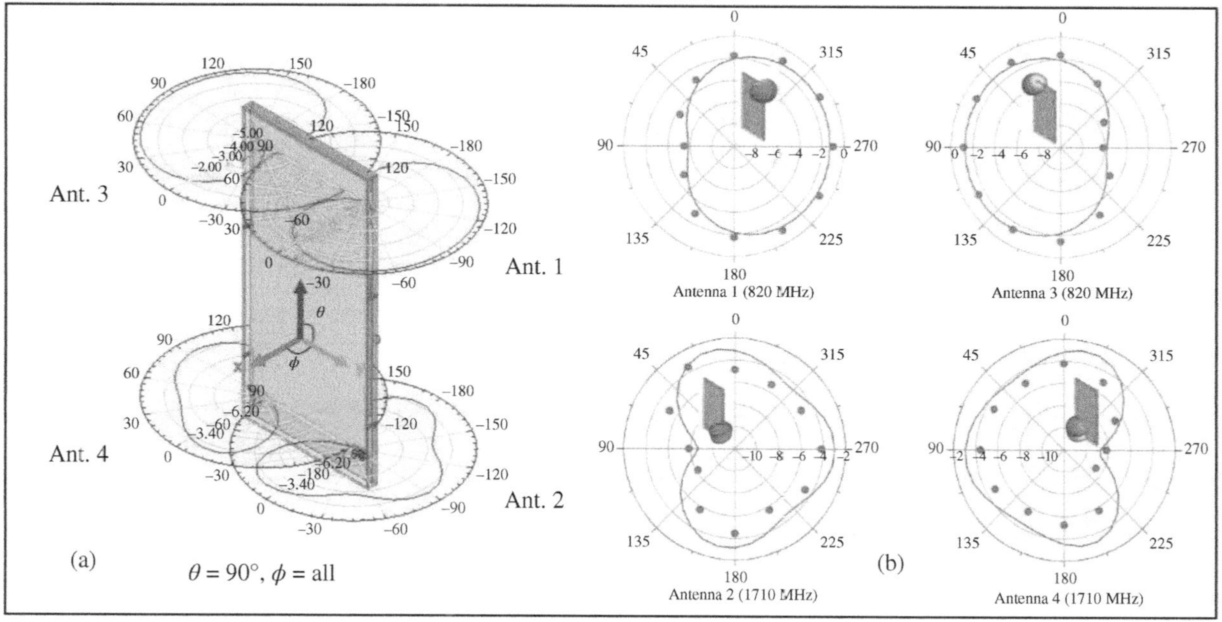

Figure 5.3 Radiation patterns of the reconfigurable MIMO antenna shown in Figure 5.2: (a) all patterns for the four elements, (b) selected patterns at 820 (elements 1 and 3) and 1710 MHz (elements 2 and 4). Source: Reproduced with permission from Ref. [6]/IEEE.

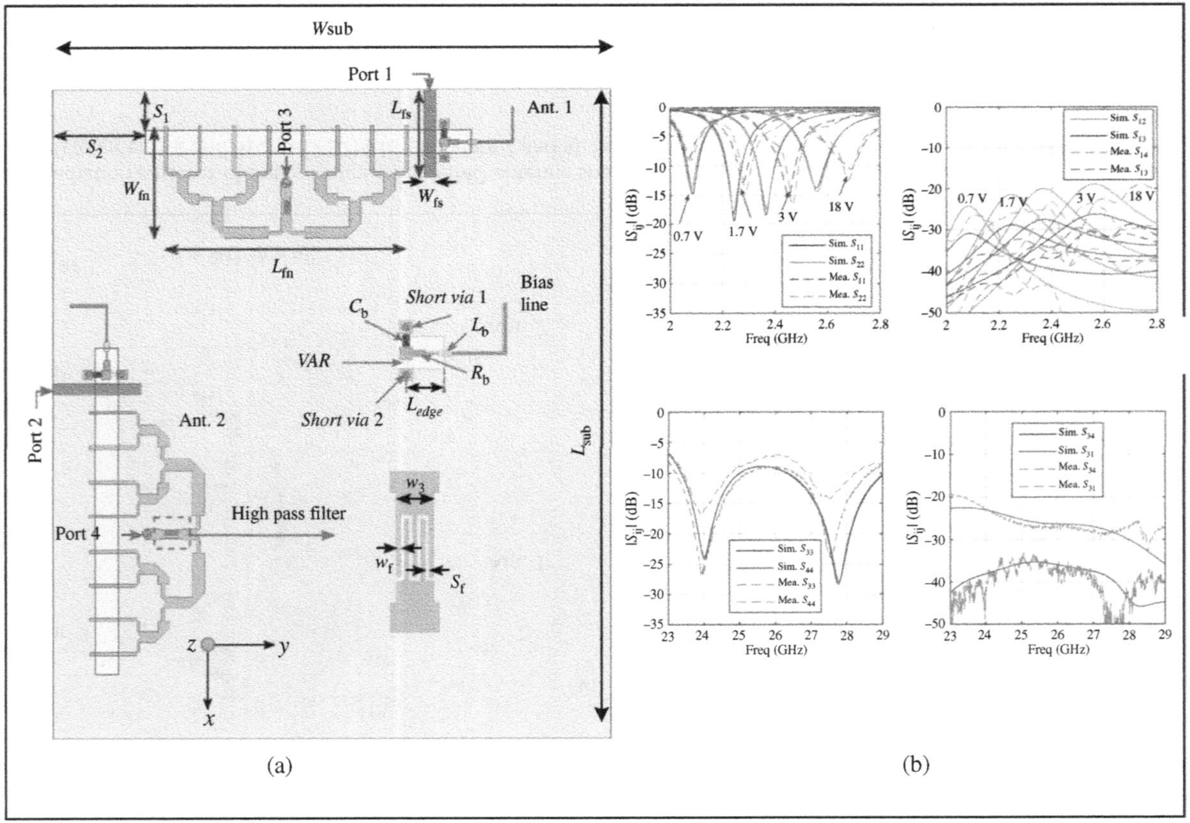

Figure 5.4 Slot-based reconfigurable MIMO antenna system integrated with a CAA, (a) geometry of the antenna system and the biasing circuit for the varactor diode, (b) S-parameter responses in the microwave and mm-wave bands. Variable values in mm are: $L_{sub} = 70$, $W_{sub} = 60$, $S_1 = 4.5$, $S_2 = 10$, $W_{fn} = 12.28$, $L_{fn} = 26.24$, $L_{fs} = 9.5$, $W_{fs} = 1.18$, Ledge = 2.69, $W_f = 1.18$, $S_f = 0.1$, $W_3 = 6.3$, $R_b = 20\,k\Omega$, $L_b = 220\,nH$, $C_b = 39\,pF$. Source: Reproduced with permission from Ref. [7]/IEEE.

proposed in [11]. Each patch (or IFA) was $20 \times 35.5\,\text{mm}^2$ in size. A middle slot was used to improve the covered bandwidth and allow multiple resonances of the patch. The antenna covered more than 100 MHz of tunable bandwidth within the bands 3300–4200 MHz via the use of varactor diodes, matching circuits, and multiple feed points. Efficiencies of more than 25% were achieved in the presence of hand phantoms. Finally, several slot-based reconfigurable MIMO antenna elements covering sub-6 GHz bands appeared in literature for mobile handset and cognitive radio applications with some integrated with ultrawideband (UWB) sensing elements such as those in [12–14]. Thus, frequency reconfigurability within MIMO implementations is widely used and with almost all types of antenna elements. Some examples of dielectric resonator antenna (DRA) reconfigurable structures are also available, where the same electrical switching or reactance loading methods are used for such reconfigurability as demonstrated in [15, 16].

5.2 Pattern Reconfigurable MIMO Antennas

The direction of the current flow within an antenna has a direct influence on the radiation pattern characteristics and shape. Thus, in order to change the pattern characteristics of an antenna, we need to manipulate the flow of the current on its surface. The use of electronic components (switches, diodes, etc.) can help in such pattern reconfigurability. We need to distinguish between two classes of pattern reconfigurable antennas that are present in literature, the first is the one that provides pattern reconfigurability via the change in the actual pattern shape via the excitation of different currents (modes) on the antenna surface (or within the antenna volume in dielectric antenna). The second type provides beam tilting or switching via the use of adjacent parasitic elements (that can be switched on and off to change the coupling mechanism with the main radiator) to the main radiator, thus tilting or shifting the radiation characteristics on such antennas. We will present several examples in this section on each category and with their MIMO configurations.

A very nice methodology to clearly view the concept of pattern reconfigurability via a change in mode excitation is via the utilization of the Theory of Characteristic modes (TCMs). CMs are obtained via the solution of an eigenvalue problem based on the surface impedance of a certain surface (usually metallic). The solution then shows the natural modes that can be supported by this structure at hand. This solution is for the structure itself without any excitation. Each structure can support certain modes. Each mode has its own current distribution and thus a certain radiation characteristic. We will not dig deeper into TCM'as there are a magnitude of references on this topic that are available in open literature. The example

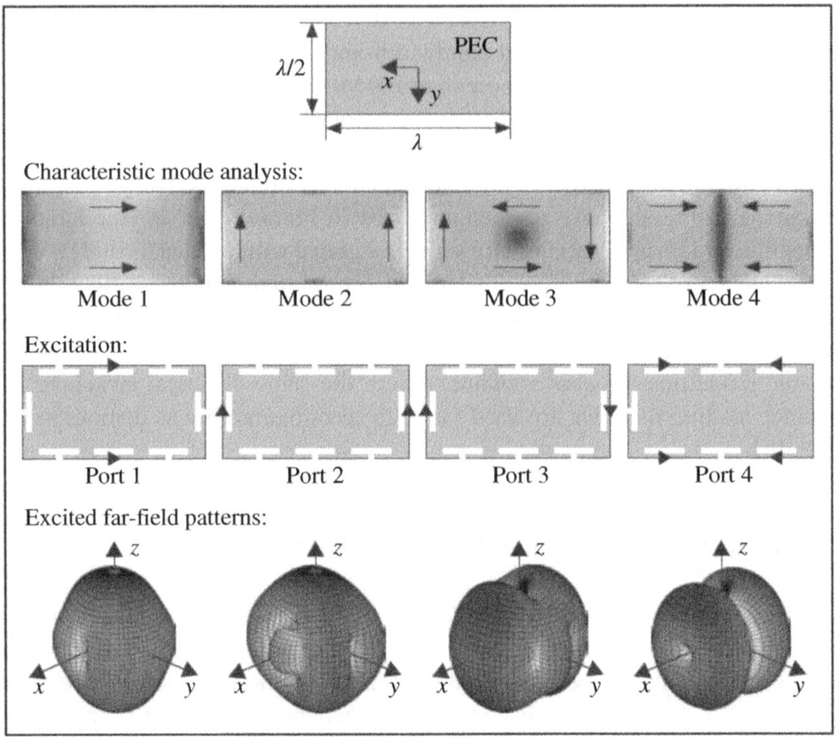

Figure 5.5 Characteristic modes of a rectangular metallic sheet (can be considered the backplane of a mobile terminal) along with the excitation locations for each mode and the resulting radiation pattern. Source: Reproduced with permission from Ref. [17]/IEEE.

we will present utilizes TCM is coming up with a pattern reconfigurable MIMO antenna system where the excitation of such CM yields various pattern shapes. For a rectangular shape, the characteristic modes of the surface current are shown in Figure 5.5 for the first four modes. Note that the locations of the excitation elements are placed at the current maxima of the specific mode from the CM solution. In order to excite a mode, all locations of current maxima should be excited together and with the same phase (current direction) relationships. Capacitively and inductively coupled elements are used for such excitations along with passive combiners. The excitation of a certain mode will yield a specific radiation pattern as shown.

The work in [18] is based on TCM and shows a MIMO antenna system that is developed using the backplane of a mobile phone and the excitation of two modes (one on each side of the backplane) to achieve pattern control via the incorporation of reconfigurable terminating loads. Two states for each port were considered and the eigen analysis was repeated in the presence of the reconfigurable loads. A direct impact on the eigencurrent (modes) and thus the radiation patterns was observed. The reconfigurable loads also helped in reducing the coupling currents and thus improving the correlation coefficient (CC) of the MIMO antenna systems when both ports are simultaneously excited. The design was optimized for operation in the 2.36 GHz band but yielded a very narrow bandwidth of less than 5%. The port isolation was marginal as well around −10 dB. Figure 5.6 shows the geometry of this pattern reconfigurable MIMO antenna system along with its two-state radiation patterns from each port. The worst case ECC achieved was 0.369.

Another example of a pattern reconfigurable MIMO antenna based on current control and yielding completely different radiation patterns is the work in [19]. In this work, a rectangular patch antenna with a u-slot is loaded with eight PIN diodes (two on each side) to be able to control the current directions and thus obtain three radiation patterns according to three states of configuration. The biasing mechanism of the diodes follows the ones presented in Figure 5.1d, where ring pads are used to separate the bias line from the GND. The geometry of the single element is shown in Figure 5.7a. The control of the PIN diode groups yields three states as shown in the configuration table in Figure 5.7b. Three orthogonal patterns are obtained, a boresight, a conical with maximum in y–z plane, and conical with maximum in x–z plane. The antenna resonates at 5.3 GHz with more than 6% of bandwidth as shown in Figure 5.7c. The MIMO configuration is shown in Figure 5.7d while the three measured radiation patterns in the x–z plane are demonstrated in Figure 5.7e. It is clear that the three states will provide good ECC values due to their orthogonality. The measured ECC of the MIMO antenna system in an indoor scenario was less than 0.33. The antenna efficiency was more than 45% for all three states with state 1 having more than 85% efficiency (the normal unloaded patch mode). An RT/Duroid 5880 substrate was used to enhance the radiation efficiency via the low-loss substrate.

The use of electrically controlled parasitic elements (coupling elements) for changing the direction or orientation of the radiation pattern is considered pattern control and reconfigurability as presented in [20–22]. The principle is simple, the presence of a parasitic element close to the current concentration can couple in a way that makes it act as a director element (as in a Yagi-Uda antenna) and thus pulls the radiation pattern maxima toward it. This was the main idea in [20, 21]. Active switching is used to turn the parasitic elements on and off to steer its beam to the left and right of the main radiator. The example in [20] presents an electronically steerable beam via the use of two parasitic elements per radiating one, and it is suggested to be used within cell phones operating at 5 GHz bands. Figure 5.8a shows this antenna solution with its MIMO configuration. It should be noted that the MIMO configuration will provide highly orthogonal patterns (completely different planes of radiation) and thus very low ECC values of less

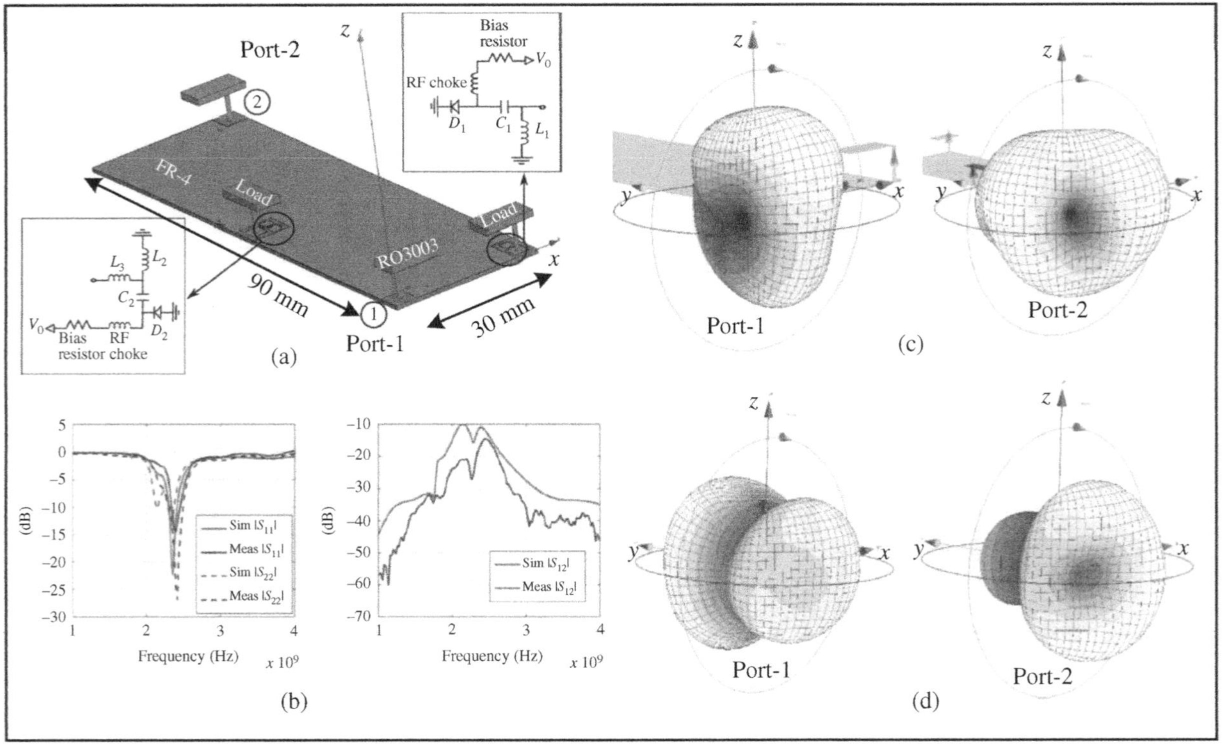

Figure 5.6 Pattern reconfigurable MIMO antenna system via mode and active termination loading: (a) geometry of the antenna system, (b) the measured and simulated *S*-parameters, (c) state 1 radiation patterns from ports 1 and 2, (d) state 2 radiation patterns from ports 1 and 2. Source: Reproduced with permission from Ref. [18]/IEEE.

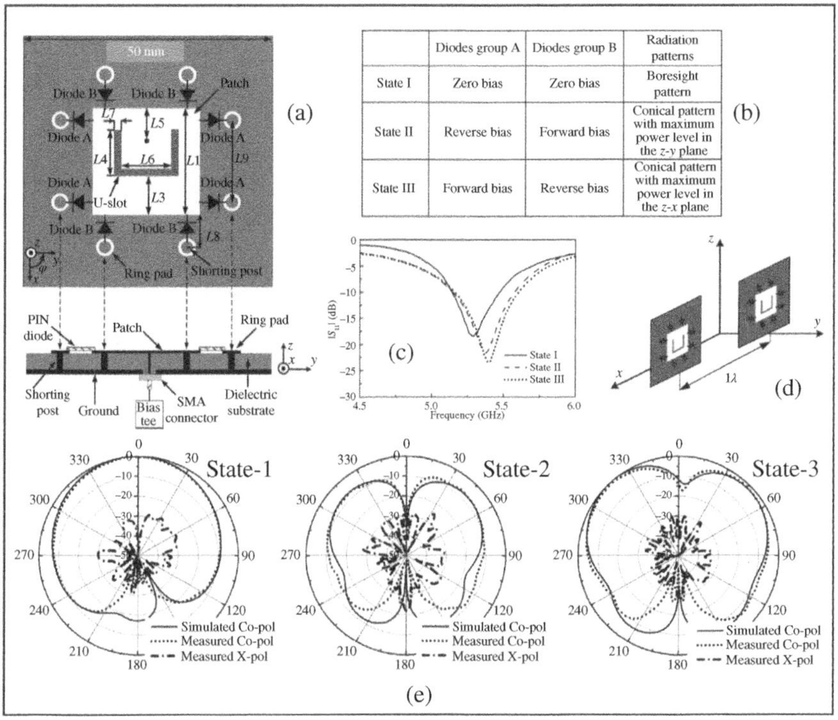

Figure 5.7 Pattern reconfigurable MIMO antenna based on a patch antenna with a u-slot: (a) geometry of the single element with eight PIN diodes, (b) the configuration table of the three states and their corresponding radiation patterns, (c) resonance curves for the three states, (d) MIMO configuration of two elements, (e) obtained radiation patterns in the x–z plane for states 1, 2, and 3 at 5.3 GHz. Source: Reproduced with permission from Ref. [19]/IEEE.

than 0.05. The port isolation will be high as well. Simple dipole elements are used as the main radiators, and dipole-like parasitic elements are used as well, but at 90° angles at each of the two sides below the main radiator surface. The patterns are located at 0, 70, and 290° in the azimuth plane as shown in the same figure.

The example in [21] presents a similar control mechanism as in [20], but having all elements on the same plane. PIN diodes are also used to activate either of the two-beam tilting or steering configurations (−35°, 0°, 35°). The design operates between 3.3 and 4.2 GHz in all states. Figure 5.8b shows the structure as well as the obtained states after reconfiguration. Two parasitic dipoles are placed on the left and right of the main dipole radiator. The activation of one switch (S) will allow the current induction on that arm and thus be coupled with the original radiator current to tilt the radiation pattern in either direction. A metasurface was added

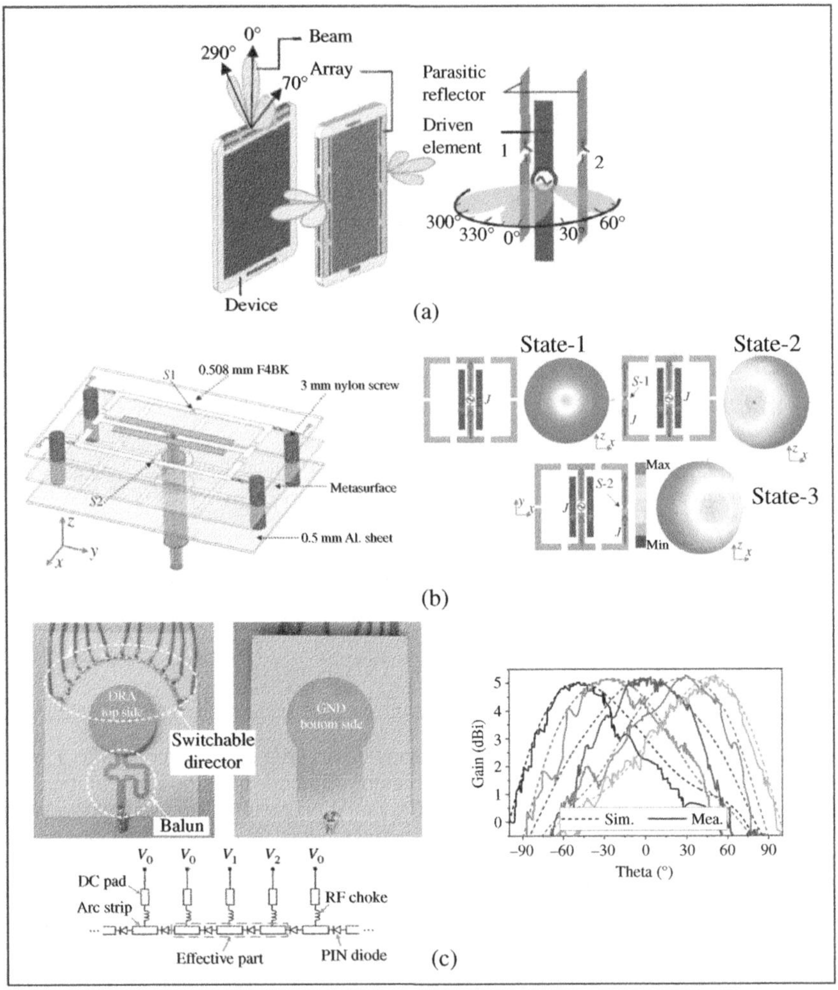

Figure 5.8 Pattern reconfigurable dipole configurations: (a) parasitic elements not on the same plane as the main dipole. Source: Reproduced with permission from Ref. [20]/IEEE. (b) Parasitic elements on the same plane as the main radiator. Source: Reproduced with permission from Ref. [21]/IEEE. (c) DRA-based pattern reconfigurable antenna element along with its diode switching mechanism and the resultant radiation patterns. Source: Reproduced with permission from Ref. [23]/IEEE.

to improve the bandwidth of the dipole, to provide end-fire radiation, and to help in the miniaturization of the elements. The antenna was not tested in a MIMO configuration, but it can be easily used in vehicular and access point MIMO systems. The example in [22] is of similar nature but uses a loop antenna as the main radiator. Two parasitic arms are placed on both of its planar sides and connected to the loop via a PIN diode one at a time to shift the location of the current maxima, thus tilting the beam $\pm 30°$. The MIMO configuration consisted of placing one element on each side of a rectangular backplane. ECC levels less than 0.015 were measured due to the orthogonality of the patterns (one on each side of a rectangle). This antenna had a very narrow bandwidth as it covered 5.15–5.35 GHz.

Finally, we present several pattern reconfigurable DRAs following the same concept of beam-tilting via parasitic element control [23, 24]. The DRA in [24] is symmetrically fed via two feeding arms to obtain a pure symmetric $TE_{01\delta}$ mode to excite a horizontally polarized omnidirectional pattern. Two director arms are added to the sides of this DRA. The activation of either director's arm will pull the fields within the DRA toward it, thus creating an end-fire radiation toward the center of that arm. The antenna was designed to operate in the 5 GHz band but with a very narrow impedance bandwidth of 1.5%. Another similar design was presented in [23] where a single director arc was used as compared to two in [24], but only portions of the arc were activated to direct the beam and cover more combinations. This design is shown in Figure 5.8c and as can be seen, the beams were tilted between $\pm 60°$. The advantage of DRA-based pattern reconfigurable antennas is that they provide high radiation efficiencies, and due to their non-metallic nature, can provide lower port coupling. The volumetric nature of DRAs makes them suitable for access point and vehicular applications in the sub-6 GHz applications, while they can be very attractive for MIMO applications in mm-wave bands.

5.3 Polarization Reconfigurable MIMO Antennas

Polarization reconfigurable antennas are usually combined with pattern (beam forming or switching) and frequency reconfigurability as shown in [25–27]. Similar to the operating or excitation principle of pattern reconfigurable antennas, the control of the radiated fields in polarization reconfigurable antennas is also dependent on the generated currents. This can be for linearly polarized or circularly polarized elements to generate any of the polarizations, namely vertical polarization (VP), horizontal polarization (HP), right-hand circular polarization (RHCP), or left-hand circular polarization (LHCP). In addition, if the mechanism of polarization reconfigurability is successful on a single element, it can be directly applied within a MIMO configuration, after which MIMO performance metrics can be checked for proper system-level performance.

The polarization agile 4×4 patch antenna array proposed in [25] was also designed to provide pattern and frequency reconfigurability. Thus, it relied on four PIN diodes for polarization and pattern control (via sectoral activation of the 4×4 array), and each element was also loaded with capacitor values to provide frequency agility between 1.35 and 2.19 GHz, with 2% impedance bandwidth in each covered band. The geometry of a single element is shown in Figure 5.9a showing the four positions of the PIN diodes (P_1–P_4) and the feeding arm. In addition, the capacitors (C_t) are located at the four edges of the patch for frequency reconfigurability. A capacitor value of 3.14 pF allows the operation at 1.35 GHz. The PIN diode was modeled according to the datasheet as a 2.1 Ω resistor in the ON state, and 0.17 pF capacitor for the OFF state. The biasing

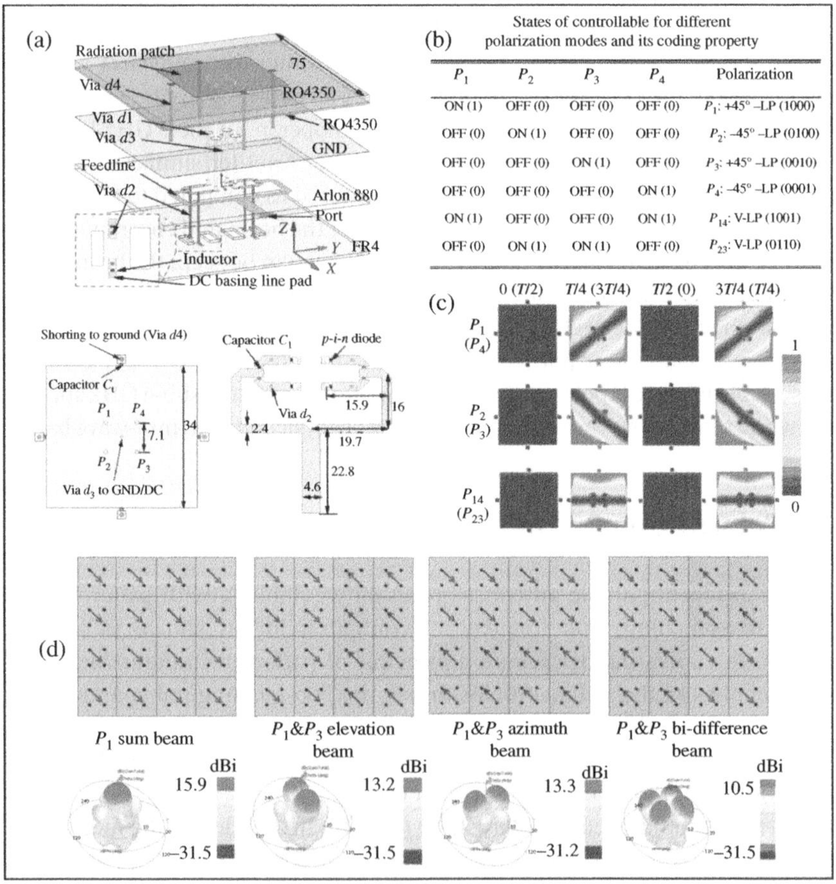

Figure 5.9 Polarization reconfigurable antenna that provides 45°-LP, −45°-LP, and V-LP: (a) geometry showing the four PIN diode posts (P_1–P_4) and frequency control capacitors (C_t), (b) polarization configuration table, (c) excited surface currents for each polarization along with excited ports, (d) 4 × 4 array excitation conditions to generate controlled pattern and polarization beams. Case shown for 45°-LP polarization. Source: Reproduced with permission from Ref. [25]/IEEE.

circuit for the PIN diodes consisted of an RF choke and a DC blocking capacitor. Three linear polarizations (LPs) were achieved via the PIN diode activation scheme shown in Figure 5.9b, a 45°–LP, a −45°–LP, and a regular V-LP. The surface currents generated on the patch are shown in Figure 5.9c for the three polarizations achieved. It is clear that the activation of the PIN diodes according to the table provided will force the currents to follow the distributions shown, and thus generate the proper polarizations. Note the ±45° cases in P_1 and P_2 and the normal patch mode in P_{14}. The pattern configuration for each polarization was generated by activating a group of elements with 1 (pink-like color) and the others with 0 (green-like color), where the difference between the excitations of a 1 and 0 is a 180° phase shift only. This allowed the generation of multiple beam shapes for each polarization. One group of beams for the 45°-LP case are shown in Figure 5.9d.

Another interesting polarization and beamforming reconfigurable array was presented in [26] for operation in the 5.8 GHz band. Polarization agility was achieved via a polarization selection module that consisted of switches and a hybrid coupler to activate two streams that are then used to generate four polarizations: VP, HP, RHCP, and LHCP. The two streams feed two Butler matrices to provide beam-switching capability. The VP and HP streams from either (or both) Butler Matrix feed each of the antennas at two points, an HP and a VP feeding point (dual-polarized elements are used). This allows for the generation of the four polarizations mentioned earlier along with beam switching into four spatial directions due to the utilization of the Butler Matrix. Figure 5.10a shows the configuration of this work. A crossed inverted V-antenna element with a director for gain and bandwidth enhancements, where each element is fed at two points corresponding to each polarization (HP or VP). For circular polarizations, the HP and VP components will have additional phase relations to generate the required circular orientation. A four-element linear antenna array is used in the final prototype, which achieved 3-dB axial ratio bandwidths of more than 22% for both LHCP and RHCP with a maximum realized gain of 13.9 dBi. Another fixed design that covered mm-wave bands around 28 GHz with very high impedance bandwidth and provided VP and HP patterns with beam steering via Butler matrix use was the one in [27]. Very wideband Butler matrices were provided with wideband couplers and phase shifters that provided 5.4 GHz of operating bandwidth.

Another example of polarization and beam steering reconfigurable antenna array is presented in [28] but is based on magneto-electric (M-E) dipole elements. The single element covered the 2–2.4 GHz band, and provided three polarizations, LP, RHCP, and LHCP via the use of four PIN diodes. The array was backed by a cavity to enhance element isolation and scanning performance of the array. The different polarizations are realized via the embedding of two switchable crossed

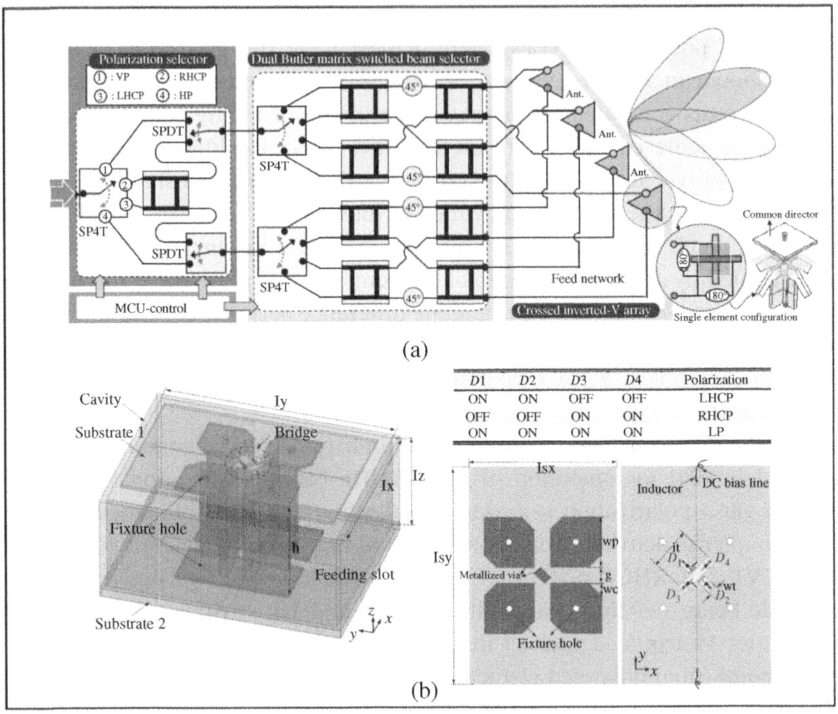

D1	D2	D3	D4	Polarization
ON	ON	OFF	OFF	LHCP
OFF	OFF	ON	ON	RHCP
ON	ON	ON	ON	LP

Figure 5.10 Polarization reconfigurable antennas: (a) based on a cross-inverted *V* with a gain enhancement director along with a polarization switching module providing four polarization modes and butler matrices for beam switching. Source: Reproduced with permission from Ref. [26]/IEEE. (b) Based on cavity-backed M-E dipoles with three polarization modes. Source: Reproduced with permission from Ref. [28]/IEEE.

strips within the M-E dipole element as shown in Figure 5.10b. The beam steering capability is achieved via the use of reflection-type phase shifters realized by varactor diodes within the feed network. The beam scanned from −40° to 40° in one plane for each of the three polarizations. Gain values around 10 dB were achieved across the band covered and for various angles and polarizations from a four-element linear array of M-E dipole elements. Finally, DRA antennas could provide polarization reconfigurable patterns if the modes generated within them are changed electronically. This was demonstrated in [29] where an electrically reconfigured dual-polarized DRA was presented. The feeding structure was controlled via PIN diodes to generate currents beneath the cylindrical DRA (cDRA) to excite the two orthogonal modes $TE_{011+\delta}$ and $TM_{01\delta}$.

Although recent polarization agile works did not consider MIMO configurations, the main ideas of polarization control can be easily applied to such configurations, and then the ECC among other MIMO metrics should be evaluated and optimized.

5.4 Challenges and Practical Considerations

There are several practical challenges that needs to be addressed when a designer is encountered with similar requirements. These challenges can be designer-driven, or market availability-driven. The designer should be aware of the consequences of each so that he/she can compensate for them when finalizing the final design.

Designer-driven challenges are those that the designer can control at the level of modeling details representing the actual final product. Thus, once the active reconfigurable antenna is designed, it needs to be put in its proper environment, i.e. casing, package, etc., to consider the various effects coming from such an environment. This is important as it will affect the resonance properties as well as the radiation properties of the designed reconfigurable antennas. The inclusion of the screen and battery within a mobile terminal model as well as the volumes and proper material properties of adjacent components to the antenna, will provide proper results that will resemble the measured ones once the antenna is put in a mock-up model. Any cuts in the GND plane will also affect the radiation and resonance properties of the antenna if it is close to its structure and if the frequency of operation considered is below 1 GHz. If the design was optimized to give a certain level of reconfigurability in an ideal environment (only antenna and biasing circuit of active element), then it needs to be re-evaluated once the battery, screen, and other parts and integrated, then the amount of shifts in the frequency bands covered or pattern orientation should be quantified and the control values re-adjusted to reflect the frequency bands to be covered within the controllability of the active part.

The consideration of MIMO antenna configurations complicates some practical aspects and trying to place the antennas further away from one another might not be feasible in a practical final product. Thus, several placement studies should be conducted in order to make sure the port isolation values and ECC levels are of acceptable nature for that application. The analysis and characterization of the effects of the user hands and head on the radiation efficiencies as well as characterization of the radiation levels for health and masking considerations is vital in the final designs and products. Such analysis can be found in various references such as [30].

Market-driven challenges can be those that are based on the information provided to the designer from the manufacturer of components, and then the amount of detail the designer will consider in his/her design. For example, the electrical properties of dielectric substrates such as the loss tangent and the dielectric constant are usually given for a specific frequency or a range of frequencies that sometimes does not cover the bands of operation of the antenna (i.e. when considering the material for mm-wave bands). Thus, the values used in the model will not be as accurate yielding inaccuracies in the resonance bands and efficiencies calculated. In addition, the data sheets of active elements do not provide complete circuit models for their components at wide frequency ranges, and thus the designer will suffer from such an accumulated error at higher frequency bands. In addition, in simplifying designs, usually active reconfigurable antennas consider a simple model of the active device at forward or reverse bias conditions, i.e. a capacitor or resistor (even though it is considered sufficient in sub-6 GHz bands). This might yield higher error values in the resonance frequencies achieved at higher bands as other parasitic effects arise along with more loss (see the example in Section 5.5 where a simple model of the varactor was used).

5.5 CAD Modeling Example

One computer-aided design (CAD) modeling example is shown to demonstrate the modeling of a frequency-reconfigurable MIMO antenna using slot elements. The design is explained in detail in [31]. The focus of this section will be on the modeling of the reactive elements to achieve frequency reconfigurability using the CAD tool HFSS. Similar steps can be followed in other modeling tools once the user knows how to set up and assign the proper elements. We present a four-element MIMO frequency-reconfigurable antenna based on slot elements where each annular slot is loaded with a single varactor diode to achieve frequency agility. The varactor diode specifications (capacitance versus applied reverse bias voltage) are taken from its datasheet. In this example, we use an SMV 1233 varactor. Figure 5.11 shows the top and bottom sides of this antenna system. The top side had only the metallic pads and biasing circuit connections as well as the feeding microstrip lines. While the bottom side contained the GND plane and the radiating elements (annular slots). A Rogers RO4350 substrate with dimensions of $60 \times 120 \times 0.76\,\text{mm}^3$ was used. The dimensions of the slots and other parts can be found in the reference.

To place the varactor diode (the small rectangle with reference designator C19), a very thin clearing ring was introduced to allow for the DC biasing and GND signals to be separated. The varactor diode was placed between the inner annular ring and the outer GND plane on the bottom layer as shown in Figure 5.12a. The active voltage (reverse bias DC) path consisting of the PAD for soldering the DC wire, the limiting resistor (i.e. R9) and the RF choke (i.e. L9) were connected through the use of a via from the top layer to the inner ring as shown in Figure 5.12b. The GND path also consisted of a similar path (i.e. R8 and L8) that connected the DC

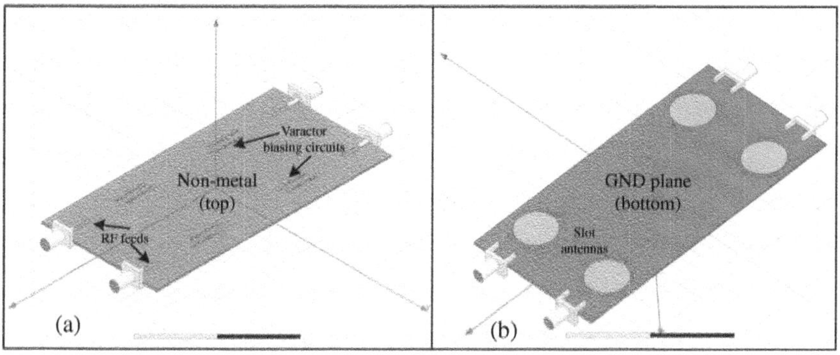

Figure 5.11 The model of a four-element annular slot frequency agile MIMO antenna system: (a) top side (non-metallic except for the biasing and RF feed lines), (b) bottom side (metallic except for the slot clearances).

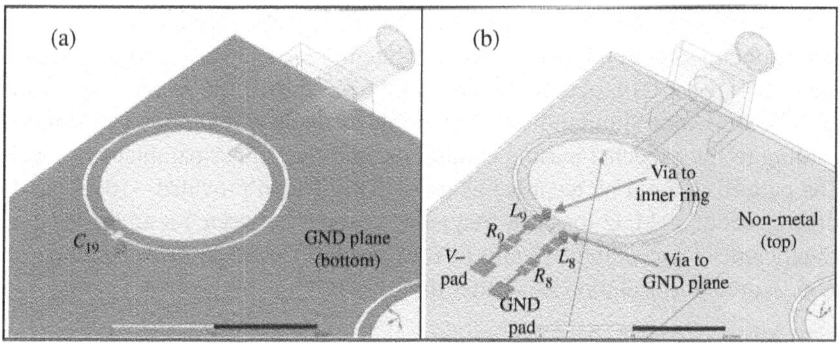

Figure 5.12 Biasing network element placement: (a) varactor diode on bottom layer, (b) biasing network on the top layer.

GND of the supply source at the top layer to the RF GND of the bottom layer via a plated via hole. The PADs for each component should be designed according to the soldering pattern of the respective surface mount components.

For each component, we need to place a rectangular sheet and then assign the proper boundary conditions for it using lumped RLC boundary conditions from HFSS. This is shown in Figure 5.13. For each component, we need to assign the proper values so that the presence of the biasing network is taken into account in the RF simulation and to obtain modeling and results as close to reality as possible. In this design, the values of all resistors (R) were 2.1 KΩ, and the inductors (L) were 1 μH. The varactor diode capacitance value when it is properly biased should be used from the datasheet. Thus, the value of C19 should reflect the value of the diode when biased at a certain voltage. For example, a value of 5.08 pF for C19,

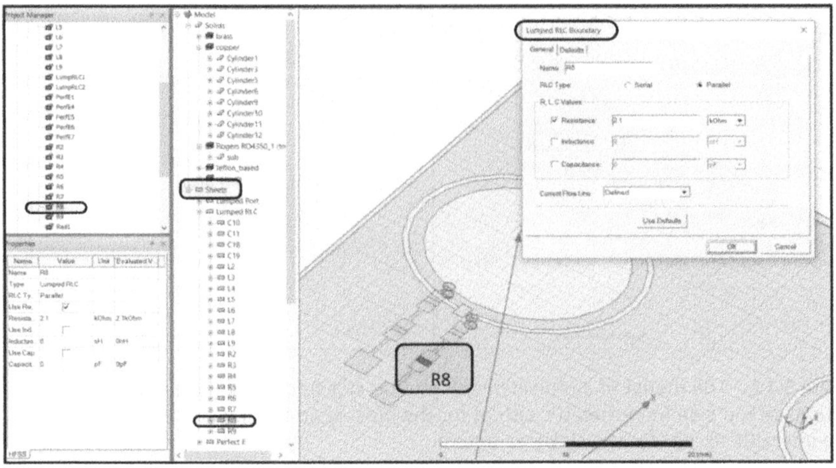

Figure 5.13 Boundary condition assignment for the RLC sheets within the model.

which is obtained with 0 V bias, made the antenna resonate at 1.8 GHz. Another value of 0.84 pF of C19, obtained with −15 V bias, shifted the resonance toward 2.45 GHz. So, several simulation sweeps should be made with different C19 values according to the expected biasing conditions, and then the *S*-parameters as well as the patterns should be recorded as usual. This design provided −10 dB bandwidths of about 40 MHz for each frequency covered between 1.8 and 2.45 GHz, with efficiencies exceeding 50% and ECC values less than 0.25 between all ports. The measured port isolation was better than 12 dB.

5.6 Summary/Conclusion

In this chapter, we have presented several methods for achieving antenna reconfigurability via electronic control. The control of the operating frequency bands, pattern types, maximum beam directions, and polarization control was demonstrated with several relevant examples from the literature. The majority of works use PIN and varactor diodes to achieve such reconfigurability features that are very useful within various applications and use cases. The use of reconfigurable antennas in MIMO antenna systems is very attractive as that might directly make them cover wider frequency bands and provide better ECC values due to beam spatial tilting capabilities. Several challenges and recommendations are provided when targeting the design of reconfigurable MIMO antennas. The chapter ends with some guidelines and use of a CAD tool example clarifying the setup of reconfigurable elements within the modeling stage to speed up the use of such models in real designs.

References

1 Christodoulou, C., Tawk, Y., Lane, S., and Erwin, S. (2012). Reconfigurable antennas for wireless and space applications. *Proceedings of the IEEE* 100 (7): 2250–2261.

2 Costantine, J., Tawk, Y., Barbin, S., and Christodoulou, C. (2015). Reconfigurable antennas: design and applications. *Proceedings of the IEEE* 103 (3): 424–437.

3 Petosa, A. (2014). *Frequency-Agile Antennas for Wireless Communications*. Artech House.

4 Hussain, R. and Sharawi, M.S. (2014). A cognitive radio reconfigurable MIMO and sensing antenna system. *IEEE Antennas and Wireless Propagation Letters* 14: 257–260.

5 Hussain, R., Sharawi, M.S., and Shamim, A. (2018). 4-Element concentric pentagonal slot-line-based ultra-wide tuning frequency reconfigurable MIMO antenna system. *IEEE Transactions on Antennas and Propagation* 66 (8): 4282–4287.

6 Choi, J., Hwang, W., You, C. et al. (2019). Four-element reconfigurable coupled loop MIMO antenna featuring LTE full-band operation for metallic-rimmed smartphone. *IEEE Transactions on Antennas and Propagation* 67 (1): 99–107.

7 Ikram, M., Abbas, E.A., Nguyen-Trong, N. et al. (2019). Integrated frequency-reconfigurable slot antenna and connected slot antenna array for 4G and 5G mobile handsets. *IEEE Transactions on Antennas and Propagation* 67 (12): 7225–7233.

8 Shamsaee-Malfajani, R., Bin-Ashraf, F., and Sharawi, M.S. (2022). A 5G enabled shared-aperture, dual-band, in-rim antenna system for wireless handsets. *IEEE Open Journal on Antennas and Propagation* 3: 1013–1024.

9 Mun, B., Jung, C., Park, M.J., and Lee, B. (2014). A compact frequency-reconfigurable multiband LTE MIMO antenna for laptop applications. *IEEE Antennas and Wireless Propagation Letters* 13: 1389–1392.

10 Padmanathan, S., Al-Hadi, A., Elshirkasi, A. et al. (2022). Compact multiband reconfigurable MIMO antenna for sub-6GHz 5G mobile terminal. *IEEE Access* 10: 60241–60252.

11 Luomaniemi, R., Lehtovuori, A., Ilvonen, J. et al. (2022). Extremely low-profile tunable multiport handset antenna. *IEEE Transactions on Antennas and Propagation* 70 (2): 911–921.

12 Hussain, R., Sharawi, M.S., and Shamim, A. (2017). An integrated four-element slot-based MIMO and a UWB sensing antenna system for CR platforms. *IEEE Transactions on Antennas and Propagation* 66 (2): 978–983.

13 Hussain, R. and Sharawi, M.S. (2019). An integrated slot-based frequency-agile and UWB multifunction MIMO antenna system. *IEEE Antennas and Wireless Propagation Letters* 18 (10): 2150–2154.

14 Hussain, R., Khan, M.U., and Sharawi, M.S. (2017). An integrated dual MIMO antenna system with dual-function GND-plane frequency-agile antenna. *IEEE Antennas and Wireless Propagation Letters* 17 (1): 142–145.

15 Yan, J.-B. and Bernhard, J.T. (2013). Implementation of a frequency-agile MIMO dielectric resonator antenna. *IEEE Transactions on Antennas and Propagation* 61 (7): 3434–3441.

16 Tang, S.-C., Wang, X.-Y., Zheng, S.Y. et al. (2021). Frequency-reconfigurable dielectric patch antenna with bandwidth enhancement. *IEEE Transactions on Antennas and Propagation* 70 (4): 2510–2519.

17 Johannsen, N.L., Peitzmeier, N., Hoeher, P.A., and Manteuffel, D. (2020). On the feasibility of multi-mode antennas in UWB and IoT applications below 10 GHz. *IEEE Communications Magazine* 58 (3): 69–75.

18 Kishor, K.K. and Hum, S.V. (2014). A pattern reconfigurable chassis-mode MIMO antenna. *IEEE Transactions on Antennas and Propagation* 62 (6): 3290–3298.

19 Qin, P.-Y., Guo, Y.J., Weily, A.R., and Liang, C.-H. (2011). A pattern reconfigurable U-slot antenna and its applications in MIMO systems. *IEEE Transactions on Antennas and Propagation* 60 (2): 516–528.

20 Chamok, N.H., Yılmaz, M.H., Arslan, H., and Ali, M. (2016). High-gain pattern reconfigurable MIMO antenna array for wireless handheld terminals. *IEEE Transactions on Antennas and Propagation* 64 (10): 4306–4315.

21 Wang, Z., Liu, S., and Dong, Y. (2022). Compact wideband pattern reconfigurable antennas inspired by end-fire structure for 5G vehicular communication. *IEEE Transactions on Vehicular Technology* 71 (5): 4655–4664.

22 Rhee, C., Kim, Y., Park, T. et al. (2014). Pattern-reconfigurable MIMO antenna for high isolation and low correlation. *IEEE Antennas and Wireless Propagation Letters* 13: 1373–1376.

23 Chen, J.-X., Ke, Y.-H., Yang, L.-L., and Yang, W.-W. (2022). Pattern-reconfigurable dielectric resonator antenna with endfire beam-scanning feature. *IEEE Antennas and Wireless Propagation Letters* 21 (7): 1398–1402.

24 Ke, Y.-H., Yang, L.-L., and Chen, J.-X. (2021). A pattern-reconfigurable dielectric resonator antenna based on switchable directors. *IEEE Antennas and Wireless Propagation Letters* 21 (3): 536–540.

25 Wu, J., Lu, X., Wang, W. et al. (2021). Design of a compact polarization-agile and frequency-tailored array antenna with digital-controllable radiation beams. *IEEE Transactions on Antennas and Propagation* 70 (2): 813–822.

26 Dong, H.-J., Kim, Y.-B., and Lee, H.L. (2022). Reconfigurable quad-polarization switched beamforming antenna with crossed inverted-V array and dual-butler matrix. *IEEE Transactions on Antennas and Propagation* 70 (4): 2708–2716.

27 Klionovski, K., Sharawi, M.S., and Shamim, A. (2019). A dual-polarization switched-beam patch antenna array for millimeter wave applications. *IEEE Transactions on Antennas and Propagation* 67 (5): 3510–3515.

28 Hu, J., Yang, X., Ge, L., and Wong, H. (2022). A polarization and beam steering reconfigurable cavity-backed magneto-electric dipole antenna array using reflection-type phase shifter. *IEEE Transactions on Antennas and Propagation* 70 (1): 296–306.

29 Liu, X., Leung, K.W., and Yang, N. (2021). Wideband horizontally polarized omnidirectional cylindrical dielectric resonator antenna for polarization reconfigurable design. *IEEE Transactions on Antennas and Propagation* 69 (11): 7333–7342.

30 Sharawi, M.S. (2014). *Printed MIMO Antenna Engineering*. Artech House.

31 Hussain, R., Ghalib, A., and Sharawi, M.S. (2017). Annular slot-based miniaturized frequency-agile MIMO antenna system. *IEEE Antennas and Wireless Propagation Letters* 16: 2489–2492.

6

Massive-MIMO (m-MIMO) Antenna Systems

6.1 Advantages of m-MIMO

With the commercialization of cellular network systems, rich application content, diverse application scenarios, and personalized service requirements continue to drive the development of mobile communications. Facing the huge data volume and system capacity, transmission rate and stability of the wireless communication networks are severely challenged. At the same time, due to the inherent scarcity of wireless communication networks, the huge data volume of user groups and services limits its own development. With imperfect resource allocation and user management, there will be interference and competition among users, which will lead to degradation of the overall system performance.

In response to the above problems, massive array antenna technology based on multiple-input multiple-output (MIMO) technology has emerged [1]. MIMO technology uses multiple transmitting and receiving antennas at the transmitter and receiver ends to enable spatial multiplexing. 4G communication systems usually employ a few MIMO antennas, which limits the system capacity. In order to provide higher communication quality for multiple users simultaneously, massive MIMO (m-MIMO) system deploying a large number of antennas at the base station has been proposed for 5G communications, which receives extensive attention from both industry and academia, as shown in Figure 6.1.

6.1.1 Spectrum Efficiency Improvement

Compared with the conventional MIMO array, the spatial resolution of m-MIMO can be significantly enhanced. With the help of the large spatial degree of freedom provided by the m-MIMO technology, multiple users in the cellular network can be served using the same time-frequency resource (i.e. simultaneously in the same band). It can significantly improve the spectrum efficiency without

MIMO Antenna Systems for 5G and Beyond, First Edition.
Xiaoming Chen, Ping Jack Soh, and Mohammad S. Sharawi.

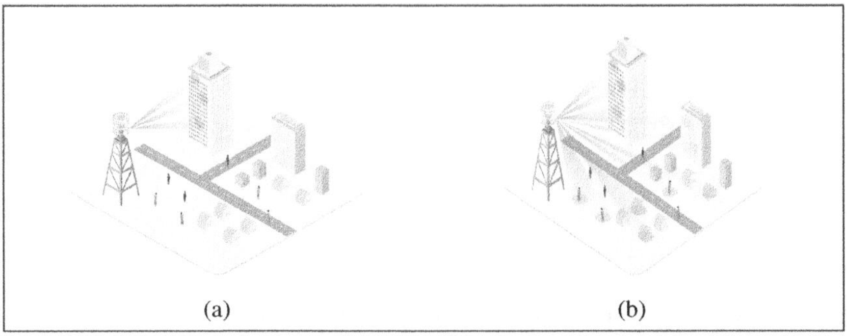

Figure 6.1 (a) Traditional MIMO method and (b) m-MIMO method.

increasing the density and bandwidth of base stations [2]. According to [3], the m-MIMO technology can improve the spectral efficiency by at least 10 times.

6.1.2 Energy Efficiency Improvement

By deploying a large number of antennas, the m-MIMO system can make full use of spatial multiplexing gain to increase system capacity. When the total transmitted energy is determined, techniques such as beamforming and precoding are used, and the BSAs radiate the energy to each target mobile terminal. This scheme can reduce the radiation power of a single antenna and the total energy loss, and the interference to other users is also reduced [4, 5]. m-MIMO can improve the system energy efficiency exponentially by forming a very narrow beam aimed at the user. The m-MIMO technology can improve energy efficiency by at least 100 times [3].

6.1.3 Air Interface Access Delay Reduction

The performance of wireless communication systems is affected by channel fading, and signal power is interfered by multipath effects over short distances. When the signal transmitted by the transmitter reaches the terminal through multiple paths, the signals with different phases and amplitudes will cancel each other. The terminal device must wait until that fading period has passed, which is the reason for transmission delay. To overcome channel fading, repetitive signals are sent at different time slots or frequencies to obtain diversity gain. Multiple antennas are used in m-MIMO technology to transmit multiple pre-coded independent signals, which reduces the impact of channel fading without using additional resources [6]. The integration and compatibility between different networks are the focus of 5G development, and the low air interface delay makes it possible for various innovative applications, such as virtual reality (VR), augmented reality (AR), autonomous driving and intelligent internet of things (IoT).

6.1.4 Signal Processing Complexity Reduction

When the number of antennas is sufficient, the channel between the base station and the user terminal will have quasi-orthogonal characteristics, so that the channels between the terminal users on the same resource are also orthogonal. Therefore, the simplest signal processing techniques can achieve near-optimal communication capacity.

6.1.5 Expansion of System Coverage

Compared with traditional BSAs or traditional integrated active antennas, m-MIMO has a large array and independent control capabilities for array units. The m-MIMO array increases the beam steering in the vertical dimension, which is beneficial for the antenna array to generate multi-beam coverage, and expanding the system coverage.

6.1.6 Insensitive to Non-Ideal Performance of Single Hardware

m-MIMO system reduces the accuracy requirement for individual antenna element, and cheaper hardware with low power consumption can be adopted. As the number of antennas increases, the power of a single antenna in the m-MIMO system, the linearity of a single amplifier, and the linearity of a radio frequency (RF) link are greatly reduced compared with traditional MIMO antennas. Using inexpensive power amplifier instead of expensive high-power amplifier helps to reduce the overall cost of the device and expands the application scenarios of the device.

6.1.7 System Reliability Improvement

For a traditional MIMO system, a failure in one module (especially the power amplifier and radio frequency link) of the system can bring down the entire system. The difference is that the failure of some modules in the m-MIMO system usually does not affect the operation of the entire system, so that the reliability and robustness of the system are improved. m-MIMO concentrates the beam in a very narrow range, which will greatly improve the robustness of purposeless human interference and intentional interference [7]. The reliability of the m-MIMO system will be improved and the user experience will be enhanced.

In addition, m-MIMO technology also has the advantages of saving energy and overcoming the impact of equipment mobility. Generally speaking, the m-MIMO technology is a transitional approach to the current mobile communication system by increasing the number of BSAs, while the terminal user still adopts the single-antenna receiving communication mode. The comprehensive performance of the system can be improved only through the modification of the base station, without updating the user's terminal equipment on a large scale.

6.2 m-MIMO Antenna Implementations

As an important part of the mobile communication system, the performance of the base station plays a vital role in the progress of communication technology. In order to achieve a more comprehensive signal coverage area and ensure system capacity requirements, BSAs have been extended from traditional MIMO antennas to m-MIMO antenna systems. Compared with traditional BSAs, the layout of m-MIMO base stations is more compact, and the available site space is becoming increasingly tight. The primary solutions to this problem are antenna miniaturization and reduced spacing between adjacent antennas in the array. This section will introduce the implementation of m-MIMO antenna from the design of antenna unit and array antenna, respectively. For the antenna unit, the antenna performance can be improved by the reasonable design of the antenna form. For the antenna array, when the spacing of the elements in the BSA system is more compact, the in-band and cross-band coupling problems need to be considered.

6.2.1 Implementation of the Antenna Unit

As the basic radiating unit of the base station, the dual-polarized antenna has two polarization modes, which correspond to the two working states of the antenna, namely transmitting and receiving. Using the dual-polarized antenna as the BSA unit can effectively increase the channel capacity. Therefore, stable radiation performance can be obtained by rationally designing the structural form of the dual-polarized antenna unit. When a large number of antenna elements are formed into an m-MIMO array, the performance of the MIMO array can be effectively improved.

Miniaturization is one of the key technologies for designing m-MIMO BSAs, which have the advantages of saving manufacturing costs and facilitating installation. When the spacing of the antenna in the array is constant, the miniaturization of the antenna elements is also conducive to the application of multiple decoupling methods. The electrical length of the antenna is inversely related to the antenna operating frequency. The lower the antenna operating frequency, the larger the electrical size of the antenna. Therefore, while satisfying the requirements of the radiation characteristics within the operating bandwidth, it is a relatively difficult challenge to effectively design the miniaturization of the antenna unit.

At present, the miniaturization design of dual-polarized BSAs is mainly achieved by loading parasitic technology, changing the antenna structure, using a substrate with high dielectric constant and loading metamaterials.

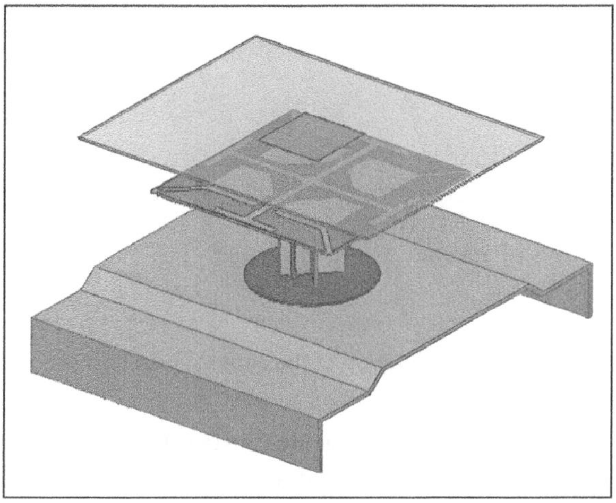

Figure 6.2 Configuration of the antenna loaded with parasitic square ring. Source: Reproduced with permission from Ref. [8]/IEEE.

6.2.1.1 Loading Parasitic Structures

By introducing parasitic resonant structures or short-circuiting branches around the antenna, the surface current of the antenna is changed and the electrical size of the antenna is extended. Miniaturization of the antenna is achieved by introducing new resonance points to expand the bandwidth.

In 2020, Ye et al. proposed a square ring scheme loaded above the folded electric dipole antenna [8], as shown in Figure 6.2. The overall dimension of the antenna is $0.37\lambda \times 0.37\lambda \times 0.27\lambda$ (λ is the wavelength of the lowest resonant frequency). The antenna consists of four folded dipoles, a parasitic patch, two feeding structures, and a metallic reflector. By loading a square parasitic patch above the antenna, a wider impedance bandwidth and a narrower beamwidth are obtained. The bandwidth of this antenna is 1.4–2.77 GHz (64.7%) with VSWR <2 and isolation between the two ports is greater than 26 dB.

6.2.1.2 Change the Structure of the Antenna

The lateral miniaturization scheme increases the electrical length of the antenna in a limited space by changing the shape of the radiation arm of the antenna.

A new miniaturized BSA element was proposed by Chu et al. in 2017 [9], as shown in Figure 6.3. The overall dimensions of the antenna are $0.31\lambda \times 0.31\lambda \times 0.19\lambda$ (λ is the wavelength of the lowest resonant frequency). The stepped square ring structure is used to widen the bandwidth, and the miniaturized design is realized through designing the feeding structure of two T-shaped branches. Impedance matching is improved by two pairs of crossed linear dipoles loaded in the middle of the coupling gap. The impedance bandwidth of the proposed antenna is 1.68–2.94 GHz (54.5%) with VSWR <1.5 and isolation between the two ports is greater than 28.5 dB.

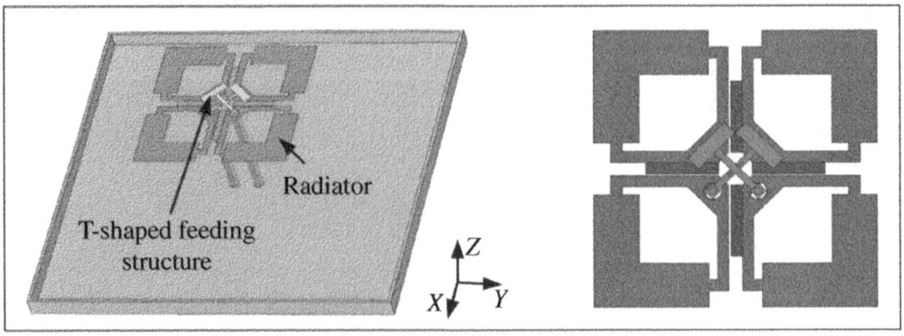

Figure 6.3 Configuration of the antenna with changed radiation arms.

6.2.1.3 Using High Dielectric Constant Dielectric

In 2019, Li et al. proposed a dielectric block-loaded miniaturized BSA [10], as shown in Figure 6.4. The overall dimensions of the antenna are $0.33\lambda \times 0.33\lambda \times 0.15\lambda$ (λ is the wavelength of the lowest resonant frequency). Part of the electromagnetic wave is confined in the dielectric block by loading a dielectric block under the antenna, which improves the isolation between the ports. The impedance bandwidth of the proposed antenna is 3.2–3.9 GHz (19.7%) with VSWR <1.5 and isolation between two ports is greater than 40 dB.

6.2.1.4 Metamaterial

Metamaterial is a new type of artificial composite structure or artificial composite material with physical properties that natural materials do not possess.

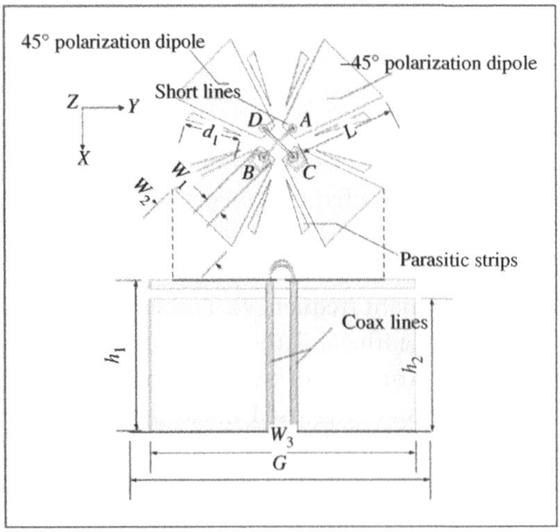

Figure 6.4 Configuration of the antenna loaded with dielectric block. Source: Reproduced with permission from Ref. [10]/IEEE.

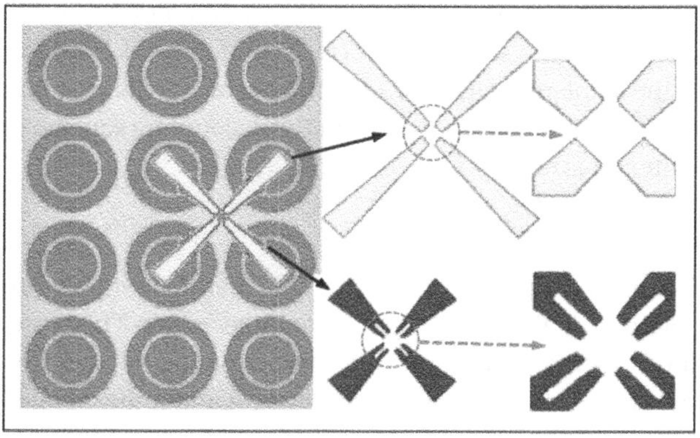

Figure 6.5 Configuration of the antenna loaded with AMC. Source: Reproduced with permission from Ref. [11]/IEEE.

Artificial magnetic conductor (AMC) is a periodic structure composed of metal sheets and dielectric substrates. It has in-phase reflection characteristics in the working frequency band, and the antenna-radiated electromagnetic wave will not produce a 180° phase difference after being reflected by AMC structure. By using a rationally designed AMC structure instead of a traditional metal reflector, the profile height of the antenna can be effectively reduced.

In 2017, Zhai et al. proposed a dual-band dual-polarized antenna based on AMC surface [11], as shown in Figure 6.5. The dual-polarized antenna consists of four pairs of orthogonal bow-tie patch dipoles, two feeding baluns, and concentric bicircular AMC surfaces. The introduction of AMC surfaces is beneficial to improve the antenna's radiation characteristics and reduce the profile of the antenna. The impedance bandwidths of the proposed antenna are 2.36–2.76 GHz (15.6%) and 5.12–5.62 GHz (9.3%) with VSWR <2. And isolation between the two ports is greater than 22 dB.

In addition, slot antennas [12] or patch antennas [13] with low-profile characteristics are also suitable as array elements for 5G m-MIMO antennas.

6.2.2 Implementation of Antenna Array

In m-MIMO systems, hundreds of antennas are usually placed on a base station platform to increase data throughput. In order to place more antennas in limited space, the spacing between antennas has to be reduced. The reduction in spacing will lead to poor coupling between antenna elements. Mutual coupling between antenna elements leads to poor antenna input impedance, deterioration of port isolation, distortion of the radiation pattern, and reduction of the MIMO system's average capacity [14]. It is foreseeable that for the future demand of multi-standard coexistence and limited site resources, antennas operating in different frequency bands need to coexist on a single base station platform. Therefore, decoupling is an inevitable problem in the implementation of m-MIMO arrays. This part will introduce the implementation of m-MIMO arrays from the decoupling scheme of in-band and cross-band antenna arrays.

6.2.2.1 In-Band Antenna Array

Due to the specific requirements of the BSA, the decoupling method not only can maintain the radiation pattern but also can be applied to large arrays. Therefore, in-band decoupling methods such as neutral line, decoupling network, defected ground structure, and electromagnetic bandgap structure cannot be applied to dual-polarized BSAs. In recent years, the in-band decoupling methods for dual-polarized BSAs can be basically divided into two kinds: neutralization methods and blocking methods. The neutralization method is to introduce a new coupling by loading the structure to cancel with the inherent coupling. The blocking method is to achieve decoupling by loading the structure to block the propagation path of the inherent coupling between the antennas. The following example is given to introduce these two methods.

In 2017, Wu et al. proposed a 2×2 high-isolation MIMO BSA array [15], as shown in Figure 6.6a. The spacing of the antenna elements in the array is 0.53λ (λ is the wavelength of 3.55 GHz). High isolation between antenna elements is achieved by loading an array decoupling surface (ADS) composed of multiple metal sheets above the antenna. By loading the ADS, the inherent coupling waves between the antenna units will cancel the reflected waves through the ADS. The impedance bandwidth of the antenna array is 3.3–3.8 GHz, and the isolation between ports is higher than 25 dB. In 2021, Da et al. proposed a 2×2 high-isolation MIMO BSA array with a dielectric structure loaded above the antenna [16], as shown in Figure 6.6b. When electromagnetic waves are radiated

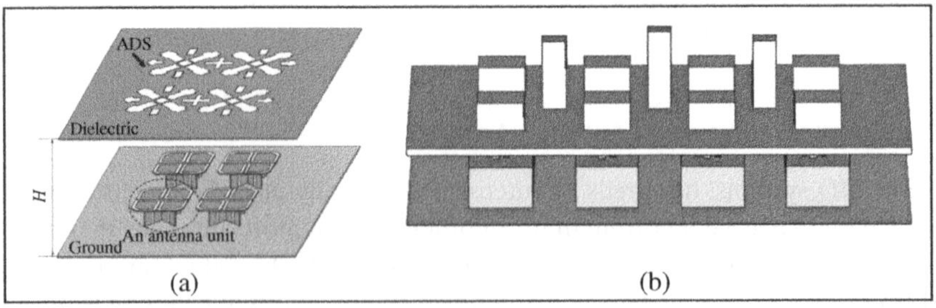

Figure 6.6 Structure of the array with (a) ADS. Source: Reproduced with permission from Ref. [15]/IEEE. (b) The dielectric superstrate. Source: Reproduced with permission from Ref. [16]/IEEE.

Figure 6.7 BSA array with SRR-loaded baffle. Source: Reproduced with permission from Ref. [17]/IEEE.

toward the dielectric, the reflected electromagnetic waves will cancel the inherent coupling waves between the antenna units. The impedance bandwidth of the proposed antenna is 3.3–3.8 GHz (14%), and the isolation between ports is higher than 25 dB.

In 2020, Li et al. proposed a 1×4 high-isolation array based on the BSA element loaded with dielectric blocks [17], as shown in Figure 6.7. The split-ring resonator structure loaded around the antenna unit can effectively suppress coupling between antenna elements in the array. The impedance bandwidth of the proposed antenna is 3.4–3.8 GHz (11%), and the isolation between ports is higher than 25 dB.

6.2.2.2 Cross-Band Antenna Array

For the future demand of multi-standard coexistence, the antennas on the base station platform need to work in different frequency bands. Compared with a single antenna covering multiple frequency bands, the multi-antenna solution will obtain more stable radiation patterns and beamwidths in different frequency bands. On a narrow base station platform, multiple antennas are often placed with shared aperture to save space and improve aperture utilization.

Multi-antenna shared-aperture placement scheme can be divided into embedded scheme, stacked scheme, and interleaved scheme [18], as shown in Figure 6.8. The embedded scheme is to place high-frequency (HF) units inside or between low-frequency (LF) units in the array. A stacked scheme and interleaved scheme is to place the antennas of different frequency bands along the vertical direction and stack them together. Depending on the coupling status between antennas in different schemes, the decoupling schemes are also different. In this subsection,

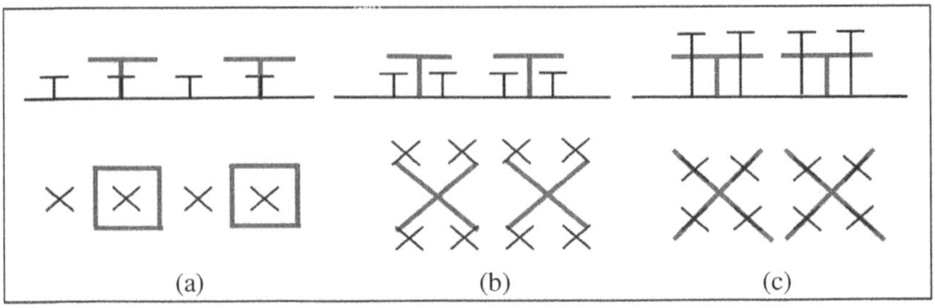

Figure 6.8 Arrangement of the (a) embedded (b) interleaved and (c) stacked BSA arrays. Source: Reproduced with permission from Ref. [18]/IEEE.

the implementation of m-MIMO arrays will be introduced based on examples of different decoupling methods.

In 2020, Wu et al. proposed a dual-band embedded shared-aperture BSA array [19], as shown in Figure 6.9. The HF antenna unit is a conventional cross dipole, while the LF antenna unit is a bowl-shaped dual-polarized unit surrounded by four dipoles. In order to improve the isolation between the HF and LF antenna units, the HF and LF units are placed on the same plane, and a secondary reflector is added to support the HF antenna unit. The impedance bandwidths of the final antenna unit with VSWR less than 1.7 are 0.698–0.96 GHz and 1.7–2.7 GHz, while the stable radiation characteristics are obtained in the frequency band.

In 2019, Zhu et al. proposed a dual-band shared-aperture interleaved BSA array structure [20], as shown in Figure 6.10. The antenna array consists of an LF antenna unit, an HF antenna unit, and a frequency-selective surface (FSS). The array structure is arranged from top to bottom as an HF antenna unit,

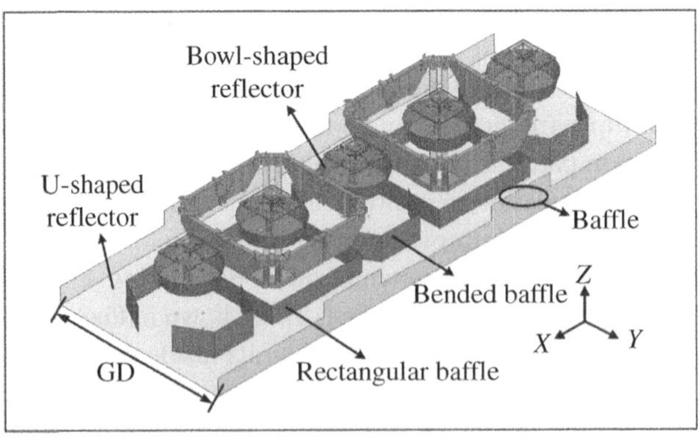

Figure 6.9 Configuration of the embedded shared-aperture BSA array. Source: Reproduced with permission from Ref. [19]/IEEE.

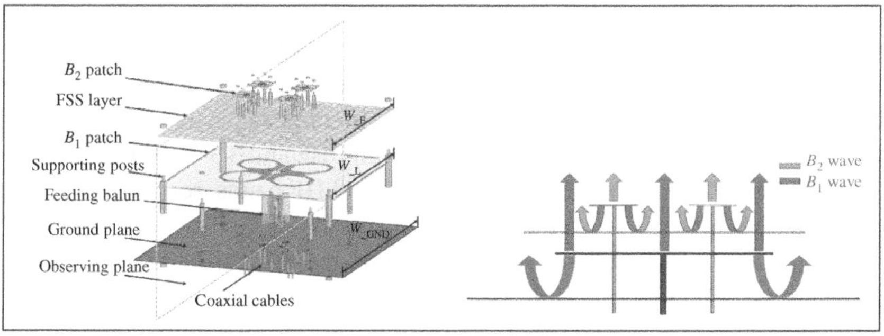

Figure 6.10 Configuration of the interleaved shared-aperture BSA array loaded with FSS. Source: Reproduced with permission from Ref. [20]/IEEE.

FSS, and LF antenna unit, respectively. For the HF antenna unit, the FSS is used as a reflector. For the LF antenna unit, the structure not only reduces the coupled waves radiated from the HF antenna unit to the LF antenna unit but also reduces the profile of the LF antenna unit by introducing a capacitive structure. The impedance bandwidths of antenna with isolation greater than 25 dB are 0.69–0.96 GHz, and 3.4–4.9 GHz, with stable radiation characteristics in both frequency bands.

In 2021, Li et al. proposed a dual-frequency shared-aperture interleaved BSA array [21], as shown in Figure 6.11. The HF and the LF antenna units are placed in the same plane, and the HF antenna unit is placed at the opening position of the ring-shaped radiating structure of the LF unit. Both HF and LF antenna elements are in the form of crossed dipole antennas. In order to obtain an independent

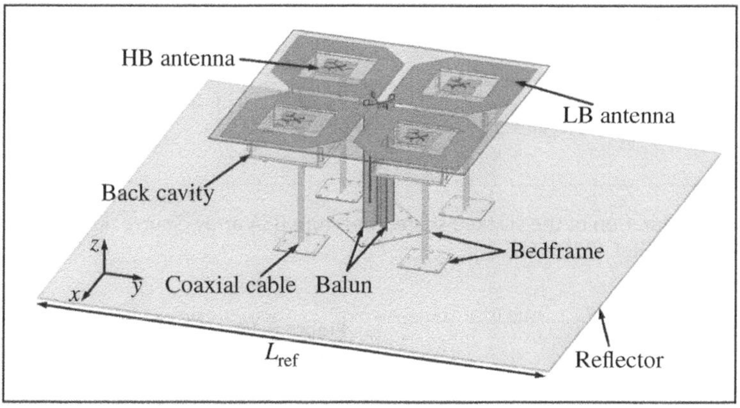

Figure 6.11 Configuration of the new interleaved shared-aperture BSA array. Source: Reproduced with permission from Ref. [21]/IEEE.

radiation environment, the reflection structure of HF and LF units is a back cavity and large reflection plate, respectively. Slots on the feeding balun structure of the LF unit can not only improve the matching characteristics of the LF unit but also can be used as an HF filter structure. The impedance bandwidths of the final antenna unit with VSWR <1.5 are 0.69–0.96 GHz and 3.3–3.8 GHz. The port isolation is greater than 38 and 25 dB in the LF and HF bands, respectively, while stable radiation patterns are obtained in both bands.

In 2021, He et al. proposed a novel dual-band shared-aperture stacked BSA array [22], as shown in Figure 6.12. Different from the antenna layout mentioned above, the HF antenna unit is placed below the LF unit. However, the bulky LF radiating unit will distort the radiation pattern of the HF unit. To reduce this effect, the radiating arm of the LF unit is replaced by a band-pass frequency selective unit. The radiation arm neither affects the normal radiation of the HF antenna nor the matching characteristics and radiation performance of the LF antenna.

Consistent with the above arrangement method, Chang et al. proposed a dual frequency shared-aperture stacked BSA array in 2021 [23], as shown in Figure 6.13. When the LF antenna element in the array is excited, the HF antenna element resonates at LF as a top-loaded monopole. This coupled resonance will deteriorate the impedance matching of the LF antenna units and distort the radiation patterns. Without affecting the performance of the HF unit, the ferrite magnet rings are loaded onto the coaxial lines of the HF unit. The common mode current on the feeding structure is weakened and the radiation patterns are

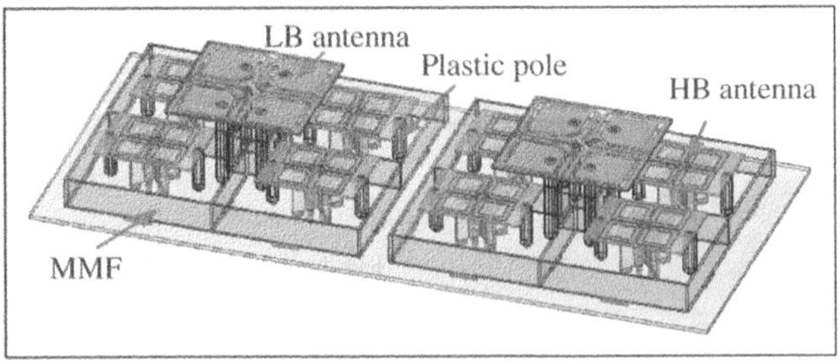

Figure 6.12 Configuration of the stacked shared-aperture BSA array. Source: Reproduced with permission from Ref. [22]/IEEE.

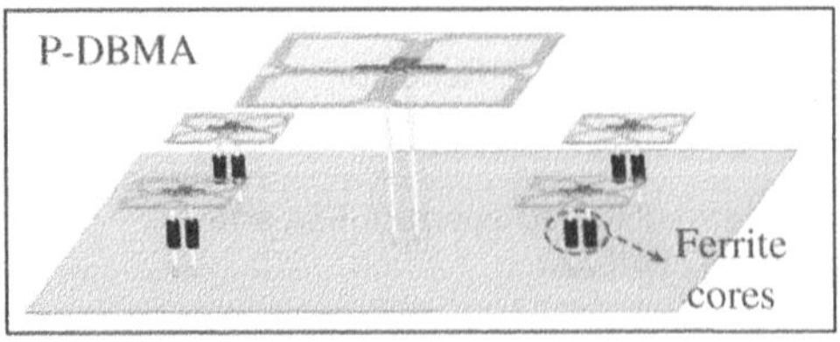

Figure 6.13 Configuration of the stacked shared-aperture BSA array loaded with ferrite magnet rings. Source: Reproduced with permission from Ref. [23]/IEEE.

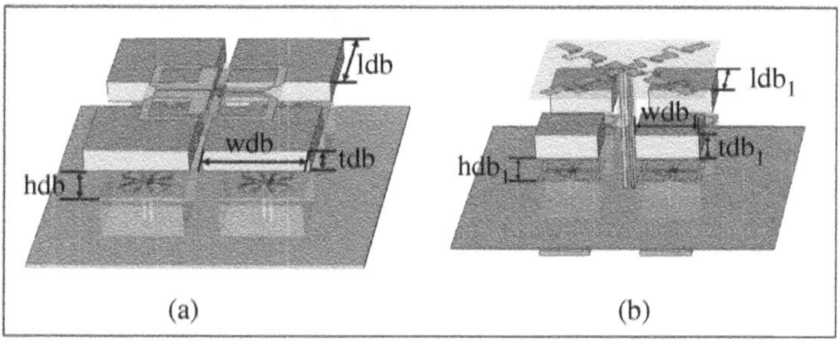

Figure 6.14 Configuration of the (a) stacked and (b) interleaved shared-aperture BSA array loaded with the dielectric superstrate. Source: Reproduced with permission from Ref. [18]/IEEE.

restored. The antenna obtains good matching characteristics and stable radiation patterns in 1.8–2.7 GHz, and its performance in 3.3–3.8 GHz is not affected.

In 2022, Da et al. proposed dual-frequency shared-aperture interleaved and stacked BSA arrays loaded with dielectric blocks [18], as shown in Figure 6.14. By loading the dielectric block above the HF antenna unit, the reflected electromagnetic wave cancels the inherent coupling wave between the antenna units. The impedance bandwidth with antenna isolation greater than 25 dB is 3.3–3.9 GHz, and the antenna performance is not affected in the frequency band of 1.9–2.5 GHz.

6.3 5G Massive-MIMO Array Example (Practical/Industry Related)

5G m-MIMO technology is an important part of 5G wireless communication network, which can improve the transmission rate and data capacity of the network. Here are some practical/industry-related examples of 5G m-MIMO arrays.

6.3.1 Huawei's 5G Blade Active Antenna Unit (AAU) Base Station

This is a small, high-performance 5G m-MIMO base station equipped with 128 antenna units, which can provide users with high-speed and stable mobile communication services, as shown in Figure 6.15.

High-quality network coverage has always been the focus of 5G network construction, and the blade AAU BSA minimalist deployment solution and indoor distributed m-MIMO solution address the key challenges of indoor and outdoor 5G deployment.

For outdoor scenarios, large bandwidth m-MIMO is the key to deliver the ultimate network performance and the core for operators to build 5G high-quality

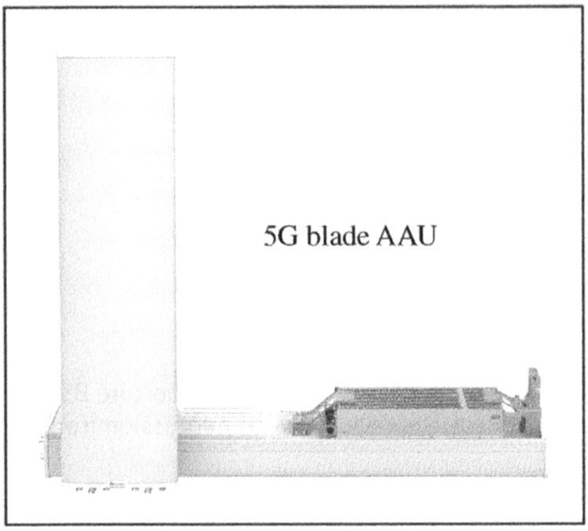

Figure 6.15 Photos of 5G blade AAU base station.

networks. The integrated blade AAU base station solves the coexistence problem of multi-band antenna modularization, and ensures optimal 2/3/4/5G network performance while deploying 5G m-MIMO.

For indoor scenarios, distributed MIMO technology not only significantly improves 5G network capacity and experience but also precisely solves the interference problem between indoor 5G cells, greatly enhancing user experience.

6.3.2 Ericsson's AIR 6488 5G m-MIMO Base Station

This base station uses an array of 64T64R (i.e. 64 transmit antennas and 64 receive antennas), as shown in Figure 6.16a. By combining antennas with a large number of steerable ports, beamforming 5G technology, m-MIMO, and multi-user MIMO (MU-MIMO) constructed based on m-MIMO can be initialized as MU-MIMO. These functions have the advantages of enhancing user experience, increasing network capacity and transmission rate, expanding network coverage, and reducing interference.

6.3.3 Ericsson's Street Macro 6701 5G m-MIMO Base Station

This is a base station designed for urban street environments with an array of 128T128R, as shown in Figure 6.16b. The base station not only provides higher coverage and capacity, but also has flexible installation and scalability.

Street Macro 6701 base station has the characteristics of small size, lightweight, low power consumption, flexible deployment, and convenient installation. At the same time, the base station adopts a laser-based technology that converts electricity into a high-intensity beam, which is then captured and converted into electricity at the base station side. This can replace the power line from the base station to the power grid, and is very suitable for deployment in hot spots such as streets and venues scenes.

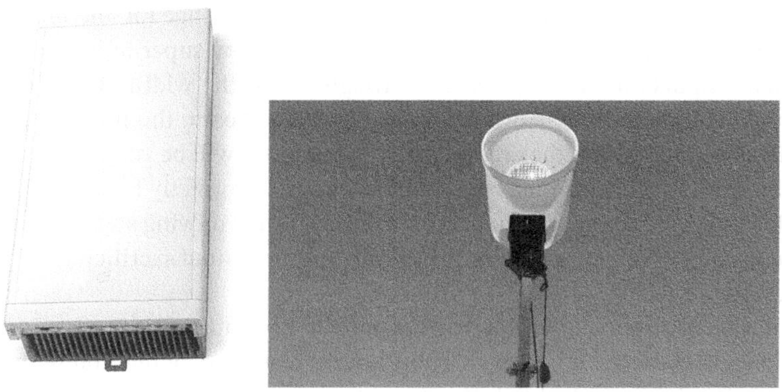

(a) (b)

Figure 6.16 Photos of Ericsson's (a) AIR 6488 and (b) Street Macro 6701 5G m-MIMO base stations.

In addition, NOKIA's Air Scale m-MIMO base station uses an array that can reach up to 128T128R, which can bring higher capacity and faster speed to the 5G network. Samsung's 5G m-MIMO base stations with 256T256R arrays are designed for higher data capacity and faster transmission speeds, making them suitable for deployment in high-density cities and populated areas.

The above are some examples of practical/industry-relevant 5G m-MIMO arrays, which can not only improve the performance of the network but also a key link in the 5G industry chain. These m-MIMO array examples are not only used in mobile communication but also can be widely used in IoT, smart home, industrial automation, and other fields to bring more convenient and efficient experience for people's life and work.

6.4 Challenges in m-MIMO Antenna Designs

It is clear that the most obvious challenge brought by m-MIMO is the increase in the number of antennas. Therefore, it is necessary to find effective theories and design methods to improve the performance of space-constrained m-MIMO antenna arrays. For the antenna element, it is necessary to achieve miniaturization while maintaining the performance of the original antenna element, such as bandwidth, power, etc., which is the first problem in the antenna design for m-MIMO.

As the number of antennas increases, the distance between the antennas becomes smaller and smaller, and it is necessary to ensure that they do not influence each other. At the same time, the mutual couplings of antennas on the frequency spectrum will also lead to the degradation of signal quality, so effectively suppressing these interferences is also a challenge for antenna design.

A key idea of m-MIMO is to increase antenna gain by superimposing multiple antenna outputs onto a single beam, through which the width of the generated beam tends to become narrower. The narrower the beam, the more directional, but this also means that the area covered by the beam will be very narrow. Beamforming and directional must be fast and accurate to correctly focus on the target user terminal, especially when the user terminal is in a moving state. Therefore, it is also a major challenge to expand the beam width without sacrificing the performance of a very large MIMO.

With the continuous development of mobile communication, green communication will become an important indicator of the future mobile communication system. As the core technology of next-generation mobile communication, m-MIMO should not only focus on spectrum efficiency but also energy efficiency. Running hundreds or even thousands of antennas in an m-MIMO system consumes a lot of energy, and how to achieve efficient energy management to reduce the total energy consumption of the system is also a challenge.

MIMO antennas need to be reconfigurable to adapt to different communication environments and application requirements. m-MIMO still has the following challenges, such as designing a flexible reconfigurable antenna system, a low-complexity system, and m-MIMO systems that are easy to install and expand. There are also design problems of low-power circuits and radio frequency modules, as well as integration issues with other technologies (such as millimeter wave technology).

6.5 Conclusion

m-MIMO antenna system is a technical scheme to meet the more comprehensive signal coverage and system capacity requirements of mobile communication systems. m-MIMO technology has many advantages in mobile communication systems, including improved spectrum and energy efficiency, reduced air interface access delay and signal processing complexity, expanded system coverage, insensitivity to single hardware non-ideality, and improved system reliability. In addition, m-MIMO technology also has the advantages of saving energy and overcoming the impact of device mobility. In the implementation of m-MIMO antenna arrays, miniaturization of antenna units, low profile, and reduction of the spacing between adjacent antennas in the array are one of the key technologies. The performance of m-MIMO arrays can be effectively improved by reasonably designing the form of antennas and using special materials. In the realization of the array antenna, it is necessary to consider the coupling problem between the in-band and cross-band antenna arrays.

The design and application of m-MIMO antenna arrays face multiple challenges, including space constraints, miniaturization, decoupling, accuracy of beamforming and orientation, improvement of energy efficiency, reconfigurability, and system complexity. These challenges need to find effective schemes and theoretical methods to promote the development of m-MIMO technology. 5G m-MIMO technology is an important part of 5G wireless communication network, which can improve the transmission rate and data capacity of the network. It has been applied in many industries, such as mobile communication, IoT, smart home, and industrial automation.

References

1 Marzetta, T.L. (2010). Noncooperative cellular wireless with unlimited numbers of base station antennas. *IEEE Transactions on Wireless Communications* 9 (11): 3590–3600.

2 Rusek, F., Persson, D., Lau, B.K. et al. (2013). Scaling up MIMO: opportunities and challenges with very large arrays. *IEEE Signal Processing Magazine* 30 (1): 40–60.

3 Larsson, E.G., Edfors, O., Tufvesson, F., and Marzetta, T.L. (2014). Massive MIMO for next generation wireless systems. *IEEE Communications Magazine* 52 (2): 186–195.

4 Ngo, H.Q., Larsson, E.G., and Marzetta, T.L. (2013). Energy and spectral efficiency of very large multiuser MIMO systems. *IEEE Transactions on Communications* 61 (4): 1436–1449.

5 Zhang, Q., Jin, S., Wong, K.-K. et al. (2014). Power scaling of uplink massive MIMO systems with arbitrary-rank channel means. *IEEE Journal of Selected Topics in Signal Processing* 8 (5): 966–981.

6 Chan, P.W.C. et al. (2006). The evolution path of 4G networks: FDD or TDD? *IEEE Communications Magazine* 44 (12): 42–50.

7 Björnson, E., Larsson, E.G., and Marzetta, T.L. (2016). Massive MIMO: ten myths and one critical question. *IEEE Communications Magazine* 54 (2): 114–123.

8 Ye, L.H., Zhang, X.Y., Gao, Y., and Xue, Q. (2020). Wideband dual-polarized four-folded-dipole antenna array with stable radiation pattern for base-station applications. *IEEE Transactions on Antennas and Propagation* 68 (6): 4428–4436.

9 Zheng, D.Z. and Chu, Q.-X. (2017). A wideband dual-polarized antenna with two independently controllable resonant modes and its array for base-station applications. *IEEE Antennas and Wireless Propagation Letters* 16: 2014–2017.

10 Li, M., Chen, X., Zhang, A., and Kishk, A.A. (2019). Dual-polarized broadband base station antenna backed with dielectric cavity for 5G communications. *IEEE Antennas and Wireless Propagation Letters* 18 (10): 2051–2055.

11 Zhai, H., Zhang, K., Yang, S., and Feng, D. (2017). A low-profile dual-band dual-polarized antenna with an AMC surface for WLAN applications. *IEEE Antennas and Wireless Propagation Letters* 16: 2692–2695.

12 Li, Q., Cheung, S.W., and Zhou, C. (2017). A low-profile dual-polarized patch antenna with stable radiation pattern using ground-slot groups and metallic ground wall. *IEEE Transactions on Antennas and Propagation* 65 (10): 5061–5068.

13 Lian, R., Wang, Z., Yin, Y. et al. (2016). Design of a low-profile dual-polarized stepped slot antenna array for base station. *IEEE Antennas and Wireless Propagation Letters* 15: 362–365.

14 Janaswamy, R. (2002). Effect of element mutual coupling on the capacity of fixed length linear arrays. *IEEE Antennas and Wireless Propagation Letters* 1: 157–160.

15 Wu, K., Wei, C., Mei, X., and Zhang, Z. (2017). Array-antenna decoupling surface. *IEEE Transactions on Antennas and Propagation* 65 (12): 6728–6738.

16 Da, Y., Zhang, Z., Chen, X., and Kishk, A. (2021). Mutual coupling reduction with dielectric superstrate for base station arrays. *IEEE Antennas and Wireless Propagation Letters* 20 (5): 843–847.

17 Li, M., Chen, X., Zhang, A. et al. (2020). Split ring resonator loaded baffles for decoupling of dual-polarized base station array. *IEEE Antennas and Wireless Propagation Letters* 19 (10): 1828–1832.

18 Da, Y., Chen, X., and Kishk, A.A. (2022). In-band mutual coupling suppression in dual-band shared-aperture base station arrays using dielectric block loading. *IEEE Transactions on Antennas and Propagation* 70 (10): 9270–9281.

19 Wu, R. and Chu, Q.-X. (2019). A compact, dual-polarized multiband array for 2G/3G/4G base stations. *IEEE Transactions on Antennas and Propagation* 67 (4): 2298–2304.

20 Zhu, Y., Chen, Y., and Yang, S. (2019). Decoupling and low-profile design of dual-band dual-polarized base station antennas using frequency-selective surface. *IEEE Transactions on Antennas and Propagation* 67 (8): 5272–5281.

21 Li, Y. and Chu, Q.-X. (2021). Coplanar dual-band base station antenna array using concept of cavity-backed antennas. *IEEE Transactions on Antennas and Propagation* 69 (11): 7343–7354.

22 He, D., Yu, Q., Chen, Y., and Yang, S. (2021). Dual-band shared-aperture base station antenna array with electromagnetic transparent antenna elements. *IEEE Transactions on Antennas and Propagation* 69 (9): 5596–5606.

23 Chang, Y.L. and Chu, Q.-X. (2021). Ferrite-loaded dual-polarized antenna for decoupling of multiband multiarray antennas. *IEEE Transactions on Antennas and Propagation* 69 (11): 7419–7426.

7

Recent Topics in MIMO Antenna Systems

This chapter discusses recent topics in MIMO antenna systems, in line with the trends of operating in the upper millimeter wave and sub-THz bands for next-generation communications. This chapter starts by discussing the aspects of on-chip and on-package antenna systems, an essential and practical aspect of the growing integration between antennas and the radio front ends. This is followed by the subtopics related to millimeter wave MIMO and sub-THz antennas, including their types, feeding methods, and recent examples. Inherent challenges arise from the use of upper millimeter wave and sub-THz bands, which include limited transmission range and narrow beamwidth (and consequently area coverage for communications). This will then potentially require the deployment of more terrestrial infrastructure, supported by non-terrestrial networks to maintain sufficient coverage of high-speed connectivity to users. In terrestrial networks, antennas are expected to be reconfigurable and capable of beamforming. This can be done by using advancing high gain antennas and MTS to enable features in beamforming, beam steering and supporting partial functionalities of radio front ends. Besides that, increasing network coverage using RISs and methods to enable reconfiguration in these components are presented. Besides that, the principles of orbital angular momentum (OAM) MIMO communications are also explained with examples, prior to the presentation of antennas in the new generation of satellite communications, an important element of NTNs. The requirements of antennas and recent development of capabilities in the area of high-speed communications based on satellites are presented. While it has coexisted for quite some years, satellite communications are expected to be the dominant candidate in complementing conventional terrestrial networks in supporting the next generation of wireless communications.

MIMO Antenna Systems for 5G and Beyond, First Edition.
Xiaoming Chen, Ping Jack Soh, and Mohammad S. Sharawi.

7.1 On-Chip and On-Package MIMO Antenna Systems

Generally, a wireless communication system consists of several functionality components, as follows [1]:

- *Digital component*: for logic operations, memory, and signal processing
- *Mixed signal component*: for signal conditioning (ADCs or DACs)
- *RF front-end*: for signal conversion (upward or downward) in frequency
- *Antennas*: receiving and transmitting signals.

As the first three components existed longer in separate or combined IC packages, the focus of this subchapter is to present several relevant aspects related to the antennas on a chip.

7.1.1 Antenna On-Chip

Antennas have been increasingly integrated into the radio front-end circuits on a silicon chip, and they are known as antennas-on-chip (AoCs). Such antenna type and its research are increasingly important due to the need to reduce interconnection losses between already small antennas required for such implementations and to result in a compact size of the RF system. Besides communications from the IC outward to free space (known as off-chip communications), AoCs can potentially further advance interchip communications, which also can be another form of MIMO communication. An example of an experimentally demonstrated wireless communication between two on-chip antennas was reported in [2] placed 5 meters apart. On the other hand, the investigation in [3] successfully designed, fabricated, and examined on-chip inverted-F and quasi-Yagi antennas at 60 GHz. The drawback of implementing on-chip antenna in silicon-based technologies is its high-dielectric constant ($\varepsilon_r \sim 11.7$) and relatively thick substrate, with a resistivity of about 10 Ω-cm. This will typically cause most of the power to be confined or attenuated in the substrate instead of being radiated into free space, further degrading the radiation efficiency [4]. To overcome this, AoC researchers developed different methods to enhance radiation efficiency, which includes proton implantation [5], micromachining [6], and the use of artificial magnetic conductors (AMCs) [7] and superstrates [8]. Due to its relevance, AoC research is focused on improving the performance and exploring millimeter and THz frequencies for imaging and advanced single-chip radios.

Figure 7.1 shows a typical RF front-end of a direct-conversion transceiver suitable for a single-chip solution. The transceiver uses an antenna by incorporating a transmit/receive (T/R) switch to connect with the low-noise amplifiers (LNA) or power amplifier (PA) [9]. In addition to that, a differential architecture which potentially improves the signal-to-noise ratio (SNR) is extensively adopted

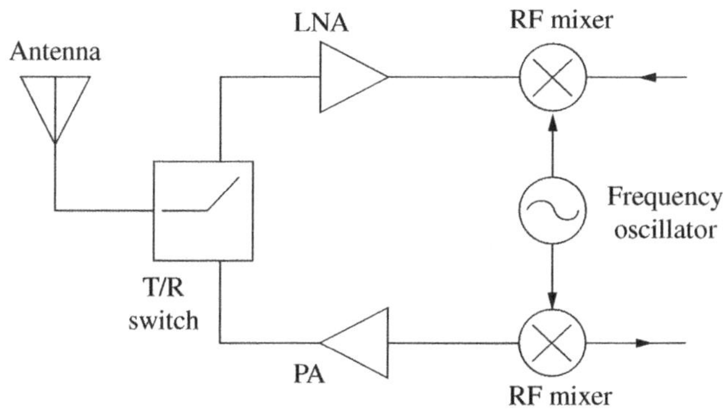

Figure 7.1 Illustrations of an AoC. Source: Reproduced with permission from [9]/Artech House.

in RF-integrated transceivers. In such cases, a differential antenna is required. Moreover, to mitigate crosstalk, placement of antenna and circuits needs to be done with care at the layout level. In general, there exist three methods in which an AoC can be used with a transceiver front-end can be summarized as follows [1]:

(1) as a receiving antenna connected to LNA in a receiver chip.
(2) as a transmitting antenna connected to PA in a transmitter chip.
(3) as a T/R antenna connected to both sides through a switch.

7.1.2 Codesign of Antennas and Circuits

An example of an on-chip antenna implementation is proposed in [10], where the antennas were integrated on top of a 100-μm thick quartz superstrate (with $\varepsilon_r \sim 3.8$ and $\tan\delta = 0.001$). It is aimed to improve radiation efficiency and suppress surface waves. To do so, a ground layer is introduced to isolate the lossy silicon substrate from the antenna. A 0.13-μm CMOS process was used to fabricate other electronic components on silicon substrate, with the metal density requirement fulfilled using additional metal squares. As illustrated in Figure 7.2, two on-chip antennas – a patch antenna and a ring slot antenna – were designed, both operating at 94 GHz with about 8 GHz of operating bandwidth. These antennas exhibited gains of 2–3 dBi and about 50% of radiation efficiency within 91–94 GHz. Figure 7.2a illustrates the two on-chip antennas in detail, whereas the layer stack-up of the chip to isolate these antennas from the silicon substrate is shown in Figure 7.2b. Their simulated and measured reflection coefficients are presented in Figure 7.2c, and the fabricated prototype of the patch on-chip antenna is shown in Figure 7.2d.

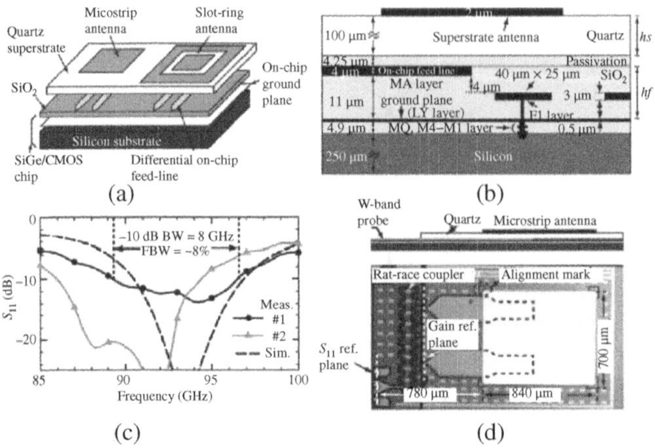

Figure 7.2 Structure of the on-chip antenna: (a) exploded side view of the two on-chip antennas, (b) layer stack-up of the chip, (c) simulated and measured reflection coefficients, and (d) fabricated prototype. Source: Ou et al. [10] © 2012 IEEE.

Another example of an in-chip/in-package antenna is presented in Figure 7.3 from [11]. To reduce the effects of the substrate on the integrated in-chip antenna, a different method is used, which is by separating the antenna from the substrate of the amplifier/active electronics and connecting them using bonding wires, as illustrated in Figure 7.3a. The amplifier is designed and fabricated using a 0.15-μm GaAs pHEMT process, with a size of $1.2 \times 2.2\,\text{mm}^2$, and producing a gain of 14.5 dB (with the bonding wire). This is shown in Figure 7.3b. The bonding wire sized at 25 μm (diameter) and 400 μm (in length) is integrated into the design of the amplifier prior to optimization to assess its effect on the overall design. The complex output impedances of the amplifier at various frequencies are recorded to match the antenna.

Two antennas were designed and studied for this purpose. The first is a quarter-wave patch, which is designed on a 0.338 mm. thick RO4350B substrate ($\varepsilon_r = 3.66$ and $\tan\delta = 0.0037$), whereas the second is a half-wave patch, designed on a 0.254 mm-thick RT6010LM substrate ($\varepsilon_r = 10.2$ and $\tan\delta = 0.0023$), as depicted in Figure 7.3c, d. These two patches were optimized in terms of impedance matching and bandwidth using additional slots. A 7 mm × 7 mm ceramic package from Kyocera ($\varepsilon_r = 8.5$ and $\tan\delta = 0.0021$) is used to host the AIA in a package. This package provided an internal cavity sized at $4 \times 4 \times 0.74\,\text{mm}^3$, as shown in Figure 7.3a. The half-half-wave patch is chosen as in the final design, with a 2.1-dB higher gain relative to the quarter-wave patch, but with 2% less bandwidth. On the other hand, the gain provided by the amplifier is 12.3–13.6 dBi, translating

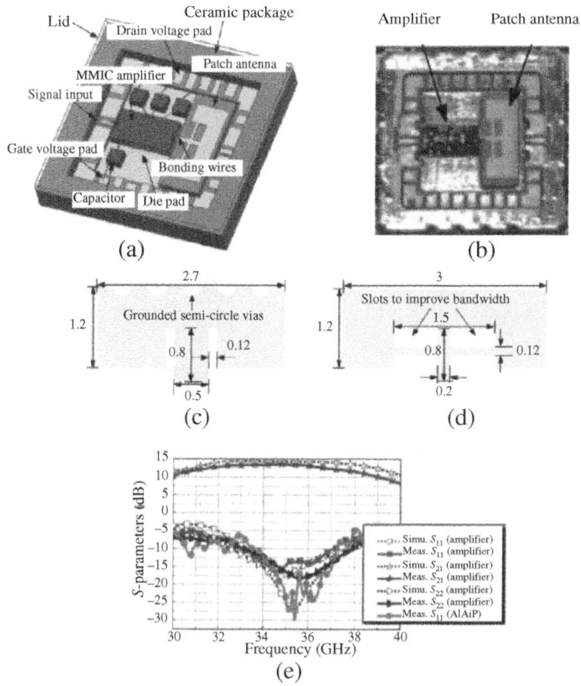

Figure 7.3 The proposed in-package AIAs: (a) the various components, active and antenna parts, (b) microphotograph of the fabricated prototype, (c) dimensions of the quarter-wave patch, (d) dimensions of the half-wave patch, and (e) amplifier gain and antenna reflection coefficients from measurements. Source: Song et al. [11]/IEEE/Pubic Domain.

to an operation from 33–37.1 GHz. Measurement results shown in Figure 7.3e indicated a maximum radiated gain of 18.9 dBi at 35 GHz.

7.1.3 Benefits of On-Chip Antennas

7.1.3.1 Non-50 Ω Co-Design

In conventional antenna-RF front-end interfaces, matching networks (MNs) are typically introduced to convert complex impedances to a 50 Ω reference,as shown in Figure 7.4a. This applies on both the receiver (Rx) and the transmitter (Tx) sides. As widely known, such matching circuits are disadvantageous as:

- Adds complexity to the design process
- Introduces additional losses in the lineup or interface
- Consumes extra space on the circuit/chip

On-chip antennas can be co-designed without the need for a 50 Ω impedance interface, as shown in Figure 7.4b. Instead, optimal matching can be directly introduced to an arbitrary impedance (of the adjacent component) via conjugate matching. This eliminates additional design efforts in matching one end of the

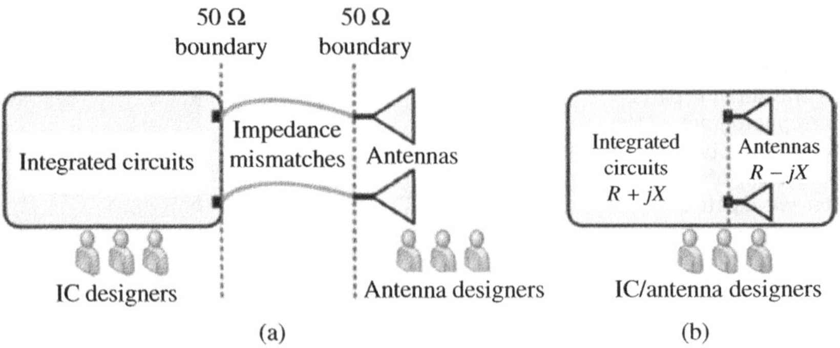

Figure 7.4 (a) Traditional 50 Ω boundary between IC and antenna designers, and (b) codesigned matching between IC and on-chip antennas. Source: Reproduced with permission from [4]/IEEE.

antenna to 50 Ω, with the same and separate effort done on the interface to the antenna. Besides that, the risks associated with non-optimal matching when a chip interconnects to PCB traces or bond wires can be reduced. Ultimately, the co-design approach provides two degrees of freedom to optimize matching between the circuit and antenna domains [12].

Examples of conjugate matching for such antennas with RF amplifiers (see Figure 7.5) are presented in [13, 14]. The main principle of the co-design is the conjugate matching of the antenna impedance, Z_{ant} and the MNs, Z_{MN} in the operating band of interest, with the design steps summarized as follows:

i) Modelling of the antenna using an electromagnetic solver to obtain its reflection Γ_{ant}.
ii) The noise and gain stability circles for the amplifier are then studied within the band of interest. The source reflection coefficient Γ_S is then selected considering the selected gain and its trade-off with noise figure.

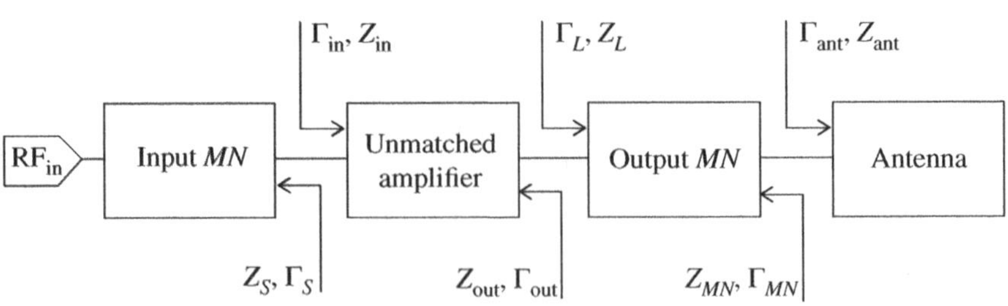

Figure 7.5 Amplifier-integrated active antenna front end block diagram. Source: Reproduced with permission from [13]/John Wiley & Sons.

iii) Next, the output reflection coefficient, Γ_{out} of the amplifier is calculated based on the required Γ_S [15] where S_{11}, S_{12}, S_{21}, and S_{22} are the S-parameters of the unmatched amplifier.

$$\Gamma_{out} = S_{22} + \frac{S_{12}S_{21}\Gamma_S}{1 - S_{11}\Gamma_S} \tag{7.1}$$

Upon determining Γ_S and Γ_{out}, the input and output MNs are then designed and optimized to convert from the 50 Ω reference to Γ_S, and from Γ_{ant} to $\Gamma_L = \Gamma^*_{out}$, respectively.

i) Measurement of the reflection coefficient from the antenna input port, Γ_{ant}(int), is not possible in the integrated system due to its inaccessibility, but can be calculated from Z_{MN} and Z_{ant}) as follows:

$$\Gamma_{ant}(int) = \frac{Z_{ant} - Z^*_{MN}}{Z_{ant} + Z_{MN}} \tag{7.2}$$

where Z^*_{MN} is the complex conjugate of Z_{MN}.

7.1.3.2 Fabrication

PCB-based antenna fabrication methods can generally involve chemical, laser etching or mechanical milling. Despite its maturity, these methods are limited in terms of accuracy, fabrication tolerances, and repeatability, as summarized in Table 7.1 [1]. Several possible factors influencing these limitations include the choice of substrates and prepregs, complexity of the structures and fabrication equipment, and expertise. On the contrary, significant improvement in fabricating antennas can be achieved by CMOS lithography process. Besides allowing better resolutions, this process also offers a good level of repeatability. A comparison of these approaches can be summarized in [1].

Table 7.1 Comparison of minimum feature sizes of antenna fabrication methods.

Fabrication technology	Minimum feature size (μm)	Tolerance	Repeatability
Mechanical milling	0.3	High	Low
Laser milling	0.3	Medium	Medium
Dry plasma etching	≈3	Medium	Medium
Wet etching	≈30	Medium	Medium
On-chip	≈0.05[a)]	Low	High

a) For Metal-1 in TSMC 5-nm technology. [1].

7.1.3.3 Size and Features

Arguably the most significant advantage of antennas' implementation on chip along with the circuit is the size reduction offered. These antennas can be also positioned at the periphery of the chip to allow more space for the placement of other components. Besides that, space for the antennas can also be further saved by folding or meandering the required electrical length of the antenna needed for radiation. These on-chip antennas typically occupy one of the top metal layers, whereas the multiple lower layers can be used to accommodate other circuits and components on chip. Note also that as the operating frequency of the chip changes from the sub-6-GHz band to the upper millimeter-wave band, the size of the antenna actually starts to become as small as (or even smaller) than most other on-chip components [16]. Several examples include a 170 GHz imaging radar transceiver with two on-chip dipole antennas that has been realized in an area less than 1 mm^2 [17], whereas the on-chip folded dipole antenna and lens combination from [18] operating from 220 to 330 GHz occupies a volume of $0.25 \times 0.29 \times 0.2\,mm^3$. Moreover, with such significant size reduction, on chip antennas can actually be combined with other components to offer additional features. An example of such is the loop-shaped antenna implemented on chip

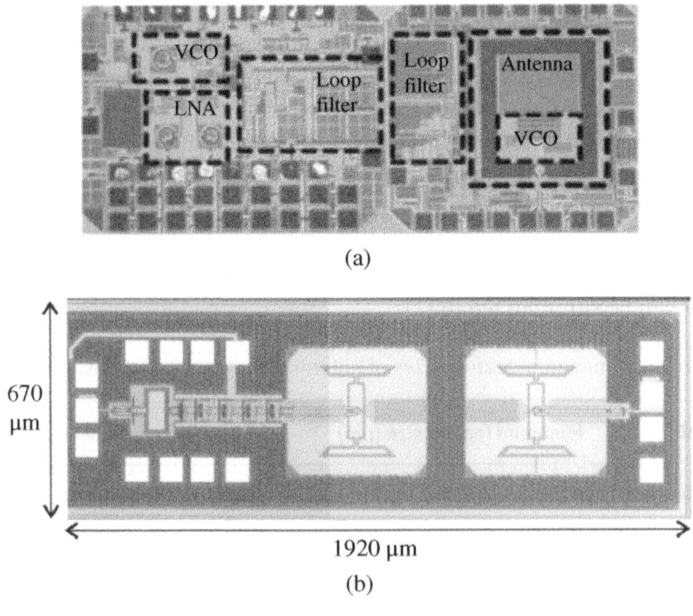

(a)

(b)

Figure 7.6 (a) A miniaturized on-chip antennas at: (a) 5.2 GHz [20] and (b) 300 GHz. Source: Reproduced with permission from [18]/IEEE.

at 5.2 GHz in [19]. This on-chip loop antenna serves a dual purpose, both as an inductor for the voltage-controlled oscillator and a radiator. Furthermore, the area enclosed within the loop is repurposed for active circuitry, resulting in efficient utilization of the chip area as shown in Figure 7.6a, resulting in a dimension of 0.5 mm^2. On the other hand, the structure shown in Figure 7.6b combines the use of a folded dipole antenna implemented on chip to feed a hemispherical lens to operate as a receiver The chip consists of square-law power detectors, an LNA and two on-chip antennas and occupies a total area of $1.92 \times 0.67\,mm^2$.

7.2 Millimeter-Wave MIMO Antenna Systems

As explained in the previous section, millimeter-wave antennas are typically compact in size and require high-precision fabrication techniques. Thus, the most practical method in realizing MIMO antenna systems in this band is either using high-precision milling/etching technologies or CMOS lithography techniques. For the former, antennas in millimeter-wave MIMO antenna systems need to be integrated with RF-front ends, which are usually in the form of chips via transmission lines or vias. On the other hand, the second method is a direct integration of antennas into millimeter-wave front ends in an on-/in-chip package. This is mainly to minimize losses due to electrically long transmission lines/fan-outs. Considering the importance of on-chip antennas in millimeter-wave MIMO antenna systems, this subsection presents two specific aspects related to such antennas, which include an example of simplified integration method and tuning method, respectively. While there are other interesting and relevant aspects of millimeter-wave MIMO antenna systems such as beamforming architectures, this has been discussed in Chapters 3 and 4, and will not be repeated here to avoid redundancy. Moreover, potential methods for multi-beam generation and control for potential implementation in MIMO communication systems will be discussed in the subsection on metasurfaces.

An example of a simple design method for the integration of on-chip patch antenna in a standard CMOS technology without post-processing is presented in [21]. To enable high antenna radiation efficiency and gain, the antenna is shielded from the substrate by a ground plane, as shown in Figure 7.7a. Next, the overall on-chip patch antenna is formed using the top metal layer (as the patch radiator) and an intermediate metal layer (as the ground plane), respectively. It occupies an area of 1220 μm by 1580 μm with carefully placed fillers and slots to adhere to the CMOS process design rules, as shown in Figure 7.7b. The electrical field distribution of this antenna at the center operating frequency of 60.51 GHz

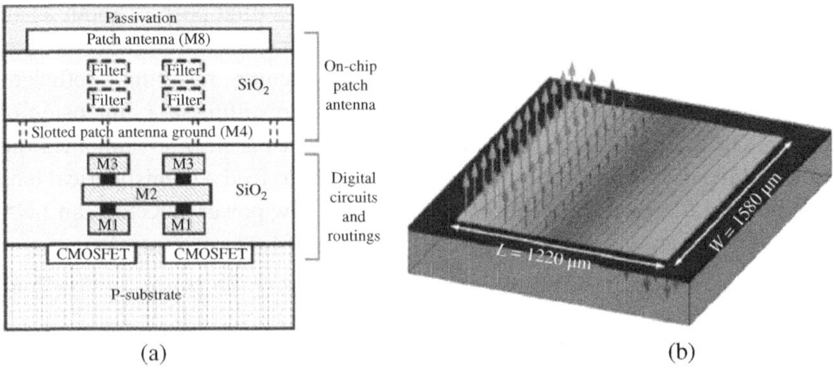

Figure 7.7 Layout cross section of the proposed integration scheme in a standard 0.13 μm CMOS process. Source: Reproduced with permission from [21]/IEEE.

is shown, indicating a typical radiation mode of a patch antenna (weaker electric fields indicated with darkened area). Overall, the antenna operated with 810 MHz bandwidth, with a peak gain and radiation efficiency of −3.32 dBi and 15.87%, respectively.

Figure 7.8a shows a practical dummy filling scheme at M6 and M7 to meet a 10% local minimum density rule over local 400 μm^2 areas stepped across the chip. The strips are composed of floating patch units, and the space between the units reduces the electrical-induced current along the length caused by the standing wave on the patch. The center frequency and efficiency of a parametric study when sweeping a strip at layers M6 and M7 along the length are shown in Figure 7.8b. Results indicate consistency with the electric field patterns, where when these strips are positioned in the center of the patch (where the electric field is null), there is insignificant effect on the antenna performance. This is despite the strip being positioned between the patch and ground. These results are consistent with the standing wave pattern, showing that if the fill is in the center of the patch area (dash-dotted lines), where the electric field null is, then the strip has little effect on the antenna even though it is in between the patch and ground. This is also validated by the spectrum of radiated electric field strength illustrated in Figure 7.8c due to 1.2 V_{pp} full-swing clock signal. Moreover, this low/null electric field area can be potentially used for the integration of additional antenna(s) for MIMO and/or multi-band communications with minimal influence on the performance of the main antenna, as validated in this study.

To achieve higher diversity gain and increased channel capacity using MIMO technique, an example of a new concept to tune on-chip antenna performance is presented in [22]. To increase the radiation resistance and consequently, antenna

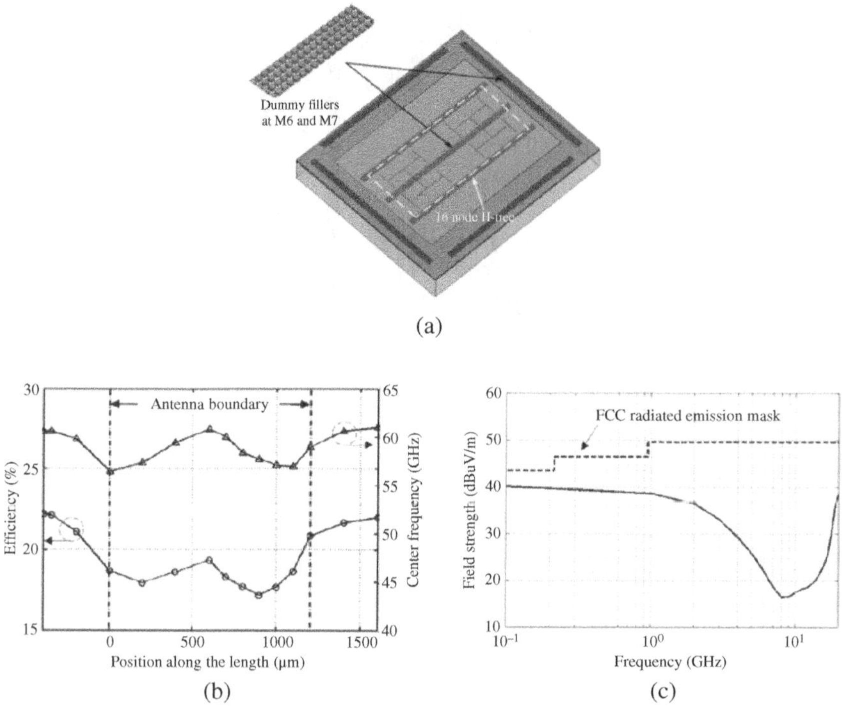

Figure 7.8 (a) A practical dummy filling scheme with a 16-node H-tree design beneath the patch antenna in standard CMOS process, (b) Center frequency and efficiency of the proposed antenna when strips are swept along length. (c) The radiated electric field strength of 1.2 V_{pp} clock signal coupled to the antenna. Source: Reproduced with permission from [21]/IEEE.

efficiency, two unbalanced antennas are used, as shown in Figure 7.9a. The main antenna is a folded monopole antenna and is located at the right edge of the chip. This is the end section of the RF front-end circuit block where no pads exist. Resonance is achieved in a compact form by meandering the monopole antenna using the same fixed length. On the other hand, a loaded IFA antenna is introduced on one side of the chip along the pad ring to operate as the diversity antenna. These antennas are a single-layer structure, residing on the pad layer of 65 nm process (M10). A capacitive load is formed by adjusting the gap between antenna and ground near the end of the antenna, whereas inductive load is introduced and controlled using a narrow meander structure close to the feeding point. These points allow the designer to tune the resonant frequency of the diversity antenna more precisely.

If the chip and package interconnections are using bond wires, the performance of the diversity antenna will drop drastically. As low resistivity silicon wafer is used

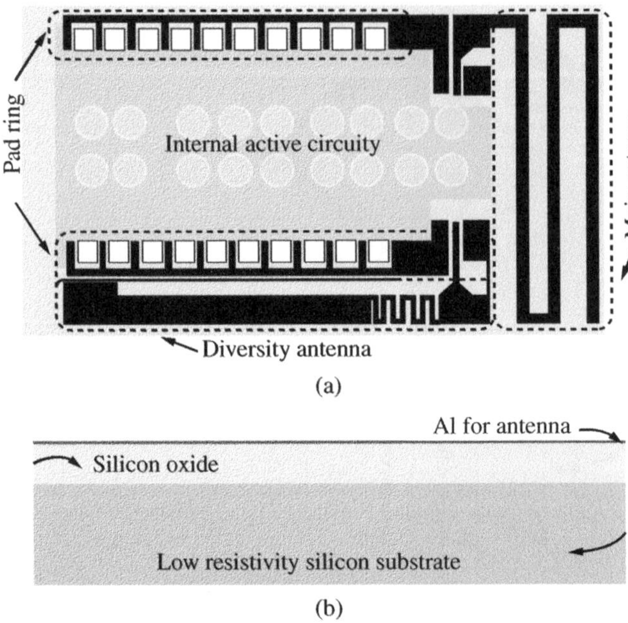

Figure 7.9 Proposed antenna tuning concepts (a) top view; (b) cross section. Source: Reproduced with permission from [22]/IEEE.

as the substrate, its resistivity measured by a four-point-probe mapping system is found to be 17 Ω-cm. Thick silicon oxide layer is deposited on top of the silicon wafer using plasma-enhanced chemical vapor deposition process. Its thickness is then validated using a reflectrometer to be 9.5 μm with ±0.2 μm non-uniformity. A layer of 1.5 μm pure aluminum is then sputtered on top of the silicon oxide. Extra aluminum is etched out using the dry etching process after completion of the photolithography process. Figure 7.9b shows the cross section of the antenna, whereas Figure 7.10 shows the die photos of four different ground patterns. *S*-parameter simulation and measurement results are provided in Figure 7.11.

The input reflection coefficient of the main antenna with ground pattern 1–3 is observed to be lower than −10 dB from 56 to 65 GHz. Measurements of ground pattern 4 indicate that the center operating frequency of the main antenna shifts to a higher frequency when more metal connects to the antenna ground. On the other hand, the diversity antenna operates from 56 to 61 GHz with ground patterns 1 and 2. Pattern 3 lowers the operating frequency of this diversity antenna, with further reduction with the use of Pattern 4. Measurements indicated improved S_{21} isolation relative to simulations, and an anti-resonance observed at 56 GHz originating from the metamaterial-like properties of Pattern 2. The peak gains of the diversity antenna and main antenna at 60 GHz are −6.1 and −3.8 dBi, respectively.

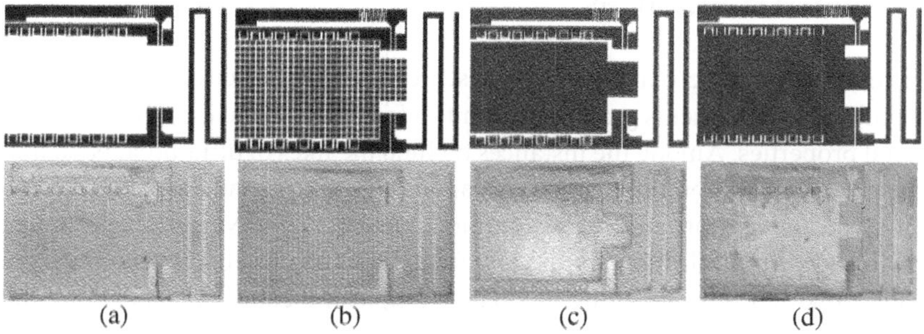

Figure 7.10 Die photos of different ground patterns (a) blank (ground pattern 1); (b) periodical metal patches (ground pattern 2); (c) solid ground plane disconnected with antenna ground (ground pattern 3); (d) solid ground plane connected with antenna ground (ground pattern 4). Source: Reproduced with permission from [22]/IEEE.

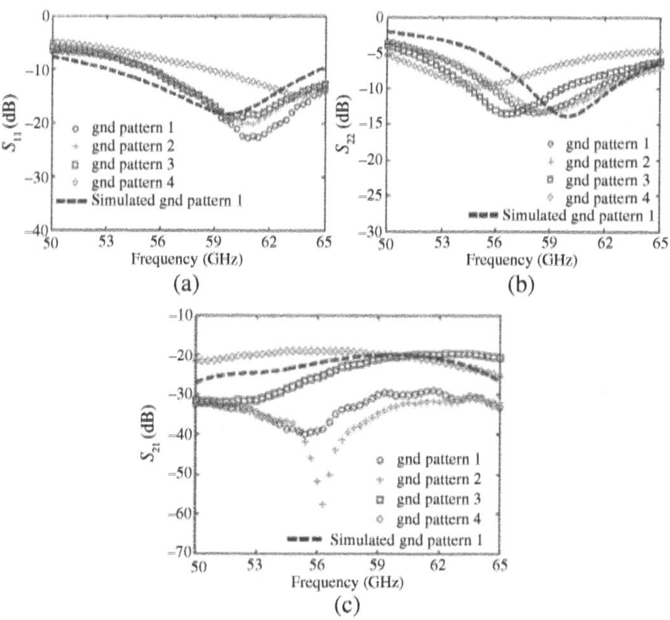

Figure 7.11 Simulated and measured S-parameters for different ground patterns of the proposed MIMO antenna; (a) S_{11}, (b) S_{22}, and (c) S_{21}. Source: Reproduced with permission from [22]/IEEE.

7.3 Sub-THz Antennas

THz technology, a new and popular research topic in electromagnetism and communications, presents great opportunities for many applications with unique spectral properties. Among the instances include the likes of ultra-broadband and high-speed wireless communications, weapons and contraband detection, disease diagnosis, intervention and high-resolution imaging [23]. Nevertheless, unlike the wireless communications in the low frequency range, sub-THz channels are usually rank-deficient, resulting in small multiplexing gain [24]. Therefore, the main focus on the design of sub-THz multi-antenna systems should be achieving high gain to compensate the high path loss instead of exploiting spatial multiplexing. While antennas play a crucial role in such wireless systems, the significant reduction in the wavelength of electromagnetic waves in the sub-THz and THz bands means that antennas operating in these bands must have extremely small physical dimensions. On the other hand, high gain and high efficiency are also necessary design specifications for high-frequency wave antennas due to the high path loss that originates from rain fading. In response to the above requirements, a variety of high-precision and low-cost processing processes suitable for sub-THz antennas have been proposed in academia and industry, and this section focuses in explaining such advancements.

The SIW technology has been able to significantly improve the high transmission loss of planar transmission line structures such as microstrip lines in the millimeter-wave (mmW) band and can be utilized to manufacture high gain, low profile antennas. However, this technology cannot avoid dielectric loss from the dielectric substrate which may be more obvious and noticeable in the sub-THz bands. In contrast to PCB-based technologies, metallic waveguides do not have dielectric loss and radiation loss, while featuring low conductor loss, and is suitable for manufacturing high-frequency antennas. However, due to the complex 3D structure in waveguide slot arrays, production costs will increase significantly, rendering it impractical for mass production. A single-layer waveguide slot array by placing the feeding and radiating waveguides in the same layer was designed in [25].

However, its use of series-feed suffers when the array size is scaled larger and the operating frequency increases, reducing the antenna bandwidth. In contrast, the corporate-feed network is not affected by this. A multi-layer structure can increase the design freedom of the feeding waveguide, which serves as a solution in implementing a corporate-feed network in the sub-THz system. Usually, conventional techniques such as machining or die-casting are incapable of manufacturing complex metal multilayer structures due to low precision [26]. The diffusion bonding process using thin metal laminates to fabricate multilayered, fully corporate-feed waveguide slot arrays is a highly accurate metal processing technique while maintaining a reasonably low cost [26]. Based on the diffusion bonding process, a broadband high-gain antenna array with a double-layer feed structure was proposed in [27]. The 32 × 32 and 64 × 64 arrays designed using this technique can achieve gains of 38 and 43 dBi in the 119–134 GHz region, with 60% and 50% efficiencies, respectively. Figure 7.12 shows the configuration of the proposed array. Diffusion bonding of thin metal laminates achieved good electrical contact by applying diffusion motion to the atoms in the adjacent metal plates, which can well protect against electromagnetic wave leakage. The process of diffusion bonding can be summarized in three steps as follows [27]:

1) etching the desired pattern on the copper plate using machining, laser cutting or electrochemical processing, with a processing accuracy of ±10 μm for this step;
2) positioning, aligning and stacking the layers of copper sheets in a vacuum, a step that produces errors of typically less than 5 μm;
3) diffusion bonding in a vacuum, high pressure, high temperature (~1000 °C) environment, a process that takes tens of hours.

The diffusion bonding technique can be machined up to a precision of ±20 μm and features very good electric contact between the metal plates. However, great care must be taken when using it to manufacture complex and large structures, as the high pressure during step 3 of the process may cause the deformation of the metal plates with large and complex patterns [27]. In addition to that, step 3 of the diffusion bonding process is time-consuming and costly, thereby affecting its commercial viability.

Low-temperature cofired ceramic (LTCC) is a multilayer substrate integration technology developed by Hughes Corporation in 1982. Fujitsu, Murata, and IBM introduced this technology into commercial communication systems in the 1990s, and it is now widely used in high-frequency antenna designs. LTCC allows the feed structure and radiator to be integrated into a multilayered, low-loss ceramic substrate in a compact package. Compared with the PCB process, it is easier to realize blind and buried vias using LTCC technology, besides featuring lightweight, low-profile, and excellent high-frequency performance. These advantages make it one of the most promising manufacturing processes for sub-THz antennas. Guo and Wong [28] proposed a dual-polarized Fabry-Perot cavity (FPC) antenna operating in the 155 GHz band. The antenna uses a new phase-shifting surface (PSS) based on LTCC and is made of a single dielectric material. For horizontal and vertical polarizations, the peak gains are 17.5 and 17.1 dBi, respectively,

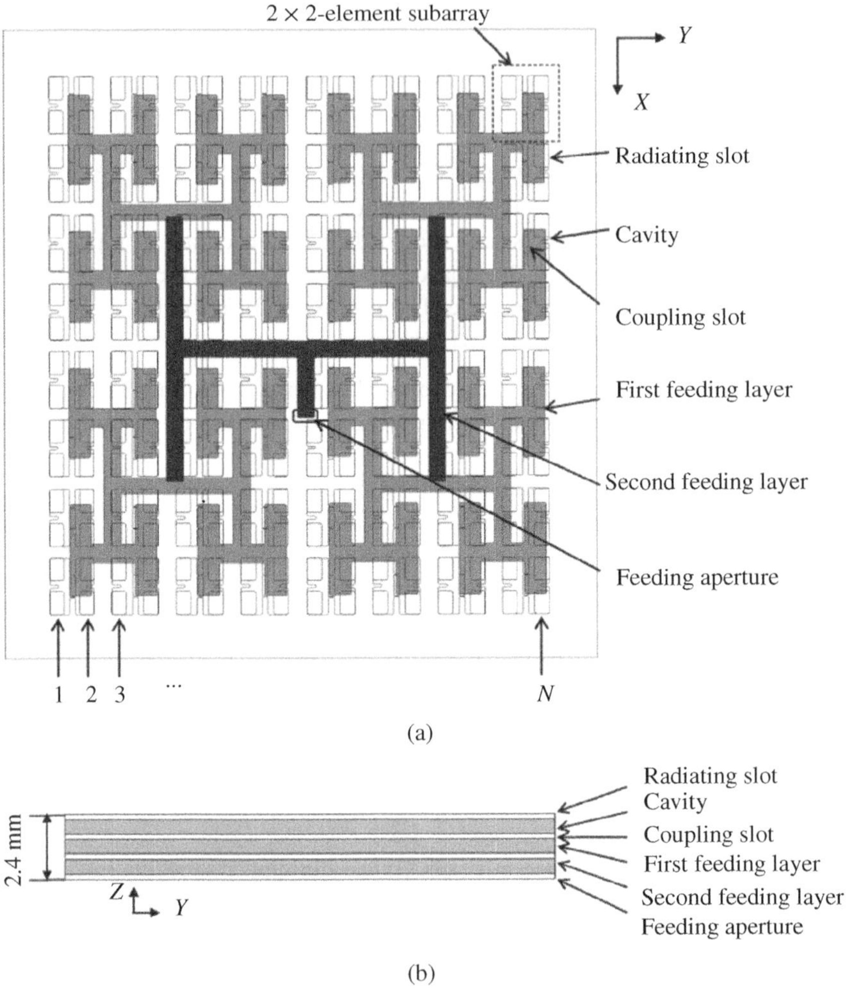

Figure 7.12 Configuration of the proposed array with corporate feed network (a) top-view and (b) side-view. Source: Reproduced with permission from [27]/IEEE.

and the impedance bandwidth is 10.4% for both polarizations. The configuration of the proposed dual-polarized feeding source is shown in Figure 7.13a, whereas Figure 7.13b–d illustrates each layer of the feeding source.

LTCC process has the following steps:

- *Slurry configuration*: making ceramic slurry from ceramic powder.
- *Tape casting*: making dense raw ceramic wafers from ceramic slurry.

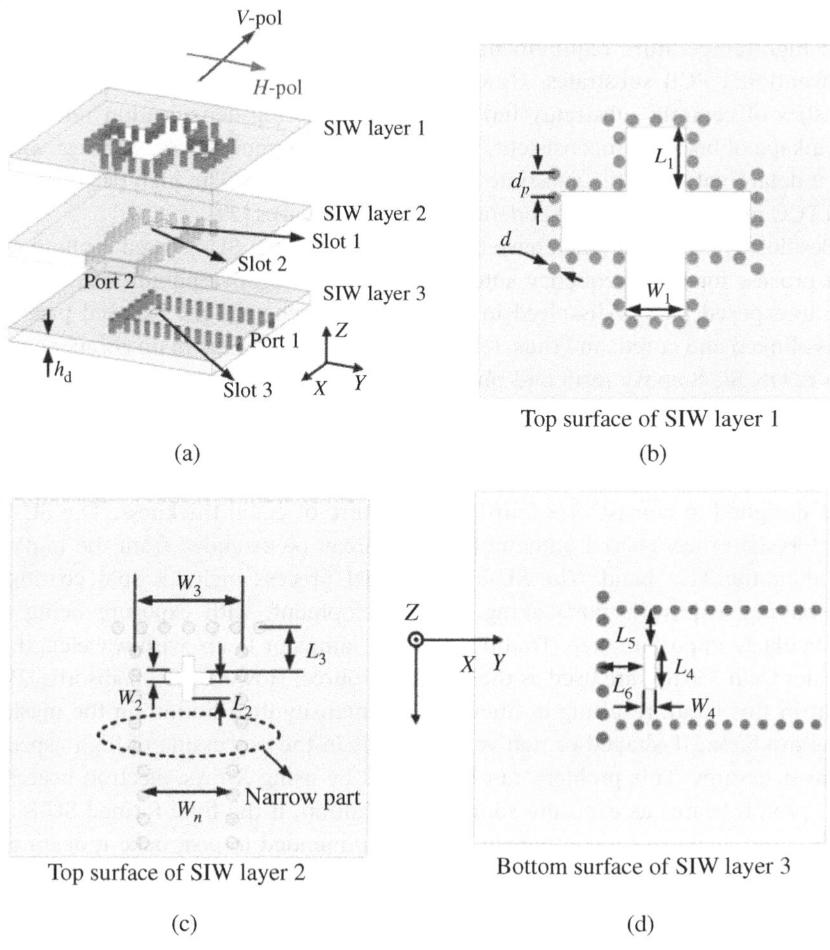

Figure 7.13 (a) Configuration of the proposed dual-polarized feeding source. (b–d) Layer-by-layer illustration of the feed source. Source: Reproduced with permission from [28]/IEEE.

- *Screen printing*: making through vias, processing metal patterns, embedding components, etc., on raw wafers.
- *Laminated sintering*: sintering all raw wafers in alignment at about 900 °C to form a high-density multilayer structure.

LTCC uses ceramic materials with high Q-value characteristics and therefore lowering losses. For instance, Ferro A6-M, a commonly used LTCC material features a relative permittivity of 5.9 and a dielectric dissipation factor is 0.002. This dielectric loss is one order of magnitude lower than FR-4 while having better insulation performance. In addition to that, LTCC can adapt to high-current and high-temperature requirements and has better thermal conductivity than conventional PCB substrates. However, due to the different sintering characteristics of ceramic substrates and pastes, the sintering densification rate and shrinkage of both are inconsistent, which can lead to unevenness, warpage, and even delamination of the substrate surface after sintering. The high permittivity of LTCC may also lead to the generation of surface waves [29].

Besides LTCC, researchers have been investigating the SU-8-based-8 photoresist process for high-frequency antennas. SU-8 is a negative photoresist, where the unexposed part is dissolved in the developer, whereas the exposed part is cross-linked and cured; and thus, retained on the substrate. Its main components are EPON SU-8 epoxy resin and photoacid-generating agent [30]. SU-8 is also a thick resist and SU-8 photoresist can be used to precisely process microstructures with high aspect ratio with vertical edges. A micromachined 300 GHz waveguide slotted antenna using SU-8 photoresist technology was presented in [31], which was designed to consist of a four-layer structure of equal thickness. The SU-8 photoresist process-based antenna fabrication can be extended from the mmW band to the THz band. The SU-8 photoresist process includes spin coating, pre-baking, exposure, post-baking, and development, with exposure being a particularly important step. Traditionally, UV lamps or lasers with wavelengths greater than 350 nm are used as the exposure source. However, SU-8 absorbs UV light in this band, resulting in uneven light intensity distribution on the mask, thus producing T-shaped or non-vertical edges in the processing of high aspect ratio structures. This problem can be avoided by using X-rays, electron beams, and proton beams as exposure sources. In addition, if the final formed SU-8 is to be used as part of a component, it is recommended to post bake it again at 150–200 °C after development to strengthen the photoresist crosslinking [30].

A new trend in high-frequency antenna design is to shift from traditional discrete designs to antenna-in-package solutions to meet the architecture requirements of System-in-Package (SiP). Considering that RF front-end circuits and signal processing circuits are generally printed on a silicon-based semiconductor substrate, the same silicon-based semiconductor substrate can be more easily integrated when designing antennas. However, the resistivity of standard silicon is only 40 Ω cm and the relative permittivity is 11.9. The low resistivity and high permittivity lead to the low radiation efficiency of the traditional silicon-based antennas. Although some related studies have used high-resistance silicon with resistivity up to 5000 Ω cm as antenna substrate [32], the high cost of high-resistance silicon makes it difficult to implement in large-scale commercial use at present. In order to satisfy high accuracy, high efficiency, and integrability at the same time, a new process combining silicon-based materials and bulk micromachining technology, i.e. silicon bulk micromachining technology, has started to receive attention in the antenna field. Chang et al. [23] fabricated the first air-substrate THz planar antenna array using bulk silicon technology. The simulation results obtained a −10 dB impedance bandwidth of 238.4–309.5 GHz, with gain and radiation efficiency higher than 10.6 dBi and 87.9%, respectively. The bandwidth and efficiency obtained for the prototype test agree well with the simulation, and the gain is higher than 8.6 dBi. Figure 7.14 shows the configuration of the proposed antenna array. The first step of the multilayer silicon bulk micromachining technology is to etch each layer of silicon to obtain the desired pattern, with an etching accuracy of 20–30 μm. After etching, all surfaces and sidewalls of the wafer are coated with a gold layer of 3.5 ± 0.5 μm thickness to avoid electromagnetic waves to be incident on the silicon, thus reducing losses and improving antenna efficiency. The final step is to align the layers and assemble the layers by wafer bonding. The aligned layers are clamped with a force of 5 kN and baked in a vacuum at 350 °C for about 10 minutes, and the gold plating on each layer will melt and join after cooling, with a tolerance of about 5 μm in this step.

Kildal et al. proposed the gap waveguide (GWG) technique in 2009 [33, 34]. The GWG is simple to manufacture, flexible to use and has almost no dielectric losses, making it a good candidate for high-frequency RF applications. GWG techniques are originated in hard and soft surfaces [35]. In the parallel-plate

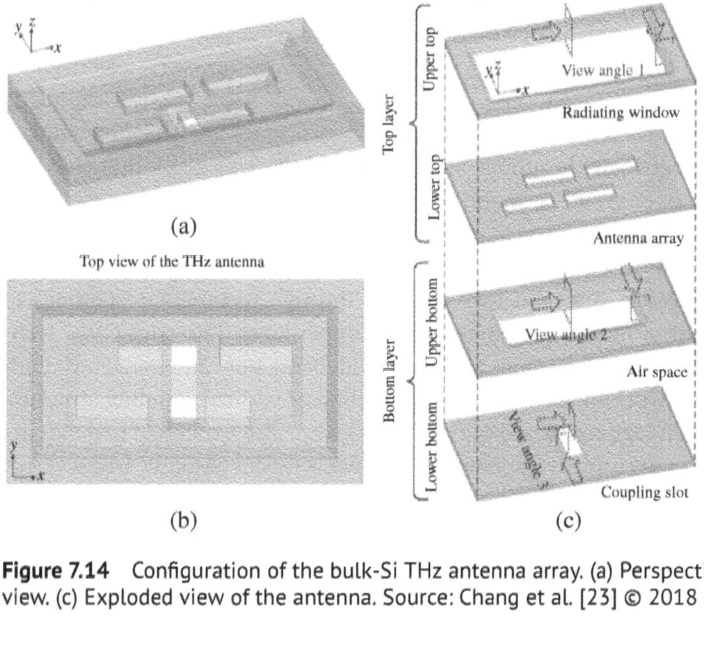

Figure 7.14 Configuration of the bulk-Si THz antenna array. (a) Perspective view. (b) Top view. (c) Exploded view of the antenna. Source: Chang et al. [23] © 2018 IEEE.

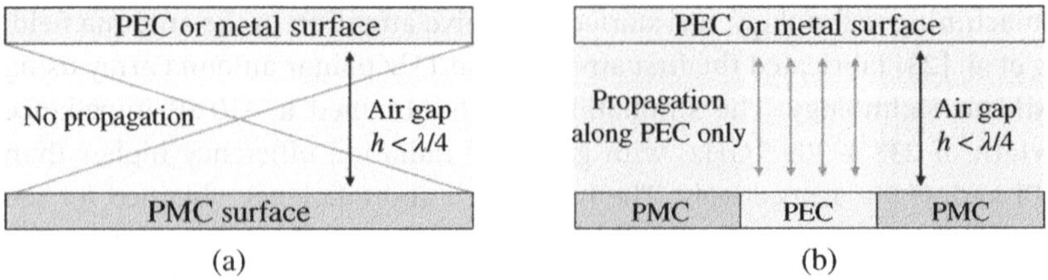

Figure 7.15 Cross section of ideal gap waveguide, (a) no wave propagation scenario, (b) wave propagation scenario.

waveguide composed of perfect magnetic conductor (PMC) plate and perfect electric conductor (PEC) plate, electromagnetic waves cannot propagate in parallel plate environment as long as the air gap between PEC and PMC plate is less than $\lambda/4$. Nevertheless, electromagnetic wave propagation can be guided by establishing a path on the PMC surface, as illustrated in Figure 7.15. Unfortunately, PMCs are not available in nature. However, in a certain frequency range, the PMC-equivalent condition can be replaced by an AMC made with periodic structures [36].

Depending on the guiding structures, GWGs can be categorized as ridge GWG, groove GWG, inverted-microstrip GWG, and microstrip-ridge GWG. The detailed geometry for each one of these configurations is shown in Figure 7.16. The propagation mode in the groove GWG is very similar to that in rectangular waveguides, so both TE/TM modes are supported. TE_{10} mode is preferred in most cases. Nevertheless, the desired modes of ridge GWG, inverted microstrip GWG and microstrip GWG are similar, and they all support quasi-transverse electromagnetic (Q-TEM) wave propagation within the stopband. Since most of the fields propagate in air, the loss is relatively small. GWG structures are designed without electric contact between the upper and lower metal surfaces so that low-loss waveguide components can be manufactured with low costs at millimeter-wave frequency bands or even higher.

GWGs are mainly used as feed networks in large antenna arrays, usually in the lower layer. Waves are then fed vertically through rectangular openings to the coupling layer or radiating elements in the upper layer. The ridge GWG technology was used for the first time to design wideband slot antenna arrays with corporate-feed network in [37]. The design method of vertical coupling structure and 3-dB power divider composed of ridge GWG were also explained in detail. Its excellent performance proves the potential of GWGs in slot antenna arrays.

Subsequent studies have focused on two-dimensional arrays with more complex feeding networks. In a two-dimensional array, the element spacing should be less than one wavelength in both orthogonal directions to avoid gating lobes, requiring a more compact feed network. An effective strategy is to add an intermediate layer

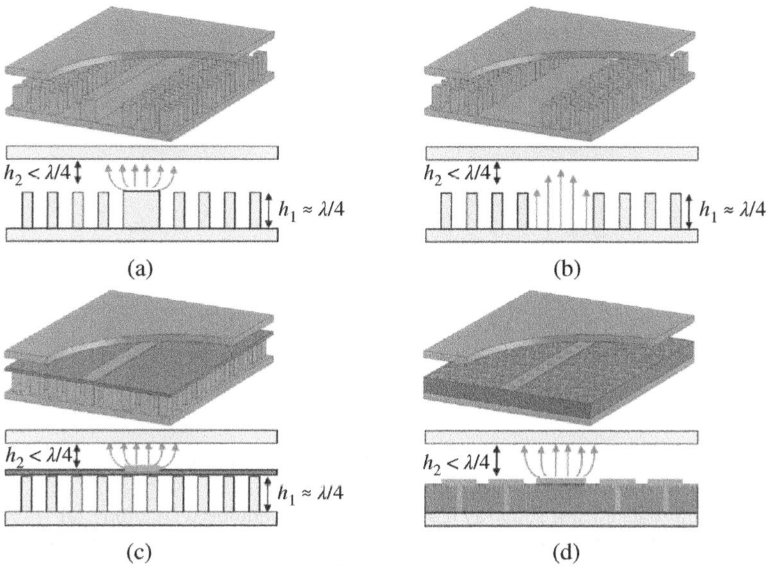

Figure 7.16 Different gap waveguide geometries and propagation modes. (a) Ridge gap waveguide. (b) Groove gap waveguide. (c) Inverted microstrip gap waveguide. (d) Microstrip-ridge gap waveguide.

between the feed layer and the radiation layer [38, 39]. The geometry of such a subarray and power divider is shown in Figure 7.17. This interlayer design is manufactured with groove GWG technique, which distributes energy coupled from the bottom layer equally to the four radiating elements. A variety of two-dimensional antenna arrays based on GWG techniques were reported, such as a patch array fed by ridge GWG [40], a patch array fed by groove GWG [41], a slot array fed by ridge GWG [42–44], a slot array fed by groove GWG [45], a slot array fed by inverted microstrip GWG [46], a slot array fed by microstrip-ridge GWG [47], a horn array fed by inverted microstrip GWG [48], a magnetoelectric dipole array fed ridge GWG [49], and a magnetoelectric dipole array fed by microstrip-ridge GWG [50–53]. Most of the above configurations can be directly applied to sub-THz bands.

Furthermore, applications of GWG in dual-polarized and circularly polarized (CP) antenna arrays have been developed. An 8 × 8 dual-polarized array antenna was designed in [54]. The antenna element had a circular radiation aperture and dual orthogonal linear polarizations provided by two feeding layers. Measured results provide an average gain of over 27 dBi and a radiation efficiency of nearly 80% within its operating bandwidth. On the other hand, CP antenna arrays based on GWGs have been reported in [55–57]. In [58], the geometry of the proposed

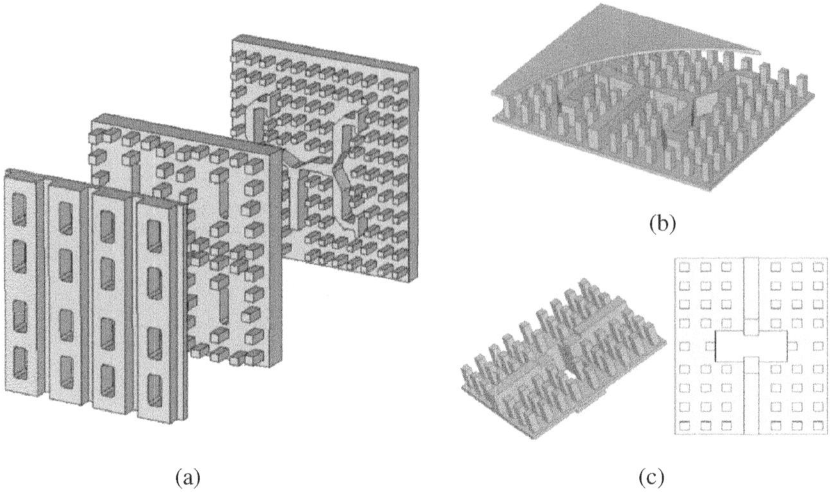

Figure 7.17 Cavity-backed slot subarray realized in gap waveguide technology. (a) Three unconnected metal layers. (b) Four-way power divider. (c) Transition from rectangular waveguide to ridge gap waveguide. Source: Reproduced with permission from [38]/IEEE.

subarray (using septum CP antenna elements) is shown in Figure 7.18a. The 8 × 8 elements array achieved 31.5% impedance bandwidth ($|S_{11}| < -10$ dB), 33.1% overlapping AR bandwidth (AR < 3 dB), 28.4 dBic maximum measured gain value, and 85% maximum aperture efficiency. A dual-CP antenna array was proposed in [57], where the radiation elements are eight U-shaped slots fed by two feeding networks (one for each CP polarization).

In slot antenna arrays with corporate-feed network, tapering with unequal power dividers is a necessary condition to achieve low sidelobe levels. Unequal power divider based on inverted microstrip GWG was first proposed in [59]. As presented in Figure 7.18c, its geometric structure is a three-port power divider with directional coupling. The proposed 16 × 16 antenna array has a −10 dB impedance bandwidth of 18.5% (in the range of 35.9–43.2 GHz), a gain of 37.5–41.8 GHz greater than 29 dBi, and a satisfactory suppression of sidelobe level (SLL) below −23 dB. Moreover, a ridge unequal power divider is fabricated in [60] and applied in the GWG feeding network. Subsequently, the sidelobe levels of x and y polarizations in E-plane and H-plane are suppressed by two-dimensional Taylor synthesis. At the same time, dual-polarization is realized through cross-shaped waveguide slots and multilayer feed networks. The detailed subarray geometry and power divider are shown in Figure 7.18b, d, respectively.

Owing to its unique structure, the GWG technology has a more expansive application prospect in the W-band [61, 62]. The losses of the ridge GWG at 100 GHz have been evaluated in [63], whereas a W-band transition from microstrip line to

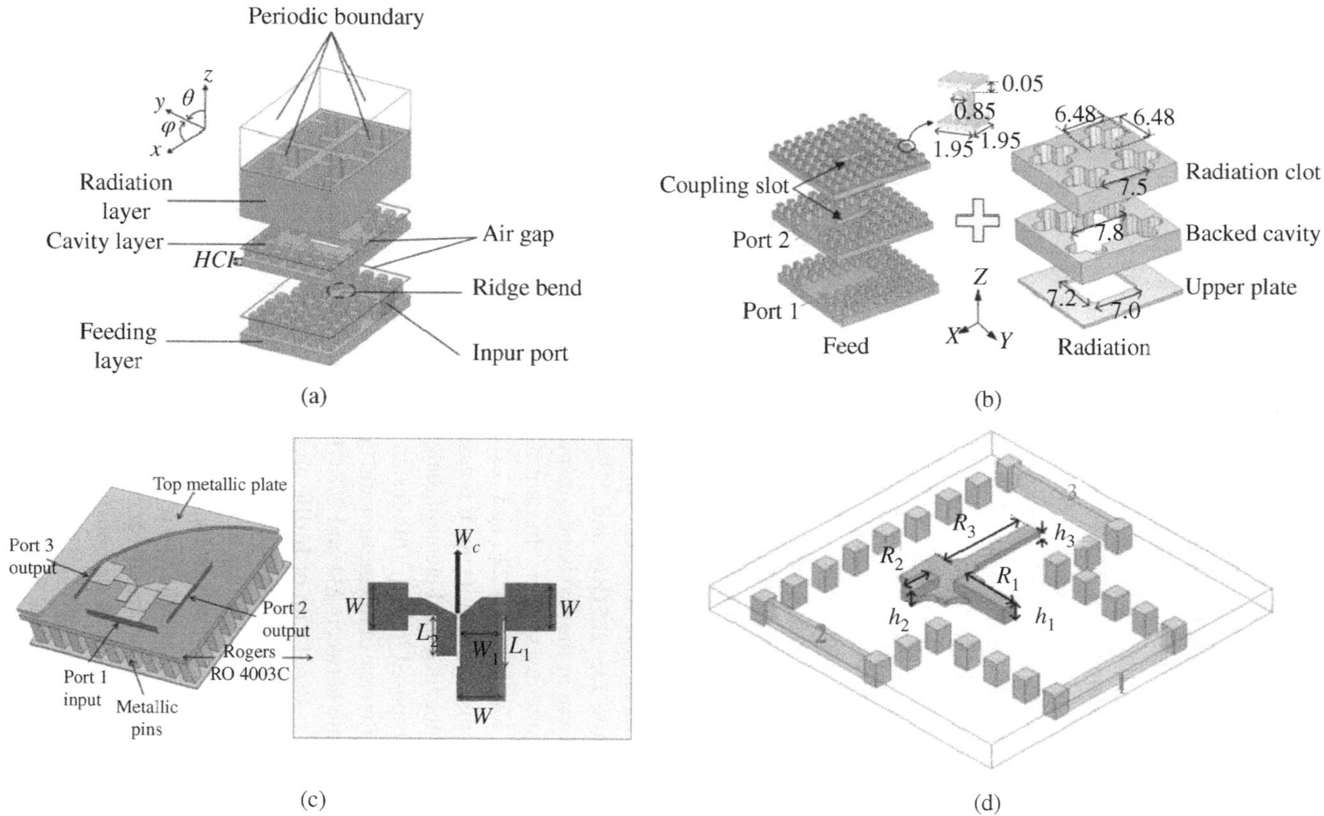

Figure 7.18 (a) CP with septum antenna elements. (b) Dual-polarized with cross-shaped aperture. (c) Unequal power divider based on inverted microstrip gap waveguide. (d) Ridge unequal power divider.

groove GWG is proposed in [64]. In [65], a compact monopole array antenna based on GWG is introduced for tracking applications in the W-band. The Magic-Tee for monopulse comparator network consists of ridge GWG and slot GWG stacked vertically, as shown in Figure 7.19. This 16 × 16 antenna array has a −10 dB impedance bandwidth of 21% (in the range of 85–105 GHz), a gain of 30 dBi in 94 GHz, and the null in the differential radiation pattern is 40 dB lower than the peak of the sum radiation pattern at 94 GHz. Another W-band monopole antenna is reported in [66]. The monopole antenna based on a radial line slot array (RLSA) fed by a Butler matrix, which is designed by GWG technology to generate sum and difference patterns simultaneously. Measurement results show that the gain is 27.8 dBi at 94 GHz, the axial ratio is less than 0.3 dB at 93–95 GHz, and a 50 dB deep null between sum and difference patterns.

Several recent works advocated the use of dielectric based waveguides at sub-THz bands for ease of integration with printed structures and systems. While several attempts to come up with low-loss dielectric waveguide structures have been documented, the concept of substrate-integrated image guide (SIIG) presented low-loss and wideband characteristics that make it a potential candidate for future sub-THz applications and systems [67]. The SIIG consists of a dielectric substrate over a ground plane. By properly designing air hole perforations and applying Floquet theory and Fourier analysis, the effective permittivity and dispersion characteristics can be extracted and then used to design the waveguide features. The works in [67, 68] demonstrated SIIG that operate in the 135–225 GHz and 215–335 GHz bands providing losses of 0.578 and 1.41 dB/cm in the two bands, respectively, opening new frontiers in sub-THz integrated antenna systems. In contrast, a grooved gap waveguide can provide 0.48 dB/cm loss between 220 and 320 GHz, but it is not easy to integrate with planar systems and circuits nor it is easy to fabricate.

A fully embedded dual-element dielectric antenna system for sub-THz applications was proposed in [69]. This design uses a variant of the dielectric waveguide technology for ease of integration and embeds two antennas that are differentially fed to operate between 234 and 278 GHz with good matching characteristics. It produced a realized gain of 16 dBi, and an efficiency exceeding 90%. The dielectric waveguide channel is controlled via perforations that are also used to provide proper impedance matching to the two antennas that are fully embedded in the substrate layer, as shown in Figure 7.20. An aluminum wall of 0.25 mm thickness is used to separate the two ports and reduce their coupling, while differential feeding allows for constructively combining of the two radiated patterns from the two antennas. This work demonstrates a very promising technique for future fully substrate-embedded antenna systems at such sub-THz bands with direct application to MIMO within 6G bands.

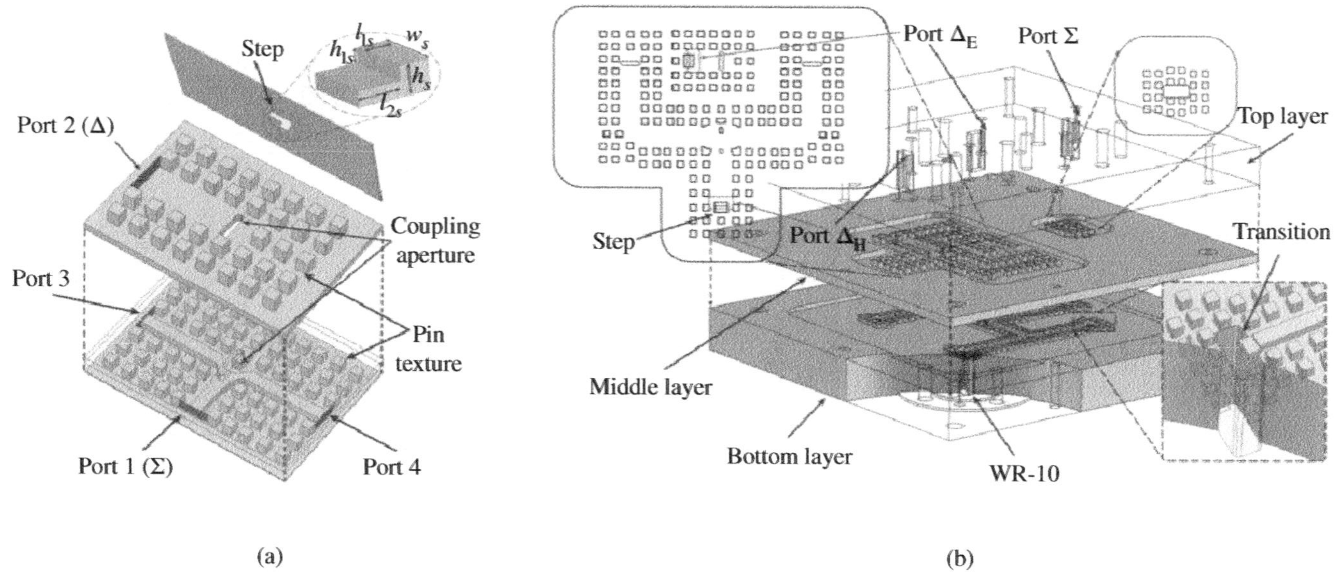

Figure 7.19 Configuration of monopulse comparator network. (a) Magic-Tee. (b) Comparator network. Source: Reproduced with permission from [65]/IEEE.

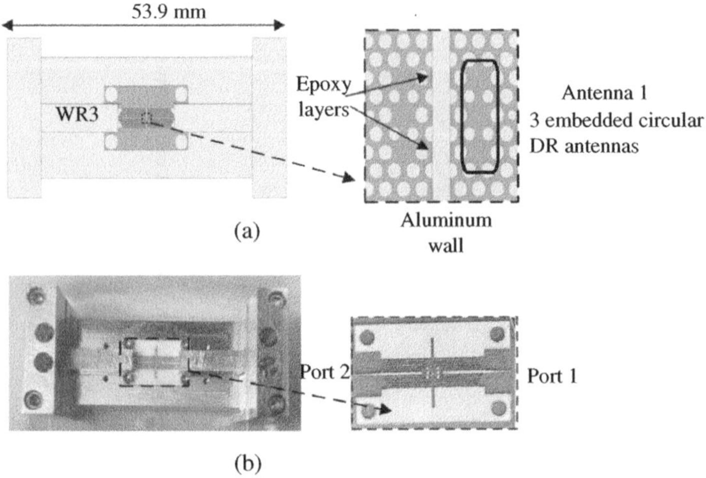

Figure 7.20 The fully integrated dual-element dielectric antenna system. (a) Design model and antenna locations. (b) Fabricated prototype placed in a custom-made waveguide and holder. Source: Reproduced with permission from [69]/IEEE.

In the context of THz antenna design, a key challenge is the rapid signal attenuation characteristic of the THz spectrum. Conversely, microwave frequencies benefit from the mitigation of fading through the use of MIMO structure. Consequently, the extension of MIMO techniques to the THz frequency range has been explored to mitigate the pronounced fading effect of THz waves. In [70], an aperture-coupled wideband antenna operating in the THz frequency range is proposed. In a 1×2 MIMO configuration, the top radiating element is coupled to the feeder via an inverted T-hole. This design achieves an impedance bandwidth of 0.3 THz (0.17 –0.47 THz) with port isolation greater than −20 dB. Gain varies from 7.8 to 12.5 dB. In addition, PBG arrays are often used in antenna design to suppress surface waves and increase gain. References [71, 72] present a 2×2 MIMO antenna using a PBG structure. In particular, Figure 7.21 shows the perforated MIMO antenna array described in [71]. To optimize gain, radiation efficiency and other performance parameters, a 7×14 array PBG structure is implemented. The resulting MIMO antenna achieves a bandwidth of 0.64 to 0.664 THz, a radiation efficiency of 73.36% and a gain of 10.68 dBi.

Reference [73] presents an ultra-wideband fractal antenna. A second-order Minkowski fractal curve is applied to both sides of the common ground plane to improve the isolation between the antenna elements. The structure is shown in Figure 7.22. Compared to a MIMO antenna without fractal, this design has a significantly extended bandwidth. The maximum gain in the frequency range reaches 8.6 dBi, while the Envelope Correlation coefficient (ECC) remains below

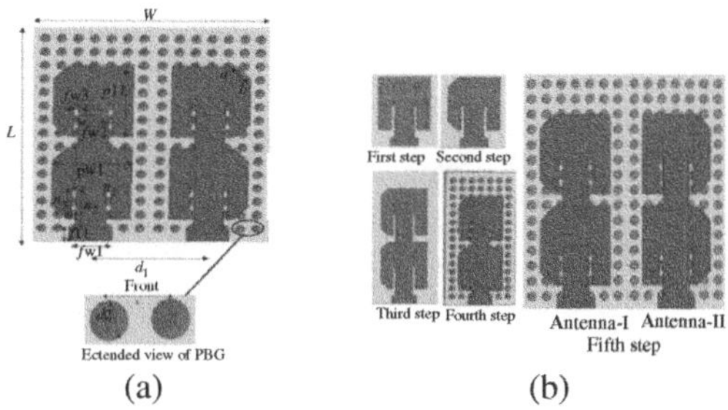

Figure 7.21 (a) Schematic views of 1 × 2 MIMO antenna array. (b) Design steps of 1 × 2 MIMO antenna array.
Source: Reproduced with permission from [71]/IEEE.

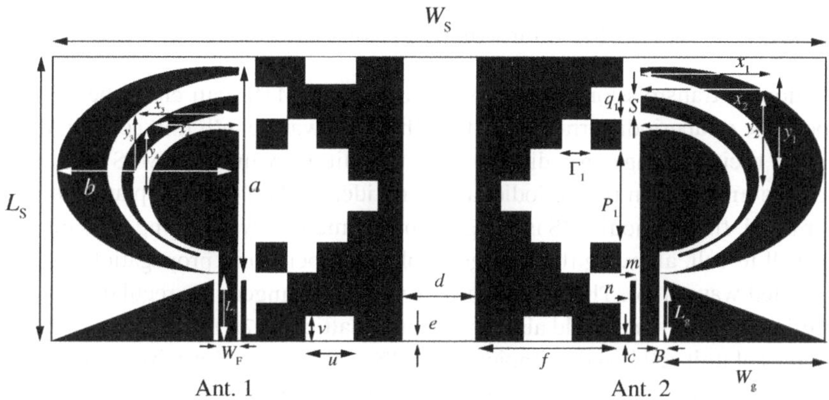

Figure 7.22 Geometry of MIMO antenna with second-order Minkowski fractal curve.
Source: Reproduced with permission from [73]/IEEE.

0.001 over the entire frequency band, resulting in a diversity gain of more than 9.995 dB. Reference [74] provides a comprehensive comparison of five THz MIMO antenna designs with different shapes from the aspects of *S*-parameter, total gain, copolar and cross-polar radiation pattern, electric field distribution and so on. At present, microstrip patch antennas dominate the THz MIMO antenna landscape, but it is expected that different forms of THz MIMO antennas will emerge in the future. At present, microstrip patch antennas dominate the THz MIMO antenna landscape, but it is expected that different forms of THz MIMO antennas will emerge in the future.

7.4 MTS

Unit cells array arranged in a plane is known as a MTS (MTS), and such surface can manipulate the electromagnetic waves impinging on it. These MTSs modify the dispersion properties of the surface or guided waves, thereby resulting in irregular

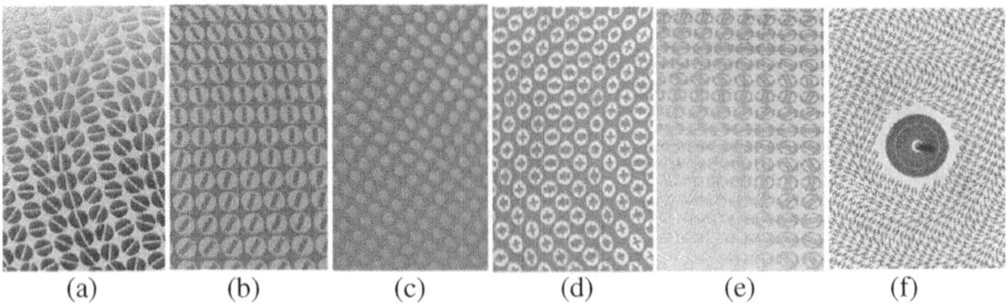

Figure 7.23 Examples of textured layout for implementing the MSs: (a) coffee bean, (b) patch with slot, (c) grain of rice, (d) patch with cross slot (e) double π (f) double anchor with a feed. Source: Reproduced with permission from [77]/IEEE.

reflection or transmission properties of waves [75]. MTSs can either be formed by periodic or aperiodic arrangements of these subwavelength unit cells (in their size and their spacing) on a dielectric host medium. A uniform MTS is formed when the unit cells in the periodic lattice are identical all over the plane. On the other hand, the aperiodic MS is designed by geometrically changing the unit cell from cell to cell, allowing the change in phase velocity and propagation path of the guided waves [76]. The unit cells are typically arranged in a regular lattice of Cartesian type, but one could also exploit other lattices [77]. Figure 7.23a–f shows examples of unit cells used to implement MTSs. Despite its discretized nature, the impedance boundary conditions (BC) formed by the patch texture can be modelled within a good approximation as a continuous sheet of anisotropic impedance BC, which supports and modifies the surface wave propagation.

7.4.1 MTS in 5G Communication Systems and Beyond

7.4.1.1 Beam Manipulation

More established methods such as the phased arrays technique have been mainly used for beam manipulation purposes in microwave bands such as beam control and steering/scanning. This can be implemented by modulating input phases on the antenna arrays. However, when moving upward in mmWave bands, individual phase shifters add to the cost and complexity of the RF front end, especially when larger arrays are needed to meet higher gain requirements. This scaling up of the array aperture also means a linear increase in the number of phase shifters required, translating to higher costs and more complex feeding networks. A viable alternative to this feature of beam focusing can be achieved by MTS by means of applying phase gradients. Moreover, in mmWave bands, the inherently shorter wavelength and higher spatial resolution enable the beam to be focused

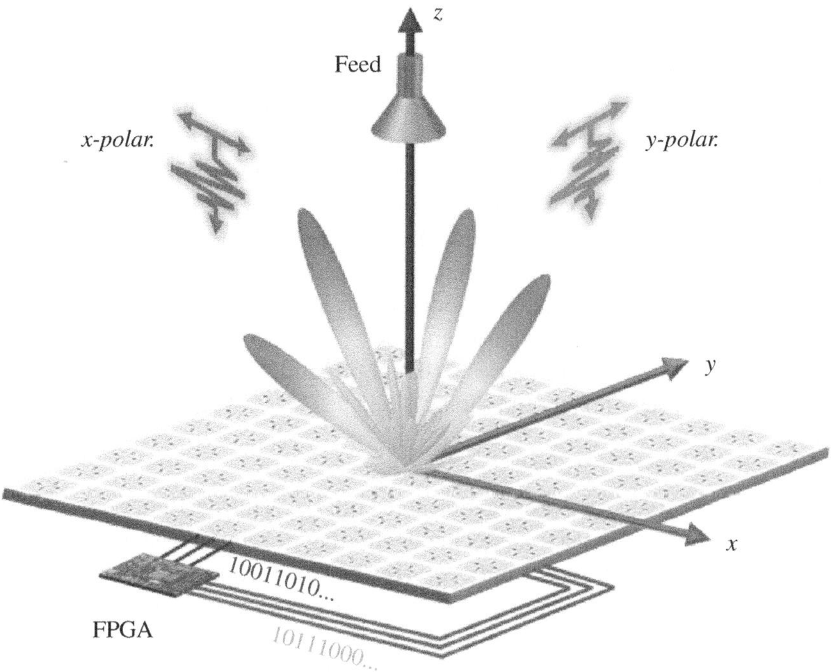

Figure 7.24 Schematic of RRA antenna with independent dual-polarized beam control in real-time controlled by FPGA-based controller. Source: Reproduced with permission from [79]/IEEE.

and steered more accurately. More importantly, in the mmWave bands, the higher losses of front-end components and transmission lines still limit their application. Examples of beam manipulation using MTSs are presented in [78, 79].

Figure 7.24 shows the schematic of a dual-polarized MTS-based reconfigurable reflectarray (RRA). Independent phase control by each element is enabled via the control of the loaded p–i–n diodes states [79]. DC biasing networks are designed on the back side of the elements, fed by multichannel DC voltage output from field programmable gate array (FPGA). As a result, dynamic beam scanning can be achieved at each dual polarization channel. When illuminated by the x-polarized wave from the feed antenna, the MTS produces well-defined beam obtained with low-side lobes. The beam direction in the upper half-space is changed by encoding different spatial phase sequences. For y-polarized excitation, beam scanning can also be realized similarly by redistributing the spatial phase states. The coding sequences in different polarization channels are encoded onto the MTS simultaneously to enable dual polarized operation of the RRA for independent information transmission with real-time reconfigurability.

The main optimization parameters in this design are high aperture efficiency and stable performance. Their optimization is done based on the following steps. Firstly, dual-polarization phase control is enabled using a dual-layer element, considering the generation of stable reflection responses under oblique incidence. Secondly, aperture efficiency should be optimized by proper choice of the focal diameter ratio. Thirdly, the spatial phase distributions of RRA need to be optimized using algorithms derived from antenna array theory. Finally, full wave simulations need to be performed using realistic elements as a final validation against phase distributions generated by the algorithm.

To radiate a beam tilted to (θ_0, φ_0), the required ideal phase distributions for element (m, n) of the MTS should satisfy [80]:

$$\varphi_{mn} = k_0(|\vec{r}_{mn} - \vec{f}| - \vec{r}_{mn}.\hat{u}) + \Delta\varphi \tag{7.3}$$

where k_0 is the free-space propagation constant; $\vec{r}_{mn}$ is the position vector, $\vec{f}$ represents the feed position vector associated with the location of feed and the focal length of RRA; $\hat{u}$ is a unit vector, indicating the beam directions; $\Delta\varphi$ represents the initial phase, which is a constant, providing further optimization of the phase distributions. For a 1-bit RRA, the continuous phase profiles on the antenna aperture should be discretized into 0° or 180°. However, different discretization methods will affect the phase distributions of the antenna, resulting in different radiation performances. As an example, two possible different discretization schemes is presented as follows:

$$\varphi^1_{1\text{-bit}} = \begin{Bmatrix} 0°, & \varphi_{mn} \notin (90°, 270°) \\ 180°, & \varphi_{mn} \notin (90°, 270°) \end{Bmatrix} \tag{7.4}$$

$$\varphi^2_{1\text{-bit}} = \begin{Bmatrix} 0°, & \varphi_{mn} \notin (0°, 180°) \\ 180°, & \varphi_{mn} \notin (0°, 180°) \end{Bmatrix} \tag{7.5}$$

As a verification, a beam tilted toward $\theta = 10°$ (*xoz*-plane) is considered, and full-wave simulation is used to evaluate the effects of different phase discretization.

An RRA with 20 × 20 elements (a total size of 320 mm × 320 mm) is combined with a spatial feeding antenna. The center feeding antenna is fixed onto a bracket and placed 160 mm away from the MTS, as seen in Figure 7.25a. Figure 7.25b shows the 800 DC-biasing ports, which have been located on both sides of the prototype. Biasing lines are then introduced on the bottom layer to enable independent addressing of each element using external DC source, see Figure 7.25c. To suppress crosstalk between the RF and DC signals, high-frequency inductors (100 nH) are connected serially between the bias line and each unit cell to form a low-pass filter. To enable high-speed dynamic beam scanning, all optimized voltage distributions for different beam scanning angles are stored in the FPGA.

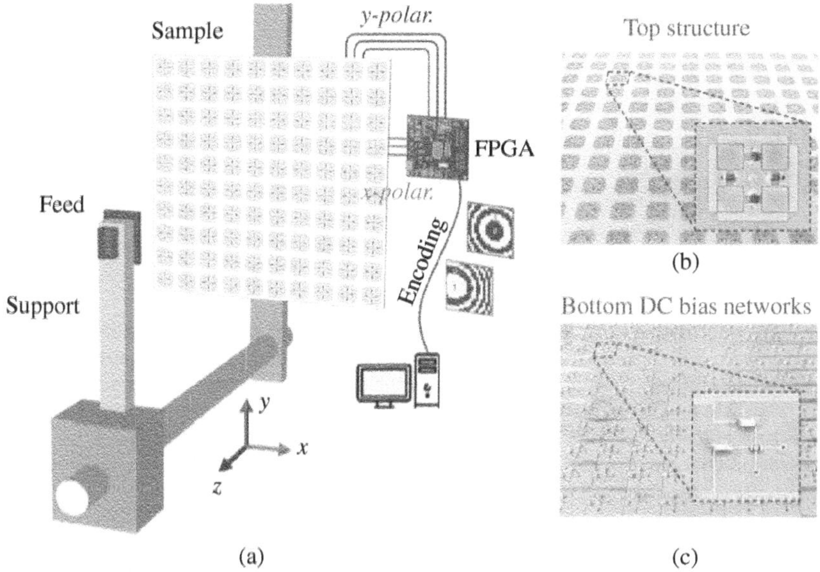

Figure 7.25 (a) The experimental setup consisting of the feed antenna, FPGA and MTS. (b) The top side of the fabricated MTS prototype. (c) The bottom DC bias network of the fabricated prototype. Series high-frequency inductors (100 nH) are used to serve as low-pass filters. Source: Reproduced with permission from [79]/IEEE.

Broadside co- and cross-polarization patterns measurement results at different polarization channels shown in Figure 7.26a, b indicated half-power beamwidths (HPBWs) of about 7°, close to the simulated 6.6°. The measured reflection coefficients plotted in Figure 7.26c (for the *x*-polarization mode) and Figure 7.26d (for the *y*-polarization mode) are lower than −14 dB around 7.45 GHz. Measured gain and aperture efficiency at the *x*-polarization channel are shown in Figure 7.26e while the *y*-polarization channel in Figure 7.26f. The measured 3 dB gain bandwidths in the *x*-polarization mode and *y*-polarization mode are 6.9–7.7 GHz and 6.9–7.75 GHz, respectively, indicating a relative bandwidth of 11.0% and 11.6%. Meanwhile, the measured maximum *x*-polarized gain is 21.13 dBi (aperture efficiency of 16.35%) and the *y*-polarized gain is 20.89 dBi (aperture efficiency of 15.47%).

In another work in [78], a reconfigurable 1-bit coding MTS operating completely independent in C-band and X-band is proposed. In the C-band, MTS is designed to enable the generation of multiple and dual beam, whereas dynamic beam scanning is featured in X band. Both features are generated by changing the coding sequences in real-time. The calculated 3D far-field scattering patterns of the MTS operating at 6 GHz is shown in Figure 7.27a shows, where colored

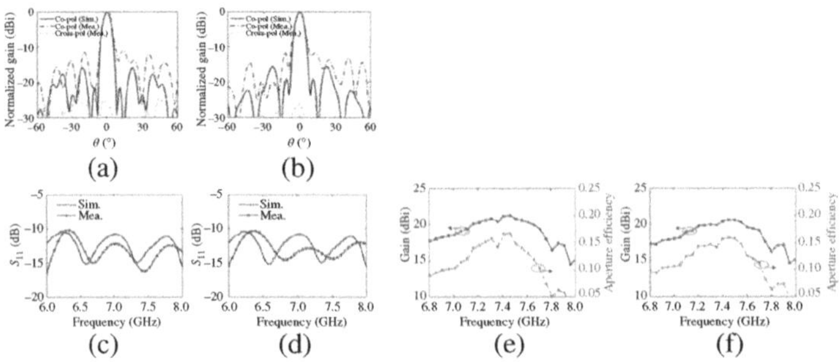

Figure 7.26 Normalized simulated and measured radiation patterns of the beam directed toward $\theta = 0°$, $\varphi = 0°$ in (a) x-polarization channel and (b) y-polarization channel. Measured and simulated reflection coefficients with the beam directed toward $\theta = 0°$, $\varphi = 0°$ in (c) x-polarization channel and (d) y-polarization channel. Measured gain and aperture efficiency of the broadside beam. (e) x-polarization channel. (f) y-polarization channel. Source: Reproduced with permission from [79].

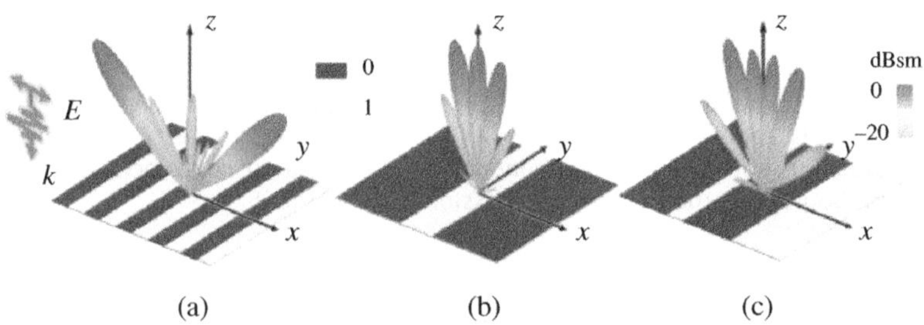

Figure 7.27 Beam patterns generated from the 1-bit coding MTS in C-band. (a) Dual-beam with the coding sequence of "00110011001100110011". (b) Three-beam with the coding sequence of "00000000111100000000". (c) Four-beam with the coding sequence of "00000011000000111111". Source: Reproduced with permission from [78]/IEEE.

blocks are used to represent the coding state of the elements. The incident wave is redirected into two directions at ±44° in the xz plane when the coding sequence is set as "00110011001100110011" along the x-axis. On the other hand, three directive beams reflected toward −11°, 0, and 11° are generated when the coding sequence is changed to "00000000111100000000", as seen in Figure 7.27b.

As shown in Figure 7.27c, four reflected beams can also be generated by the MTS toward ±5° and ±16.2° along the x-axis. In general, versatile and more complex multibeam conditions can be generated by only changing the coding sequences on the MTS. When a number of n (n is an integer) scattered beams is desired, the electric field $\mathbf{E}_n^{x_i\,y_i}$ reflected from the element in the position of (x_i, y_i) from the MTS plane can be first calculated to produce the desired n-th beam. Then, the total field on the MTS aperture to obtain the desired "n" beams can be calculated by $\mathbf{E}_{\text{total}}^{x_i\, y_i} = E_1^{x_i\, y_i} + \ldots + E_n^{x_i\, y_i} = \sum_{c=1}^{c=n} E_C^{x_i\, y_i}$. The continuous phase profile of the total field can be finally calculated, discretized into binary phases and represented by real MTS elements. The phase of the MTS can also be optimized using algorithm to obtain better scattering performance, as the total field may have a non-uniform amplitude distribution.

The coding elements should be arranged symmetrically in a spatially interleaved form to achieve dual-beam functionality in far-field scattering pattern. This will cause destructive interference between the secondary scattering waves from each element and the surface normal direction while being constructive in the two deflected directions. The total number of the element "0" and "1" is usually designed with equal values that result in a symmetric dual beam with equal scattering strength. Code 0 and Code 1 from Figure 7.27a can be made into grating-like distributions with interval set as two columns, leading to reflection angles of $\pm 44°$, which is consistent with the theoretical results of $\pm 43.9°$ calculated by the following equation [81]:

$$\theta = \pm \sin^{-1}(\lambda/\Gamma), \tag{7.6}$$

where λ is the free-space wavelength, and Γ is periodicity of the coding sequence.

When the grating-like spatial coding profiles are changed with interval of 4, 5, and 10 columns, (a period of 144, 180, and 360 mm), as illustrated in Figure 7.28a–c, the dual beams will then deviate by θ of 19°, 15°, and 7°, respectively. These deviation angles are consistent with those predicted theoretically for all cases, indicating the effectiveness of the MTS in generating the derivatives of the dual scattered beams, as shown in Figure 7.28d–f.

In theory, the beam from a reflect array consisting of $M \times N$ elements distributed in xy plane (see Figure 7.29) can be directed to a specific angle (θ_b, φ_b). This requires the ideal phase for each unit element to be [82]

$$\varphi_{ij} = k_0(|\vec{r}_{ij} - \vec{S}| - \vec{r}_{ij} \cdot \hat{m}) + \Delta\varphi \tag{7.7}$$

where k_0 is the propagation constant in free space, $\vec{r}_{ij}$ is the position vector of the element (i, j), $\vec{S}$ is the position vector of the feed, $\hat{m}$ is the unit vector of the main beam direction, and $\Delta\varphi$ is the initial phase constant for all elements. The progressive reflection phase distribution based on the MTS in Figure 7.29 can be designed as [83]

$$\varphi_{\text{reflection}}(i,j) = -k_0 x_i \sin\theta_b \cos\varphi_b - k_0 y_j \sin\theta_b \sin\varphi_b + \Delta\varphi \tag{7.8}$$

where (x_i, y_j) is the position of the element (i, j), $\Delta\varphi$ is the initial phase constant. Assuming that an open-ended waveguide antenna is used as the feed source, the phase compensation for the out-of-phase illumination on the MTS by the feed

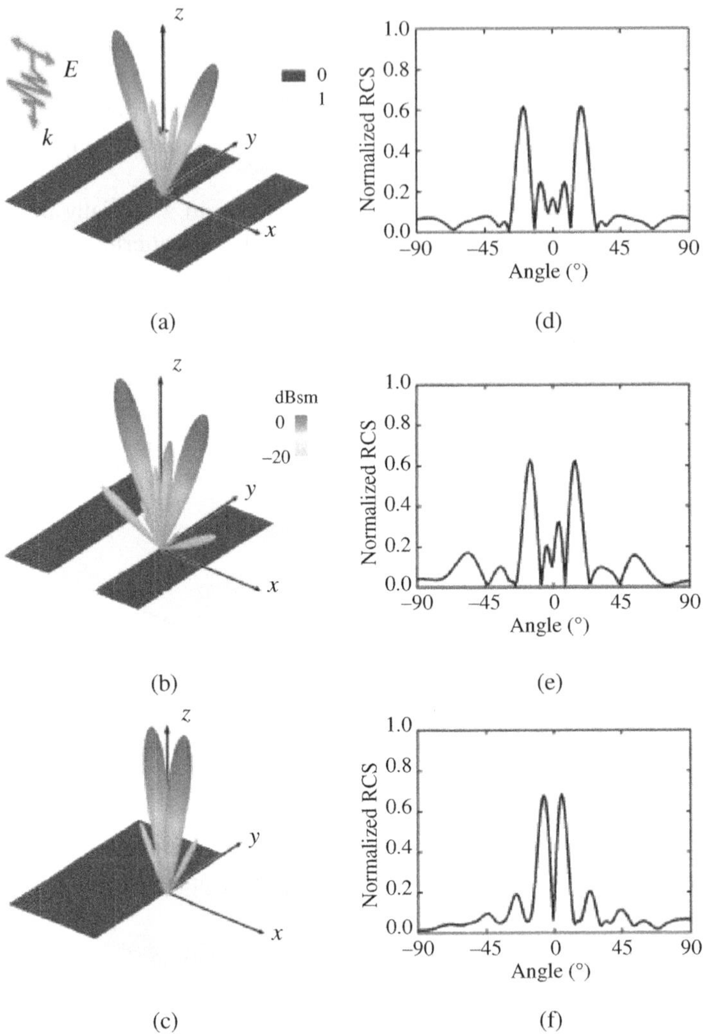

Figure 7.28 Simulated scattered beams from the 1-bit dual-band programmable MTS in C-band. (a–c) The far-field scattering from MTS with different spatial coding patterns, and (d–f) the corresponding 2D simulated results in *xz* plane. Source: Reproduced with permission from [78].

source can be calculated by

$$\varphi_{\text{feed}}(i,j) = k_0 d_{i,j}, \tag{7.9}$$

where $d_{i,j}$, is the distance between the feed and the element (i, j). Given that the feed is located at the geometric center of the MTS and the focal length is F, the

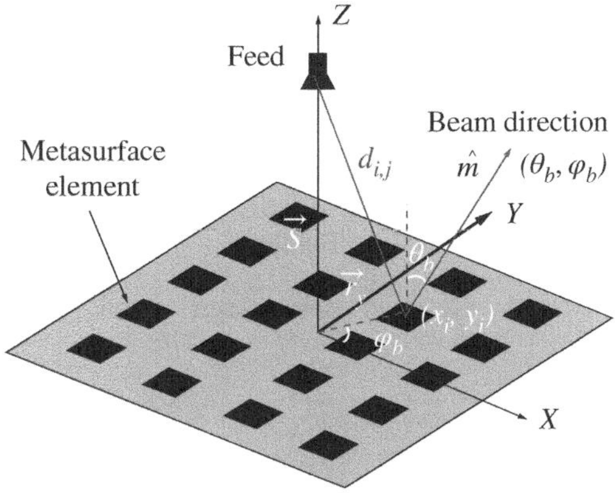

Figure 7.29 The MTS for 3D beam-scanning. Source: Reproduced with permission from [78]/IEEE.

distance can then be calculated as

$$d_{i,j} = \sqrt{x_i^2 + y_j^2 + F^2} \tag{7.10}$$

Via the combination of Eqs. (7.7)–(7.9), the required phase of the elements can be obtained as

$$\varphi_{\text{element}}(i,j) = k_0 d_{i,j,} - k_0 x_i \sin\theta_b \cos\varphi_b - k_0 y_j \sin\theta_b \sin\varphi_b + \Delta\varphi \tag{7.11}$$

Figure 7.30a depicts the programmable MTS system which consists of the printed circuit boards and the FPGA. The fabricated prototype of the MTS is shown in Figure 7.30b–c, with an enlarged view of the elements shown inset. The MTS is formed by an array of 20 × 20 digitally reconfigurable unit cells which can each be controlled dynamically by the FPGA in terms of reflection properties in a programmable manner.

Dual beams can be observed at 6 GHz from Figure 7.31a with deflected angle, θ of ±19° (top panel), ±15° (middle panel), and ±7° (bottom panel), respectively. On the other hand, the measured co- and cross-polarized radiation patterns of the MTS at 9.8 GHz shown in Figure 7.31b validated its beam-scanning functionality. The coding sequences applied to the MTS resulted in reflected angles of 10° (top panel), 20° (middle panel), and 30° (bottom panel). Overall, the programmable MTS can be simply programmed and controlled using the FPGA using specific coding patterns to dynamically and independently generate various beams in two separate frequency bands.

7.4.1.2 Time-Modulated MTS

While one of the main enablers in 5G is the use of MIMO technology, the increasing sizes of arrays in realizing its implementation are becoming prohibitive. The higher number of required independent MIMO channels increases the needed number of RF chains and antennas. Consequently, this need also adds to

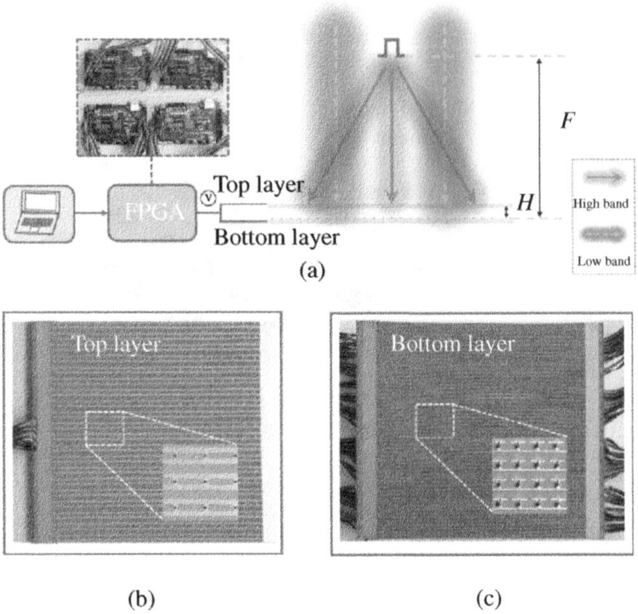

Figure 7.30 (a) The prototype of the dual-band programmable coding MTS. (b) The top layer and (c) the bottom layer. Source: Reproduced with permission from [78]/IEEE.

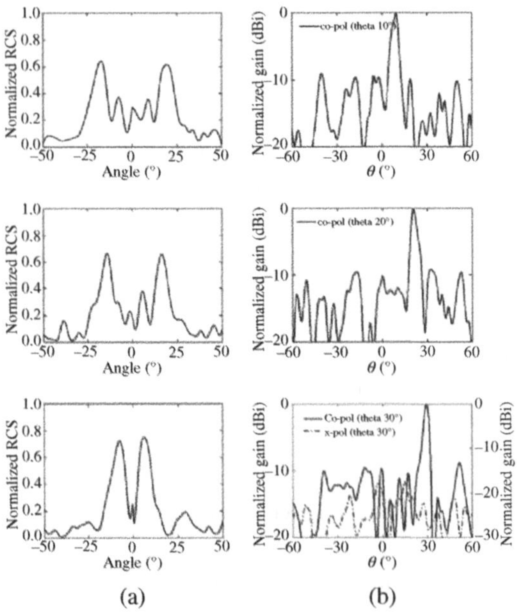

Figure 7.31 Measurement results of the programmable MTS in (a) C-band (6 GHz) and (b) X-band (9.8 GHz). Source: Reproduced with permission from [78]/IEEE.

the cost, complexity, and space requirements of the antennas. MSs can potentially overcome these issues by modulating incident waves in the frequency domain directly. This is done by applying control signals to the MSs while allowing them to perform time modulation. Such a method potentially replaces the need for modulators and transmitters in the RF chain. For instance in [84], a flexible and simple method to realize digital modulation by regulating the biasing voltages of time-domain digital coding MTS (TDCM) in a nonlinear fashion. This leads to a time-varying reflection coefficient at the interface. At the incidence of the wave, the carrier wave will shift up to a higher frequency with excellent energy conversion efficiency. Furthermore, by introducing a cycle of time shift in the regulation voltage signal, an extra phase shift will be added to the desired frequency without any influence on its amplitude, so that the two parameters are totally decoupled from each other. Therefore, it is possible to construct standardized constellations and implement high-order modulation schemes through this novel method.

To explain this, an incident EM wave can be modulated on the MTS with the time-varying reflection coefficient $\Gamma(t)$ [85], and the reflected wave is written as:

$$E_r(t) = \Gamma(t) \cdot E_i(t) \tag{7.12}$$

Note that although the reflection coefficient is usually defined in a linear time-invariant (LTI) system, it is still approximately valid in time-varying situations when the modulation frequency is much slower than the carrier frequency [86]. A monochromic incidence at the carrier frequency f_c with $E_i(t) = e^{j2\pi f_c t}$ is assumed. Based on the Fourier transform theory, the expression of $E_r(t)$ in the frequency domain is as follows:

$$E_r(f) = \Gamma(f) * E_i(f) = \Gamma(f) * [\delta(f - f_c)] = \Gamma(f - f_c) \tag{7.13}$$

where * denotes convolution operation, and $\delta(f - f_c)$ represents the Dirac delta function at $f = f_c$. The effect of spatial modulation is described in Eqs. (7.12) and (7.13). With the assumption that the $\Gamma(t)$ can be expressed as the superposition of a series of scaled and shifted pulses

$$\Gamma(t) = \Gamma_m(t) \cdot g(t), \quad 0 \leq t \leq T, \quad \Gamma_m(t) \in M \tag{7.14}$$

where the gate function $g(t)$ serves as the pulse shaping function, T is the period of message symbol, M is a set of constellation points with the cardinal number $|M|$, and $\Gamma_m(t)$ is the reflection coefficient in the message symbol, representing $\log_2 |M|$ bits of digital information during message transmission. To implement modulation schemes ranging from QPSK, 8PSK to 16QAM, the amplitude and phase waveforms of $\Gamma_m(t)$ in one period (Figure 7.32c) can be expressed as

$$\Gamma_m(t) = A \cdot e^{j\phi}, \quad 0 \leq t \leq T \tag{7.15}$$

with the reflection coefficient of the TDCM (phase and amplitude) to be ideally adjusted independently in different time slots. However, the generally strong coupling of the reflection amplitude and phase of the unit cell at its operating

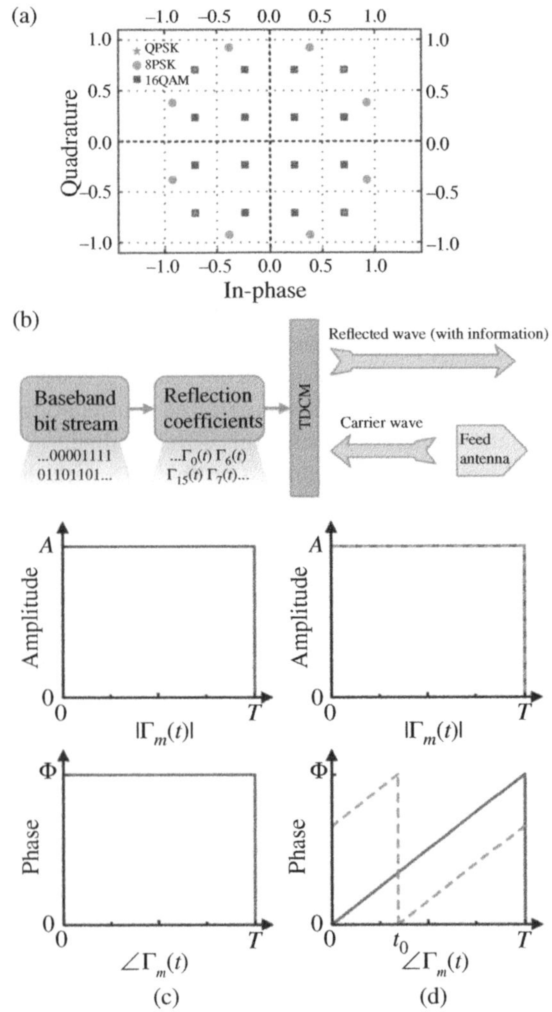

Figure 7.32 (a) Constellation diagrams of QPSK, 8PSK, and 16QAM. (b) Block diagram of the TCDM-based transmitter. Waveforms of $\Gamma_m(t)$ in the period T. (c) Constant amplitude and phase. (d) Time-varying amplitude and phase. The original waveform is denoted by the solid line, whereas the waveform after applying the cycle time shift t_0 is denoted by the dashed line. Source: Reproduced with permission from [84]/IEEE.

frequency makes such TDCM challenging in practice. A method to overcome this is to tune the harmonic phase separately regardless of the corresponding harmonic amplitude via the insertion of additional time delay to the reflection function [87]. Thus, a new reflection coefficient $\Gamma_m(t)$ is defined to implement high-order modulation schemes:

$$\Gamma_m(t) = A \cdot e^{j\left(\frac{\phi}{T}t\right)}, \quad 0 \leq t \leq T \tag{7.16}$$

where the phase variation of $\Gamma_m(t)$ is linear with a constant amplitude A from 0 to ϕ within period T. The amplitude and phase waveforms of $\Gamma_m(t)$ in one period is shown in Figure 7.32d (solid line). With A set to 1, the waveform of $\Gamma'_m(t)$ with the cycle time shift t_0 is marked by the dash line in the same figure.

To validate this TDCM principle further, an MTS with 8 × 16-unit cells is designed and fabricated using PCB technology. Its overall size is 126 × 384 mm^2 and is illustrated in Figure 7.33a, with the enlarged picture of the element (including varactor diodes, chip capacitance and biasing lines) inset. The

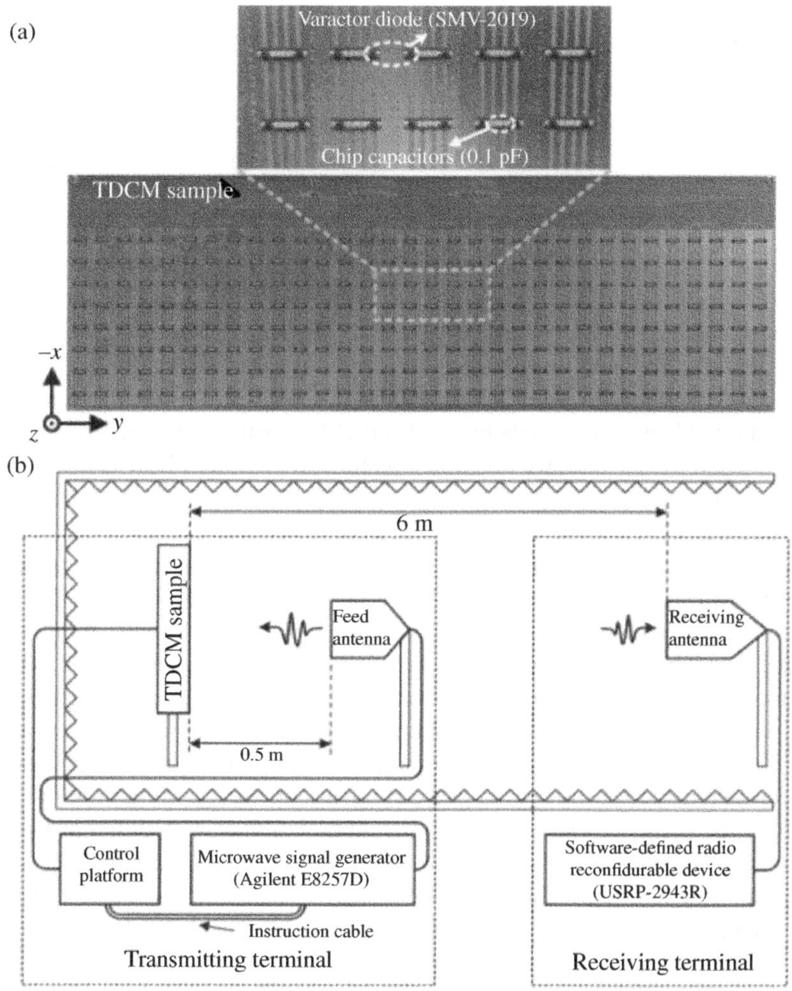

Figure 7.33 (a) The fabricated TDCM. (b) Experimental setup. (c) The reflection properties and the biasing voltages of the TDCM at 4.25 GHz. Source: Reproduced with permission from [84]/IEEE.

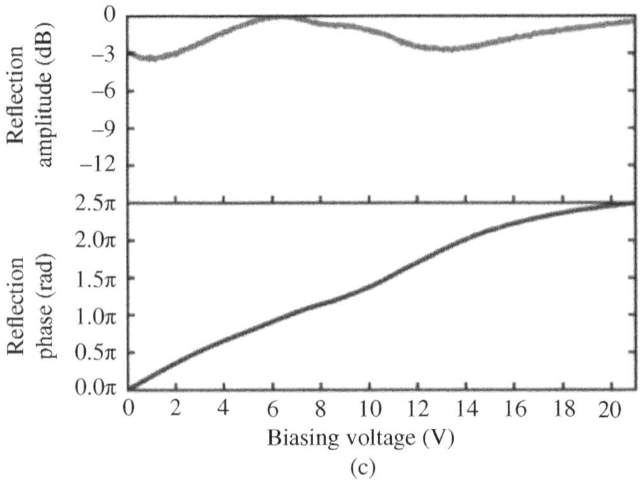

Figure 7.33 (*Continued*)

measurement setup of the system is depicted in Figure 7.33b whereas Figure 7.33c relates the reflection amplitude/phase distributions of the TDCM and the biasing voltages at 4.25 GHz. The phasing range of the TDCM is measured to be 2.5π with the variation of biasing voltage between 0 and 21 V. Despite being slightly smaller than predicted by simulations, this range is still sufficient for phase modulation. It is also found in the experiments that there is a maximum amplitude fluctuation of 3.5 dB.

In summary, a 1-bit high-efficiency programmable MTS can achieve completely independent functions with real-time beam control by FPGA hardware system in both C-band and X-band. Specifically, the MTS is designed with abilities to generate multi-beam or twin-beam scanning in the low-frequency band while dynamic beam-scanning in high-frequency band. Such features are especially useful in MIMO communications considering the increasingly scarce available spectrum. Available (but typically narrow) bands which are spaced apart can be used contiguously for MIMO to improve spectrum efficiency. More importantly, the ability of such MTS to generate and control multibeam at different bands makes it more attractive as a new MIMO communication method. In this specific work, experimental verification was performed by designing twin-beam in the lower frequency band, whereas beamscanning feature is enabled in the high band. By regulating the reflection phases of the TDCM in a nonlinear way, the limitations of the TDCM such as low conversion efficiency, spectrum pollution, and hard-to-implement high-order modulation schemes can be solved. To reduce the leakage of spectral energy, two ways can be taken toward that goal. One is to optimize the EM responses of the TDCM element design to meet the requirement of the pulse shaping function. The other is to introduce the frequency-selective surface above the MTS as a bandpass filter to reject the signals out of band. Based on the integration of a tunable surface and a digital controlling module, the nonlinear nonreciprocal process is unfolded into successive active and digital stages, thus producing a series of magnet less nonreciprocal functions for linear-polarized EM waves using the same structure.

7.4.1.3 Non-Reciprocal Transmit-Receive MTS

Besides time modulation, MSs can also be time-space modulated to overcome the fundamental limitations of reciprocity. Besides in communication systems, the lifting of this limit potentially opens various possibilities in heat management and energy harvesting. While such reciprocity can be implemented using magnetic or nonlinear materials, in practice, such implementations have been limited to circulators and antennas. Besides breaking the reciprocity, it is also imperative that the nonreciprocity can be accurately controlled dynamically. This can be done by introducing spatiotemporal phase gradients by modulating the phase responses of MTS elements, as proposed in [88]. More specifically, space-time coding (STC) operates in a way that when a plane wave with a frequency f_1 is incident on a surface at an angle of θ_1, it will be anomalously reflected toward an angle θ_2 and frequency f_2. However, in time reversal scenario, when an incident plane wave with a frequency f_2 is incident on a surface with an angle θ_2, the reflected wave is directed toward an angle of $\theta_3(\neq\theta_1)$, with a frequency f_3 $(\neq f_1)$. Such a phenomenon breaks the time-reversal symmetry and the Lorentz reciprocity [89]. This is summarized in Figure 7.34 illustrates the non-reciprocal effects from this principle.

For instance, a STC MTS comprising 16 space-coding elements can be represented by the spacetime-coding matrix in Figure 7.35a. It features periodic time-coding sequences of length $L = 4$, and each coding element can produce a temporal phase gradient with steps of 90°. Shifting of the temporal sequences along the structure is done progressively with an interval of $T_0/4$ to enable two adjacent MTS to produce a 90° phase difference in every modulation interval. A spatiotemporal phase gradient distribution is then generated from this simple 2-bit space-time-coding scheme. This enables the conversion of an incident plane wave operating at a center frequency of f_c into the +1st harmonic frequency f_c+f_0, as shown in Figure 7.35b. Besides that, phase gradient at the +1st harmonic frequency shown in Figure 7.35c will enable steering of the reflected wave toward a specific direction. The incident wave can be converted to the +1st harmonic frequency with a reflection amplitude of 0.9, as seen from the calculated resulting spectrum distribution in Figure 7.35d. The gray bars in the same figure indicate

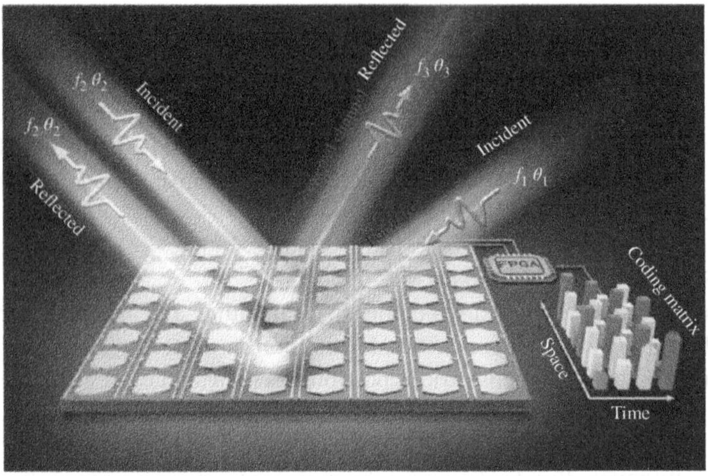

Figure 7.34 The non-reciprocal reflection based on a space-time-coding MTS. The forward and time reversal reflection scenarios. Source: Reproduced with permission from [88]/John Wiley & Sons.

results obtained by assuming an imperfect phase distribution, with phase steps among adjacent 2-bit coding states which are not exactly 90°. Nonetheless, these results are similar to the ideal 90°-step phase distribution, with weakly excited spurious harmonics.

To understand this more in-depth, some theoretical background is presented as follows. A time modulation of the reflection coefficient is applied onto a generic pth coding element with a periodic function $\Gamma_p(t) = \sum_{n=1,\ldots,L} \Gamma_p^n U_p^n(t)$, with $U_p^n(t)$ denoting a periodic impulse function with modulation period T_0 [90]. Therefore, each element is characterized by a time-coding sequence of length L, and the entire structure can be represented in terms of a space-time-coding matrix [90]. According to previously developed theory of space-time-coding MTS for modulation frequencies $f_0 = 1/T_0$ much smaller than the plane-wave frequency f_c, the time-domain scattered far-field can be approximated as

$$f(\theta, t) = \sum_{p=1}^{N} E_p(\theta)\Gamma_p(t)\exp\left[j\frac{2\pi}{\lambda_c}(p-1)d(\sin\theta + \sin\theta_i)\right] \tag{7.17}$$

where $E_p(\theta) = \cos\theta$ is the scattering pattern (approximated by a cosine function) of the pth coding element at the center frequency f_c, $\lambda_c = c/f_c$ is the center wavelength (with c denoting the speed of light in vacuum), and d is the period of the coding elements. By Fourier-series expansion of the reflection coefficient $\Gamma_p(t)$, the scattered

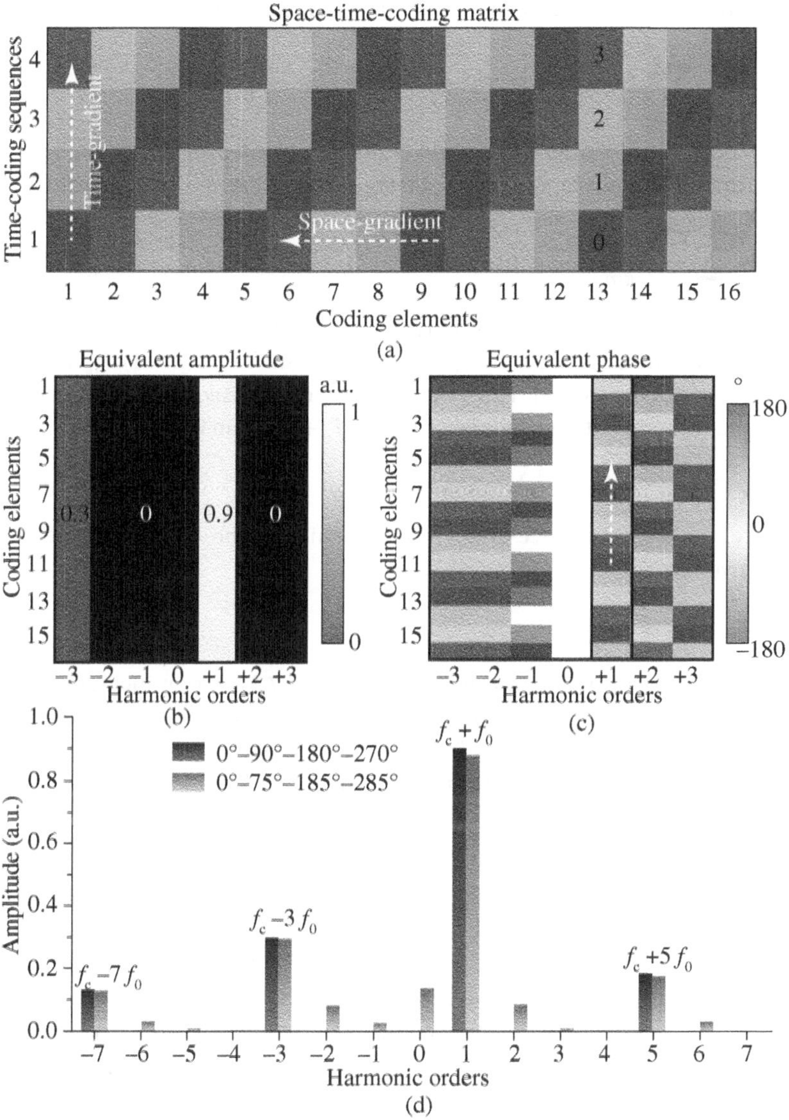

Figure 7.35 Space-time-coding matrix for realizing non-reciprocal reflection. (a) 2D space-time-coding matrix of dimension 16 × 4, in which the different grayscales represent the "0," "1," "2," and "3" digits, respectively, as marked in the plot. (b–c) Equivalent amplitude and phase distributions, respectively, for the coding elements at different harmonic frequencies. (d) Corresponding theoretical predictions for the spectral distributions of the harmonic orders (black bars). Gray bars are the corresponding values pertaining to an imperfect phase distribution with steps not exactly 90°, as reference. Source: Reproduced with permission from [88]/John Wiley & Sons.

far-field at the mth harmonic frequency $f_c + mf_0$ can be written as

$$F_m(\theta) = \sum_{p=1}^{N} E_p(\theta) a_p^m \exp\left[j2\pi(p-1)d\left(\frac{\sin\theta}{\lambda_r} + \frac{\sin\theta}{\lambda_c}\right)\right] \tag{7.18}$$

where $\lambda_r = c/(f_c + mf_0,)$ is the wavelength of the reflected waves corresponding to the mth harmonic frequency, and a_p^m are the Fourier coefficients of $\Gamma_p(t)$. From [90]

$$a_p^m = \sum_{n=1}^{L} \frac{\Gamma_p^n \sin(\pi m/L)}{\pi m} \exp[-j\pi m(2n-1)/L] \tag{7.19}$$

The time-coding sequence of the generic pth element in Figure 7.35a is derived from the periodic function $\Gamma_p(t)$ by applying a time shift $t_p = (p-1)\,T_0/4$, from basic Fourier-series theory [90]. Thus, the phase difference $\Delta\psi_m$ between adjacent coding elements at the mth harmonic frequency $f_c + mf_0$ can be written as

$$\Delta\psi_m = -2\pi m f_0 (t_{p+1} - t_p) = -\frac{m\pi}{2} \tag{7.20}$$

Therefore, the effective space-gradient phase shift for the +1st harmonic frequency is given by

$$\frac{\partial\psi}{\partial x} = \frac{\Delta\psi_1}{d} = -\frac{\pi}{2d} \tag{7.21}$$

Recalling the forward-incidence scenario in Figure 7.34, for an incident plane wave with frequency f_c and wavenumber $k = 2\pi f_c/c$, the dominant reflected order is at frequency $f_c + f_0$ and wavenumber $k + \Delta k = 2\pi(f_c + f_0)/c$. The relationship between the incidence (θ_1) and reflection (θ_2) angles is given by [89]

$$(k + \Delta k)\sin\theta_2 = k\sin\theta_1 + \frac{\partial\psi}{\partial x} \tag{7.22}$$

For the time-reversal scenario in Figure 7.34, by assuming an incident wave with frequency $f_c + f_0$ and wavenumber $k + \Delta k = 2\pi(f_c + f_0)/c$, the dominant reflected order is now at frequency $f_c + 2f_0$ and wavenumber $k + 2\Delta k = 2\pi(f_c + 2f_0)/c$, and the relationship between the incidence (θ_2) and reflection (θ_3) angles becomes

$$(k + 2\Delta k)\sin\theta_3 = (k + \Delta k)\sin\theta_2 - \frac{\partial\psi}{\partial x} \tag{7.23}$$

From Eqs. (7.22) and (7.23), the reflection angles in the forward and time-reversal scenarios (θ_2 and θ_3, respectively) are found to satisfy the following relationships

$$\sin\theta_2 = \frac{k\sin\theta_1 + (\partial\psi/\partial x)}{k + \Delta k} = \frac{\sin\theta_1 - (\lambda_c/4d)}{1 + (f_0/f_c)} \tag{7.24}$$

$$\sin\theta_3 = \frac{k}{k + 2\Delta k}\sin\theta_1 = \frac{\sin\theta_1}{1 + (2f_0/f_c)} \tag{7.25}$$

To quantify the angular separation between the time-reversed reflected and incident waves, a deviation factor is defined as follows:

$$\delta = |\sin\theta_3 - \sin\theta_1| = \frac{\sin\theta_1}{1 + (f_c/2f_0)} \tag{7.26}$$

which is always nonzero for oblique incidence. In summary, the reflected wave in the time-reversal scenario does not propagate along the same direction as the original oblique incident wave. This also comes along with a frequency shift of $2f_0$, thereby breaking the time-reversal symmetry and Lorentz reciprocity in both the space and frequency domains. Eq. (7.26) indicates that the angular separation increases with the ratio f_0/f_c and the incidence angle θ_1. If the modulation frequency f_0 is much smaller than the center operational frequency f_c, this spatial separation may not be observable in practical experiments.

By increasing the frequency ratio f_0/f_c (where modulation frequency is $f_0 = 1/T_0$) and the incidence angle θ_1 to suitable levels, the angular separation may become in the order of few degrees. It is worth noting that when the space-time-coding MTS is illuminated by a normally incident wave ($\theta_1 = 0°$), in the forward scenario, the non-reciprocal effect only exists in the frequency domain. The reflection angle θ_3 in the time-reversal scenarios is then $\theta_3 = \theta_1 = 0°$, indicating the nonexistence of spatial separation between the time-reversed reflected and incident waves. In summary, the proposed concept enriches the features of wave manipulation in digital MTS. The completely reconfigurable and programmable feature using FPGA enables new functionalities in software-defined metamaterials, which can be applicable in nonreciprocal wireless and radar systems with separated transmitting and receiving channels.

Another example [91] is proposed as a digital nonlinear MTS with non-reciprocal properties which can be customized, as illustrated in Figure 7.36. The center figure indicates that the proposed non-reciprocal MTS consists of a tunable transmissive surface integrated with two EM detectors and a digital controlling module. The digital module consists of an FPGA and two analog–digital–analog (ADA) modules. Each of the latter includes an analog-to-digital (AD) submodule and a digital-to-analog (DA) submodule. Figure 7.36f divides the non-reciprocal process into several successive active and digital steps which enables it to be defined digitally and customized flexibly. Conventional non-reciprocal process of nonlinear metamaterial is difficult to reconfigure. The EM waves with opposite directions can be captured by the two detectors, respectively, which intensities are then converted into analog DC signals (front DC and back DC in the figure). These DC signals are then channeled into the front and back AD submodules, respectively, which generate the digital signals that carry the front and back intensity information.

The transmission features of five cases of customized nonlinear nonreciprocal can be considered for linearly-polarized EM waves: EM diode functions with

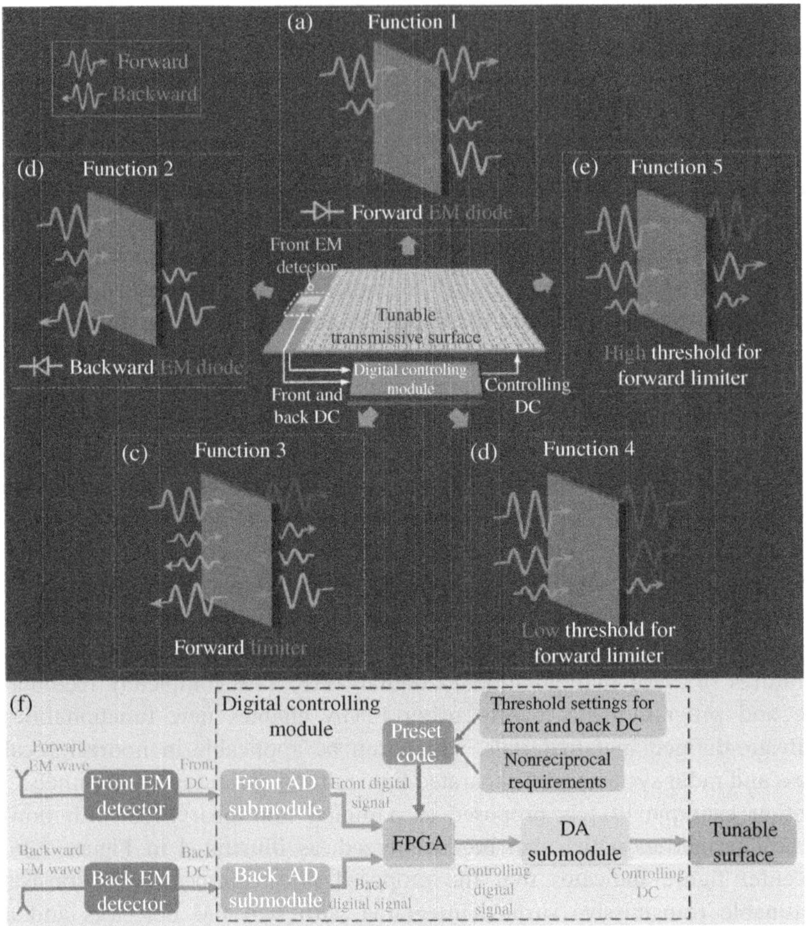

Figure 7.36 (a–e) Five customized nonreciprocal functions, which are preloaded in the module and can be dynamically switched. EM diode functions are plotted with (a) forward and (b) backward directions. (c) EM unidirectional limiting function for forward waves. Threshold tunability is proved in insets (d) and (e), where backward transmissive propagations are omitted. (f) The digital non-reciprocal process. Source: Reproduced with permission from [91]/John Wiley& Sons.

reversible directions Figure 7.36a, b, unidirectional limiting function Figure 7.36c, and tunable thresholds Figure 7.36d, e, which are integrated into the shared MTS and is dynamically switchable. The possible hysteresis issue faced by some nonlinear nonreciprocal MTS is also avoided due to the use of the digital nonlinear principle. This results in predictable and precise EM responses. A prototype consisting of 29 × 29 unit cells and the EM detectors is fabricated on the same

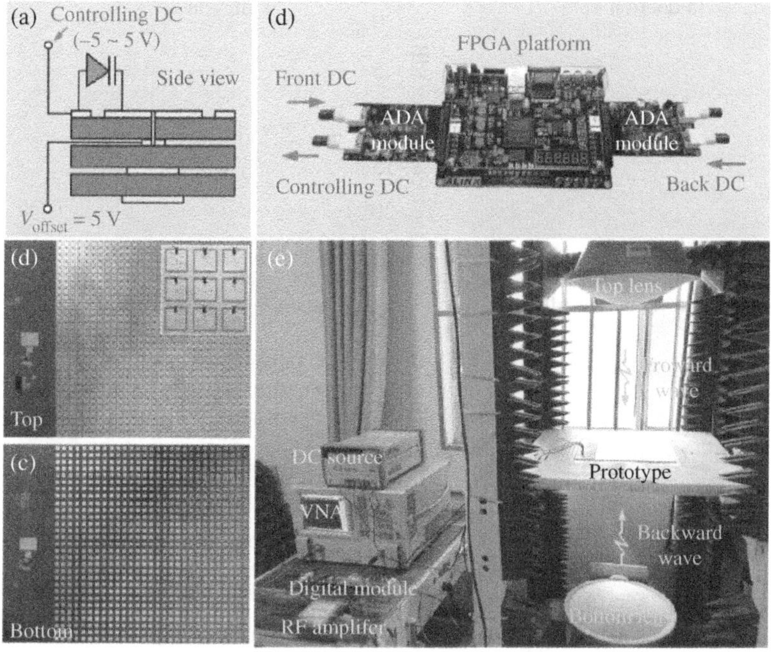

Figure 7.37 (a) Side view of a particle of the tunable surface, indicating the connections of the constant offset voltage, V_{offset} and the controlling DC. (b, c) The MTS prototype with zoom-in particles in the upper right inset. (d) Digital controlling module. (e) Measurement setup. Source: Reproduced with permission from [91]/John Wiley & Sons.

panel, as shown in Figure 7.37b, c (enlarged view of the layer inset, upper right). Measurement of the nonreciprocal transmission coefficients is performed in the focusing lens antenna system, as shown in Figure 7.37e. The control of the nonreciprocal transmission coefficients is done according to the incident intensities and nonreciprocal function requirements via the digital module.

RF amplifier is not used in the linear measurement; thus, the transmission coefficient is obtained using

$$T = S_{21p} - S_{21\text{air}} \tag{7.27}$$

where S_{21p} is the transmission coefficient in the presence of the sample, and $S_{21\text{air}}$ is the coefficient when the sample is removed (through air).

The normalized measurement results from 5.6 to 5.8 GHz are depicted in Figure 7.38a. It is observed that the transmission coefficients are below −17 dB at 5.75 GHz when the prototype is illuminated by backward waves only. With the excitation switched to the front side, the MTS still operates in opaque mode. The transmission starts upon approaching an input intensity of about 12 dBm,

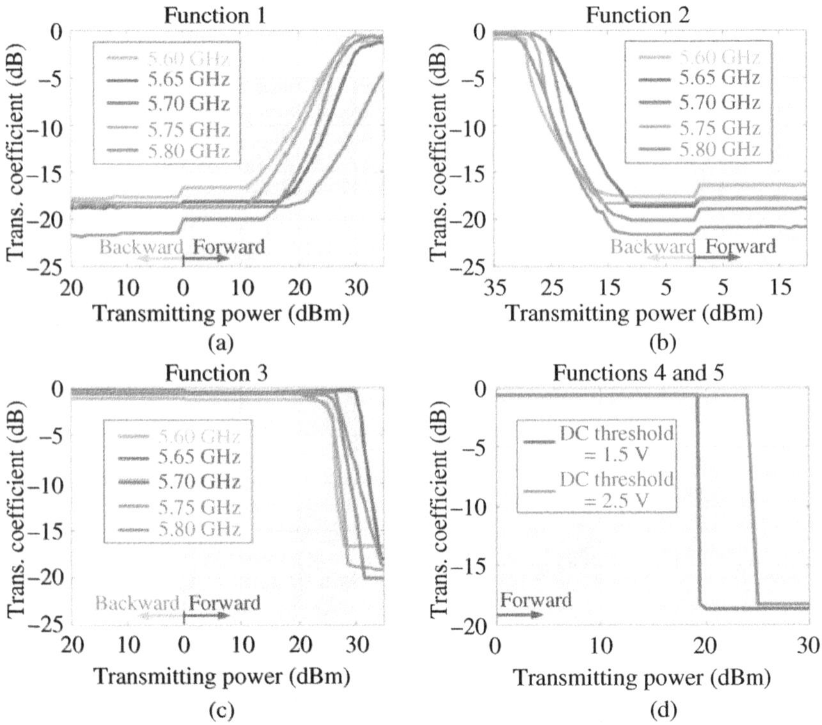

Figure 7.38 Normalized measured transmission performances of the MTS under forward or backward illuminations with varying intensities. (a) Function 1: forward EM diode. Backward waves are blocked, while forward waves experience rising transmission coefficients when the intensity is larger than the threshold. (b) Function 2: backward EM diode. The mirrored symmetry of Functions 1 and 2 proves the direction reversibility of the nonreciprocity. (c) Function 3: EM unidirectional limiting for forward waves. For backward waves and low-intensity forward waves, the MTS works in transparent mode. The transmission reduces when the forward waves are stronger than the threshold. (d) Functions 4 and 5 demonstrate the ability to tune the threshold that triggers the nonlinear and nonreciprocity. Two thresholds are set, which can be clearly observed as the different tuning points of the transmission curves. Backward transparent performances are omitted here. The above functions are preloaded in the digital module and can be switched simply by pressing keys on the FPGA board. Source: Reproduced with permission from [91]/John Wiley & Sons.

with a maximum of −0.6 dB when the intensity is higher than 31 dBm. This result validated the one-way transmission property of the EM diode. Some discrepancies are noticed between the curves, which are caused by the performance variations of the EM detectors and tunable surfaces at different frequencies. On the other hand, a mirrored symmetry can be seen when comparing the normalized transmission results presented in Figure 7.38a, b. At 5.7 GHz, small differences can be observed, which can be attributed to inconsistency between the two detectors during the fabrication.

Figure 7.38c shows the normalized measurement results under forward or backward illumination at a time. It can be observed that the transmission coefficients for backward waves are close to 0 dB. On the contrary, the forward transmission remains high until it abruptly drops when the intensity approaches 20 dBm. This is caused by the nonlinear feature of the varactors, whose capacitance change becomes increasingly sharper as the reverse biasing voltage decreases. Conversely, based on the reversibility, the limiting function can also be realized for the backward waves by switching the roles of the front and back DC in the code. The measured forward transmissions at 5.75 GHz illustrated in Figure 7.38d show the weak transparent performance for forward waves. The abrupt changes in the transmissions happen at different points of forward intensity. This reflects the different preset thresholds that trigger its nonreciprocity. On the other hand, the transparent transmissions for backward waves are omitted due to the identical function with the transparent transmission for the forward waves.

In summary, the nonreciprocity of the receiving and transmitting waves (with angle and frequency) generated by such spatiotemporal modulation scheme of MTS elements can potentially further improve MIMO communication channels in terms of diversity and spectral efficiency. This is especially true if multiplexing of different MIMO channels can be done within the same (narrow) frequency band by proper segmentation of the T/R paths in the frequency and angular domain. Despite that, this method comes with more complex design considerations and similarly more complex hardware, with the potential benefit of realizing a new type of MIMO communication scheme with a reduced number of RF chains.

7.5 Reflective Intelligent Surfaces

7.5.1 Introduction

RISs, also known as intelligent reflecting surfaces (IRS) are referred to as two-dimensional structures that can dynamically manipulate the impinging electromagnetic waves. RIS has unique features to potentially be incorporated into the infrastructure of the current communication systems or deployed into the wireless environments' without requiring change to its workflow. Since RIS composes many low-cost, passively reflecting elements, it can manipulate the electromagnetic propagation environment according to its dynamics. This is by properly adjusting phase shifts and/or amplitudes for each reflecting element via an external control channel. Advantages for RIS implementation include improvements to the signal-to-interference-noise ratio (SINR), enhanced

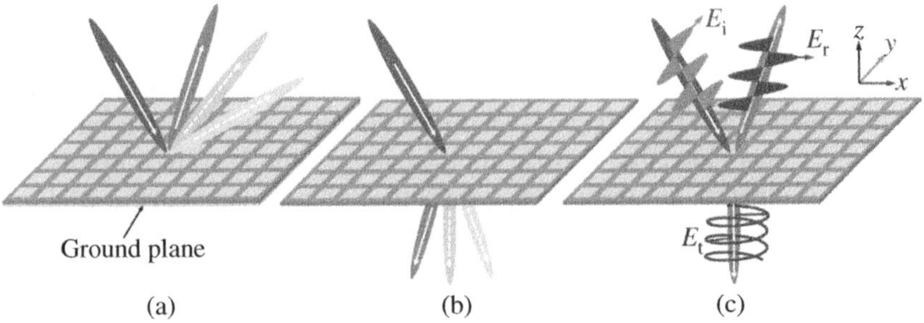

Figure 7.39 Schematic drawings of (a) reflecting RIS, (b) transmitting RIS, and (c) RIS working as a polarization converter. Source: [16]/IEEE/Public Domain.

quality of service (QoS) of the link, and reduced energy consumption and cost of additional base stations/repeaters [92]. Besides hardware, it is important to develop a channel estimation algorithm for the RIS-assisted wireless communication system. As reliable beamforming requires accurate channel state information (CSI), the amount of phase shifts can be intelligently updated by the RIS controller to achieve beamforming [93]. To achieve phase/amplitude tunability in unit cells to control the incident waves, various mechanisms can be implemented on RISs to enable phase modulation. These unit cells can be modified and programmed in response to external stimuli originating from voltages, thermal, optical, or mechanical. Most importantly, reconfigurable reflecting, transmitting, and polarization conversion surfaces (illustrated in Figure 7.39), can be considered an RIS if the external stimuli can be controlled using artificial intelligence to improve overall system performance.

7.5.2 System Architecture

Joint RIS reflection coefficient and transmit covariance matrix optimization can be used in maximizing the capacity of a point-to-point IRS-aided MIMO system illustrated in Figure 7.40 [94]. A MIMO communication system with $N_t \geq 1$ antenna at the transmitter and $N_r \geq 1$ antenna at the receiver can be integrated with an RIS/IRS equipped with M passive reflecting elements deployed to enhance the MIMO communication performance. Under frequency-flat channels, an alternating optimization algorithm can be used to find a locally optimal solution. This is by iteratively optimizing one optimization variable with the others being fixed so that the optimal solutions are derived in closed form. By designing the IRS reflection coefficients, the IRS-aided MIMO channel can be significantly improved in terms of channel power, rank, or condition number, thereby leading to enhanced capacity. These intelligent radio surfaces implemented with antenna

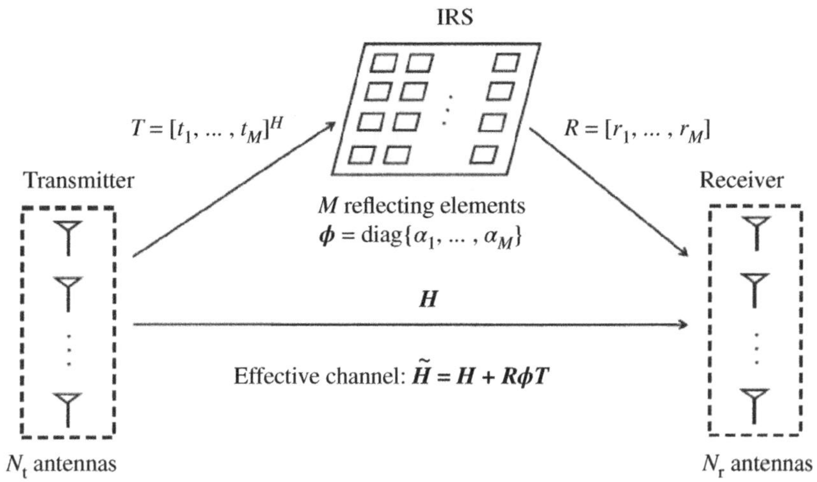

Figure 7.40 Diagram of an IRS-aided MIMO communication system. Source: Zhang and Zhang [94]. © 2020 IEEE.

arrays or MTS can be validated further via channel measurements to evaluate the potential contribution in terms of coverage extension and channel capacity improvement.

An IRS-assisted mmWave MIMO system containing a single BS, a single IRS and a single user is shown in Figure 7.41. More specifically, the IRS is deployed to provide a *virtual* LOS link for the user who is blocked by the obstacles. Both BS and user are assumed to be equipped with an analog antenna architecture consisting of a full-connection uniform linear array, where the numbers of antennas are N_{BS} and N_{UE}, respectively. The IRS is equipped with a planar array consisting of N_I passive reflective elements whose amplitude and phase can be adjusted dynamically via the IRS controller. The geometric channel model is adopted to characterize the channels for the considered IRS-assisted mmWave MIMO communications, and the channel matrix $H \in \mathbb{C}^{N_R \times N_T}$ is expressed as [95].

$$H = \sqrt{\frac{N_T N_R}{L}} \sum_{l=1}^{L} \alpha_l a_R(N_R, \phi_l) a_T(N_T, \Omega_l)^H \tag{7.28}$$

where N_T and N_R denotes the number of antennas at transmitter and receiver, respectively. L is the number of paths, and α_l is the channel complex gain of the lth path. For convenience, $\phi = \cos(\theta)$ and $\Omega = \cos(\phi)$ are called angle-of-arrival (AoA) and angle-of departure (AoD), respectively, $a_R(\cdot)$ and $a_T(\cdot)$ are the steering vectors at receiver and transmitter, respectively. For the considered IRS-assisted

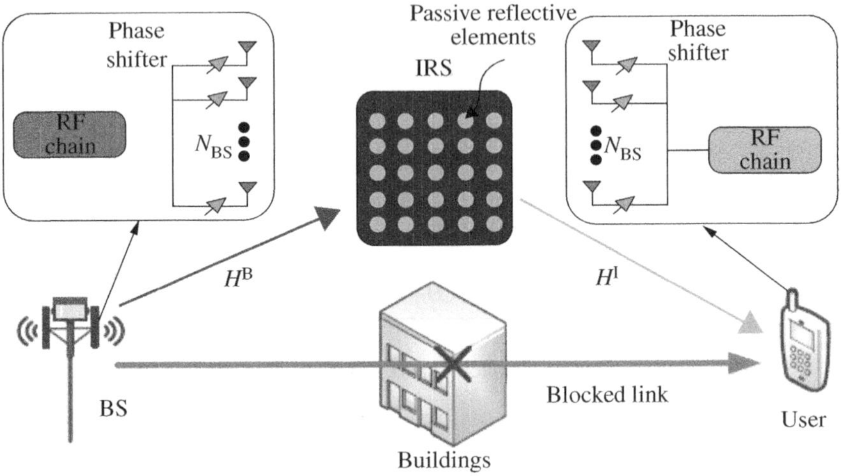

Figure 7.41 Illustration of the considered IRS-assisted mmWave MIMO system. Source: [94]/IEEE/Public Domain.

mmWave system, $H^B \in \mathbb{C}^{N_I \times N_{BS}}$ denotes the channel from BS to IRS, $H^I \in \mathbb{C}^{N_{UE} \times N_I}$ denotes the channel from IRS to UE. Assuming that the direct link from BS to UE is blocked, the cascaded channel is obtained [96].

Furthermore, a cell-free RIS-assisted massive MIMO framework can also be used for secure communications, as proposed in [97]. The access point (AP) was established with low-resolution ADCs. On the other hand, the work in [98] combines IRS and MIMO technologies, and develops a cooperative beam training scheme to facilitate the channel estimation in a multiuser analog digital hybrid IRS-assisted THz MIMO system. In particular, two different hierarchical codebooks for the proposed training procedure are designed. Based on the training results, two hybrid beamforming (HBF) designs for both single-user and multiuser scenarios were also proposed. To reduce the computational complexity of the real-time implementations, an efficient ternary-tree search is proposed for the BS and users, which is proved to be more efficient compared to the binary-tree search. The results are presented in terms of the sum rate showing a small difference between the system performance with perfect and imperfect CSI when employing the proposed procedure for channel estimation.

7.5.3 RIS-Based Wireless Communication Prototypes

For the hardware, two main practical implementations of RIS structures can be either in the form of arrays, with unit cells spaced at least a quarter-wavelength apart or in the form of MTS (unit cells spaced about a tenth of a wavelength apart). In essence, RIS should incorporate controllers which can dynamically tune phase and/or amplitude profile in each reflecting unit cell, thus the term "reconfigurable." These unit cells can be tuned using different methods, which are also explained in the following subsections.

In terms of hardware performance, one of the most significant parameters to evaluate an RIS is the efficiency (η), defined as the ratio between the power reflected/transmitted to a desired direction and the total incident power. Other key parameters of an RIS are its steering range and angular resolution. The steering range depends greatly on the technique: e.g. mechanical steering with (partially) movable components may provide a wide steering range up to 360°, whereas an electronic approach typically has a more limited steering range at its operating band, e.g. around 120° [99]. Meanwhile, the angular resolution mainly depends on the main lobe beamwidth, which depends on the number of phase quantization bits and the number of unit cells. The quantization depends on the number of integrated switches. On the other hand, the number of unit cells has an impact on the aperture size and periodicity of the RIS: with the same aperture, switching the periodicity from sub-wavelength ($\lambda/8$, similar to MTS) to $\lambda/2$ (similar to reflect arrays) greatly reduces directivity [100]. Finally, the power consumption of a RIS is also an important consideration. This includes power to the controller and power consumed for each actuator on a unit cell (e.g. transistors, diodes, heaters, pumps, and motors). For active RISs, significant amount of power may also be needed to amplify the incident signal prior to reconfiguration and reflection/transmission. In this section, RIS-based communication prototypes are described, with an overview of the important tuning technologies to enable phase and/or amplitude changes in the unit cell (and consequently the full reflector array). A more comprehensive review regarding RIS hardware in the sub-THz band is also available from [16].

7.5.3.1 Electronic-Based RIS Prototypes

The first example presented in [101] demonstrates a large-scale programmable MTS using arrays of CMOS-based chip tiles at 300 GHz. Each unit cell is a split-ring resonator which consists of 8 transistors working as switches that enable programming the unit cell to realize 84 different structures, thereby changing the amplitude and phase response. Moreover, eight sub-wavelength inductive loops are added to each unit cell, enabling the switches to operate beyond their cutoff frequency, reaching up to 5 GHz. A single chip tile features 12 × 12 unit cells. The final MTS structure consists of a 2 × 2 array of tiled chips with a total of 576-unit cells fabricated in a 65-nm industry-standard CMOS process. The unit cells are individually addressable and are digitally programmable with 8 bits of control at GHz

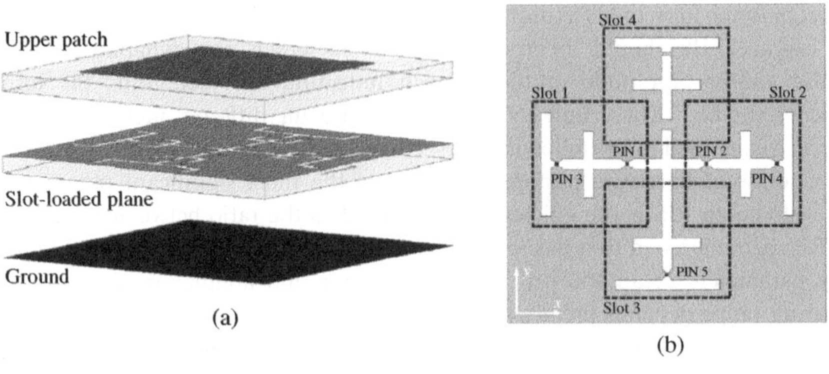

Figure 7.42 Structure of the proposed 2-bit RIS element: (a) exploded view; (b) detailed view of the slot-loaded plane. Source: [102]/IEEE/Public Domain.

speed. In addition to that, both the amplitude and phase response of each unit cell can be independently controlled, allowing a beam steering angle of 30° on each side.

The next RIS with 2-bit elements operating in the S band with a center frequency of 2.3 GHz is depicted in Figure 7.42a, with each antenna element consisting of a square-shaped upper patch, a slot-loaded plane and a ground plane [102]. The upper patch receives and radiates energy, while the ground plane suppresses the back radiation. Radiation occurs through the slots in the slot-loaded plane, the key component controlling the RIS's phase state, as shown in Figure 7.42b. The four sets of slots are positioned symmetrically and integrated with five PIN diodes to ideally provide four quantized phase states (2-bits) with a 90° phase increment. The PIN diode states are combined to control the RF current paths and the resonant lengths of the slots. The tunable phase shifts for the RIS element design are explained specifically as follows. Slots 1, 2 and 3 are arranged in a T-shaped configuration, being complemented by a dummy Slot 4 for maintaining a symmetric element structure. The states of PIN 1 and PIN 2 are alternatively turned ON or OFF, so that the currents induced may be reversed in the x-direction, corresponding to a 180° phase shift. The other three PIN diodes are turned ON or OFF simultaneously to change the resonant lengths of the slots, allowing an additional 90° phase shift, hence realizing the 2-bit phase resolution.

The upper patch has a size of 37 mm × 37 mm etched on a 1-mm thick FR4 substrate. The slot-loaded plane is etched on another FR4 substrate placed 6 mm below the upper patch. The ground is made of an aluminum sheet, placed 12 mm below the slot-loaded plane. The type of the commercial PIN diode is SMP1340-040LF from Skyworks. The ON state of the PIN diode is modeled as a series of $R = 0.8\,\Omega$ lumped resistors and $L = 780\,\text{pH}$ inductors. The OFF state can be modeled as a series of $C = 202\,\text{fF}$ lumped capacitors, $R = 10\,\Omega$ resistors

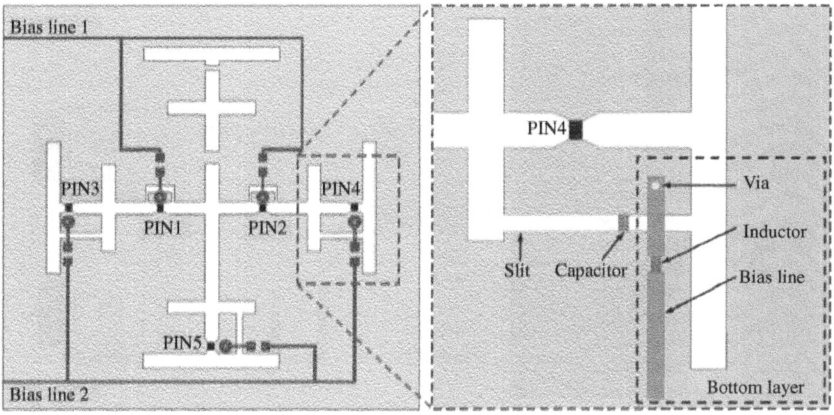

Figure 7.43 Layout of the DC bias network of the 2-bit RIS element. Source: [102]/IEEE/Public Domain.

and $L = 780\,\text{pH}$ inductors. The DC bias network of the element is depicted in Figure 7.43. The five PIN diodes are divided into two groups, which are then independently controlled by a pair of bias lines. The forward and reverse DC voltages are +0.9 and −0.9 V, respectively.

Figure 7.44a shows the design and fabrication of the RIS bearing 16 × 16 2-bit elements. The size of the surface is 800 mm × 800 mm, and the distance between the primary feed and the surface is 720 mm. Upon being illuminated by the primary feed, these 16 × 16 elements can be dynamically reconfigured to convert the spherical wavefront impinging from the feed into a planar wavefront in the desired direction. Hence, the designed RIS produces a focused high gain beam capable of promptly switching its direction within a two-dimensional ±60° angular range. As shown in Figure 7.44b, the RIS-based wireless communication system prototype consists of the base station side, including the transmitter host, the USRP at the transmitter and the RIS having 256 2-bit elements. The user side includes the receiver antenna, the USRP at the receiver and the receiver host. At the base station side, a graphical interface is realized at the transmitter host. The interface is responsible for controlling the parameters at the transmitter, including the carrier frequency, transmit power, modulation and encoding modes, and more. The signals are then delivered to the USRP at the transmitter via the transmitter host. Once the signals from the transmitter host have been received, the USRP at the transmitter carries out the signal processing, which aims to transform the signals into a form suitable for transmission in wireless channels. After that, the processed signals are forwarded to the RIS having 256 2-bit elements. At the user side, the signals are received from the wireless channel, which are then processed by the USRP at the receiver. The USRP at the

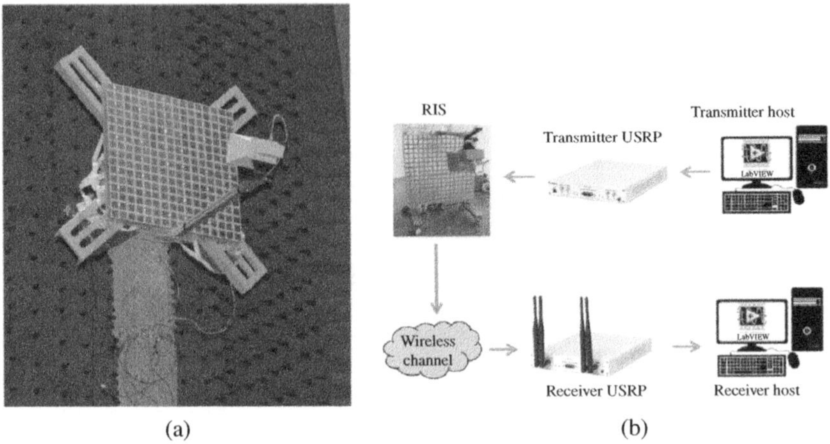

Figure 7.44 (a) The fabricated RIS with 16 × 16 2-bit elements. (b) The RIS-based wireless communication prototype. Source: Reproduced with permission from [102]/IEEE.

receiver processes the signal for recovery, which is an inverse process of those in the USRP at the transmitter. Finally, the recovered signals and the corresponding parameters, such as the received signal power, constellation, bit-error-rate (BER), data rates, and so on, are displayed by the graphical interface at the receiver host.

The USRP at the transmitter consists of four modules such as the embedded processor, the high-speed FPGA processor, the DAC module, and the RF module. The embedded processor and the high-speed FPGA processor jointly perform the baseband signal processing. The embedded processor handles the media access control (MAC) layer process, such as data framing. The high-speed FPGA module handles the physical layer signal processing, such as channel coding and orthogonal frequency division multiplexing (OFDM) modulation. The DAC module converts the digital signals output by the FPGA processor to analog form. The RF module up-converts the required for RF signal transmission. Similarly, the USRP at the receiver also consists of four modules: the embedded processor, the high-speed FPGA processor, an ADC module and a RF module. The baseband signal processing at the receiver is the inverse procedure of that at the transmitter. Down-conversion of the received RF signal is done by the RF module, whereas the ADC module convers the analog signals obtained from down-conversion into digital format. The signals can be obtained from diverse data sources, such as text, images, videos and so on. To achieve efficient and robust wireless communications, a series of signal preprocessing operations have to be done.

Figure 7.45a shows the normalized measured radiation patterns of the broadside beam at 2.3 GHz. High-gain pencil beams are formed by controlling the phase shifts of the 2-bit RIS elements, with half-power beamwidths of 9.1° and 8.8° in

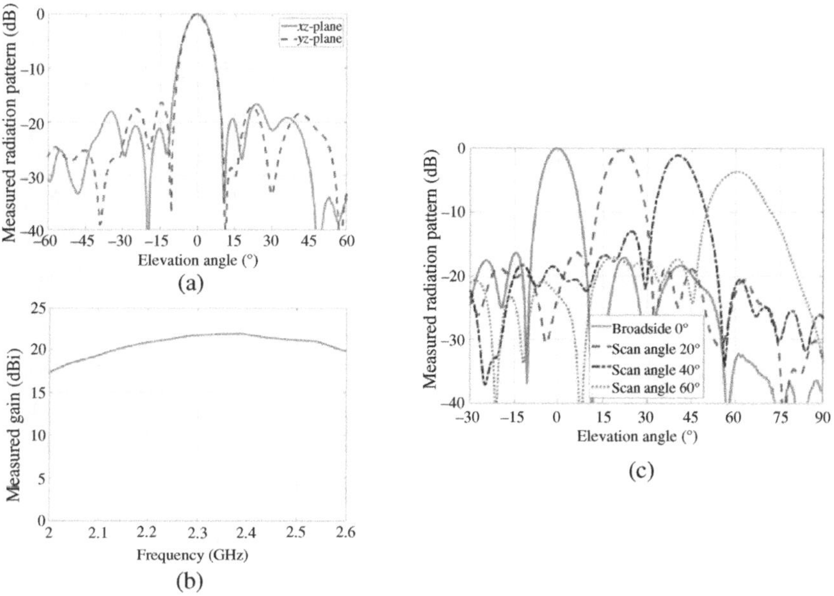

Figure 7.45 Measured performance of the broadside beam: (a) Normalized radiation patterns at 2.3 GHz. (b) Antenna gains within the frequency band of interest (from 2 to 2.6 GHz). (c) Measured radiation patterns of the scanned beams. They are normalized to the measured gain of the broadside beam. Source: [102]/IEEE/Public Domain.

the two principal planes, respectively. The measured antenna gains within the frequency band of interest are illustrated in Figure 7.45b, with measured sidelobe levels of −16.7 and −16.4 dB. The measured gain is between 21.7 and 21.9 dBi, which corresponds to an aperture efficiency of about 31.3%, whereas the 1-dB gain bandwidth is 350 MHz (15.2% at 2.3 GHz). Beamscanning from 0° to 60° can be achieved by the control board, as seen in the measured patterns plotted in Figure 7.45c (normalized to broadside). The increasing scanning angle decreases the measured gain while broadening the main beam. For instance, a measured gain of 18.0 dBi is produced when scanned at 60°, indicating a scanning gain reduction of 3.7 dB. It is worth noting that a 21.7 dBi antenna gain can be achieved using a conventional phased array with 64 elements.

7.5.3.2 Phase Change Material-Based RIS Prototypes

Another proven method for phase modulation in RISs is by using phase change materials (PCMs). Examples of such materials include liquid crystal, vanadium oxide (VO_2) and Germanium telluride (GeTe). Suitable control methods have shown capabilities in modulating the responses of these surfaces through thermal, optical, or electrical stimuli, with their own respective advantages. A chalcogenide

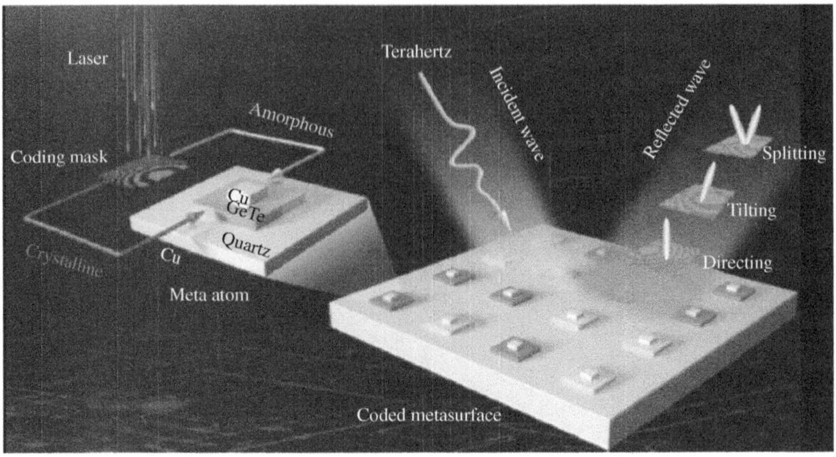

Figure 7.46 The optical-based coding MTS for multifunctional control of terahertz beam: The coding MTS is composed of meta-atoms based on *GeTe* phase change material. Source: Reproduced with permission from [103]/John Wiley & Sons.

PCM is able to perform conversion reversibly between two stable binary states: amorphous (insulator) and crystalline (conductive). Their fast phase change response, broadband responses, semiconductor-compatibility integration, and nonvolatility with zero hold power are suitable features to realize these reflectors [103]. This is schematically represented in Figure 7.46.

Particularly for GeTe compositions, these demonstrations include efficient manipulation of THz waves using split-ring resonators-based metamaterials [104] or the use of multilevel nonvolatile resonant states in metamaterial structures [105]. Phase change of the material from the amorphous to its crystalline state can either be done using thermally or electrically. Selective laser activation using a properly designed coding mask enables selective reconfiguration of the between the crystalline (conductive) and amorphous (insulating) states. This then allows multifunctional control of waves via the specifically patterned MTS by changing the permittivity (63) and conductivity (10 S/m) of the amorphous *GeTe* to its crystalline (permittivity of 1 and conductivity of 410,000 S/m), as shown in Figure 7.47a.

Each meta-atom then can change between two-state coding element: "0" for the *GeTe* in the amorphous phase and "1" for *GeTe* in the crystalline state (Figure 7.47b). In particular, the dimensions and design of the meta-atom are chosen for providing distinct responses for the "0" and "1" states (reflected amplitude and phase) to an incident wave (Figure 7.47c, d) to obtain coding states with 180° (π) phase difference.

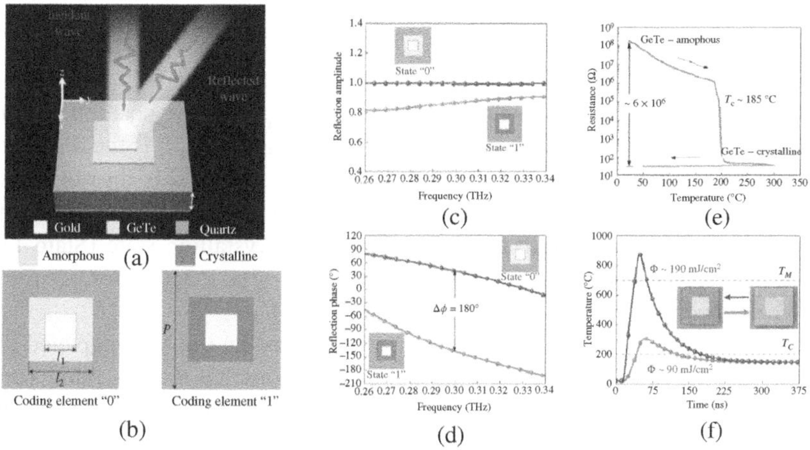

Figure 7.47 (a) Structure of the meta-atom coding element with a 1-μm thick gold squares on top of a 1-μm thick square *GeTe* built on a SiO_2 substrates. (b) The two states of the coding element: state "0" indicates the amorphous (insulating) state, state "1" indicates the crystalline (conductive) state. (c) Amplitude variation and (d) phase variation of the coding element in the two states. (e) Variation of the surface resistance of a *GeTe* layer vs temperature (2-probe measurement) between the two states and the non-volatile behavior of the crystalline state upon cooling to room temperature. (f) Finite-element simulations of the temperature rise of the *GeTe* layer upon single-pulse UV laser irradiation with different energy densities. T_C, crystallization temperature; T_M, melting temperature. Source: Reproduced with permission from [103]/John Wiley & Sons.

The MTS consists of *GeTe* square unit cells, each sized at 229 μm × 229 μm which are 1 μm thick. They are fabricated on a quartz substrate ($\varepsilon_r = 3.8$, thickness = 100 μm). On top of each *GeTe* patches, square metal patches made using titanium/gold bilayer (40/1000 nm-thick) sized at 107 μm × 107 μm are placed. A 1-μm thick metal Ti/Au bilayer is fabricated beneath the quartz layer and is acting as the MTS ground plane. Figure 7.47e shows the change in DC resistance between the amorphous (insulating) and crystalline (conductive) states of *GeTe* layer, which is more than six orders of magnitude. Besides that, a transition temperature of around 185 °C to the crystalline state is observed. Note that the material retains its conductive behavior (non-volatility of states) upon cooling to room temperature. The *GeTe* is irradiated with optical energies using single laser pulses (248 nm and ~35 ns of pulse lengths). Simulations of laser-beam and GeTe material interactions illustrated in Figure 7.47f show that a 90 mJ/cm^2 single laser pulse increases the temperature of the *GeTe* layer above its crystallization temperature, T_C. This will turn the material from its amorphous to its conducting state. Conversely, a subsequent 190-mJ/cm^2 laser pulse will change the crystalline *GeTe* material back to the amorphous state on a fast timescale. This is done above its local melting temperature, T_M, preventing its re-crystallization. The needed phase compensation at the unit cell level is determined by $\varphi_e = k_0(|\vec{r}_e - \vec{r}_f| - F - \hat{u}\vec{r}_e)$, where k_0 is the free-space wavenumber and $\vec{r}_e = (x_e, y_e, 0)$ is the position vector of the unit cell, $\vec{r}_f$ is the position vector of the feeding antenna, F is the distance between the incident feeding source and the center of MTS, $\hat{u}$ is the vector of the reflected main beam direction [106]. Therefore, to generate phases between $131.5° \leq \varphi_e < 311.5°$, the *GeTe* film must be in the crystalline state (State 1), whereas for phases between $0 \leq \varphi_e < 131.5°$ and $311.5° \leq \varphi_e < 360°$, the *GeTe* film of unit cell needs to be in the amorphous state (State 0).

The fabricated device in its initial state is represented by the MTS depicted in Figure 7.48a, where all *GeTe* patterns are in the amorphous State 0. The coding pattern/function of the MTS was determined by the laser irradiation patterns based on Figure 7.47f using specific proximity contact mask on the *GeTe* layer. Figure 7.48 illustrates the simulated and measured radiation patterns of different functions for each specific coding MTS. The simulated EM excitation, its associated coded pattern, the fabricated prototype, and the measured radiation patterns in the *yz*-plane are represented from left to right of this figure. Meanwhile, each row represents different coding states. Figure 7.48a indicates that when the unit cells are all in the amorphous State 0), the incident wave is reflected at $\varphi = 90°$ when radiated at $\theta_{rad} = 30°$ using a horn antenna. The placement of the coding mask on top of this MTS followed by laser irradiation onto specific *GeTe*-based unit cells result in the change of the phase center to π or their crystalline State 1.

This phase control technique enables the manipulation of an incident beam within a wide bandwidth, as illustrated in Figure 7.49. Measurements indicated that the reflected beam angle using function 1 varies from $-3°$ to $+2°$ from 0.275 to 0.315 THz, whereas for function 2, the measured reflected varies from $+38°$ to $+42°$. On the other hand, the reflected beam angle for function 3 varies from $+30°$ to $+34°$ and the two reflected beam angles resulting from function 4 varies from $+28°$ to $+32°$ and $-34°$ to $-30°$. This indicates stability within the measured radiation patterns and gains within the bandwidth of interest, validating the capability of the proposed MTS functionality using coding pattern reconfiguration.

Similar to the principles of *GeTe*, a digitally coded MTS can also be used to manipulate incident waves using electrically tuned liquid crystals (LC) molecules [107]. This time, contrast to the reflective type presented previously, the MTS operate in the transmissive mode, as schematically illustrated in Figure 7.50. Different types of beams can be generated based on different coding of the MTS using the LC principles shown in the bottom right corner. The orientation of the LC molecules (represented by elliptical rods) is changed by applying different voltages to the two metallic layers. Similar to the examples presented in the previous section on space and/or time coded MTS, the ability of generate multiple beams and more

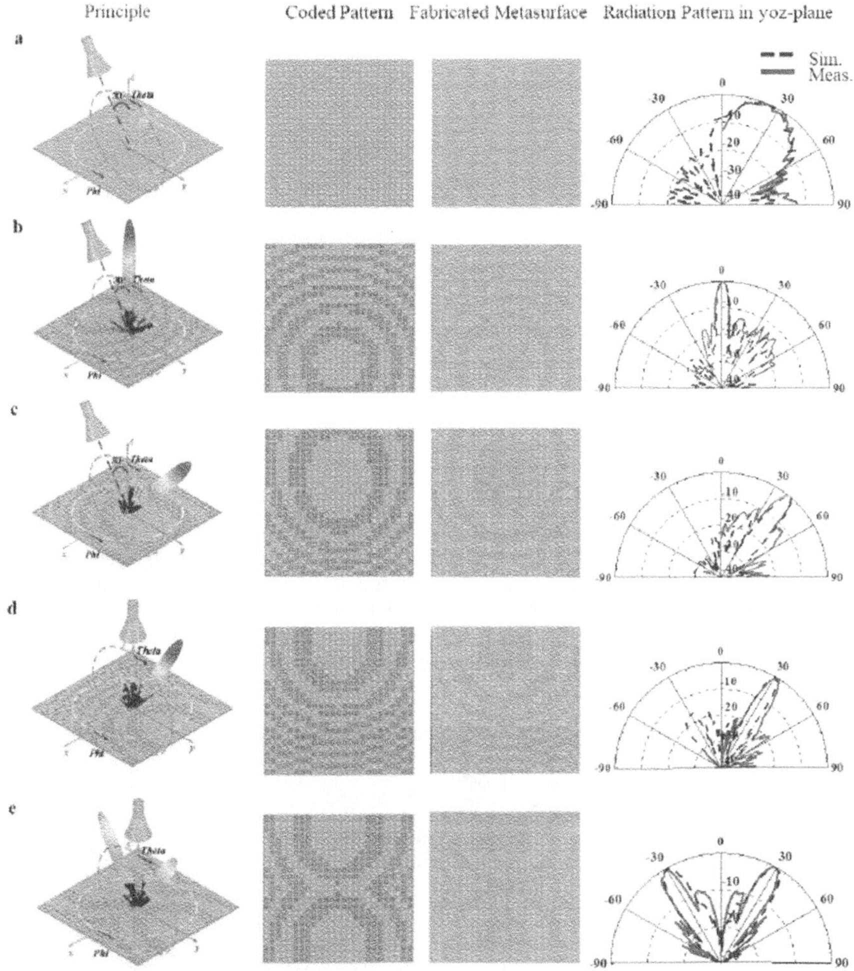

Figure 7.48 Simulated and measured results of the reflected beam of the coding MTS. The left row shows the MTS excitation by a horn antenna and the simulated reflected beam, the middle row shows the specific coding MTS (simulated and fabricated), and the right row compares the simulated and measured normalized far field of the reflected wave. (a) Code function 0 (uncoded MTS), (b) code function 1 (reflected beam at $\varphi = 0°$ when fed at $\theta_{rad} = 30°$), (c) code function 2 (reflected beam at $\varphi = 40°$ when fed at $\theta_{rad} = 30°$), (d) code function 3 (reflected beam at $\varphi = 30°$ when fed at $\theta_{rad} = 0°$), (e) code function 4 (dual reflected beams at $\varphi = \pm\ 30°$ when fed at $\theta_{rad} = 0°$). Source: Reproduced with permission from [103]/John Wiley & Sons.

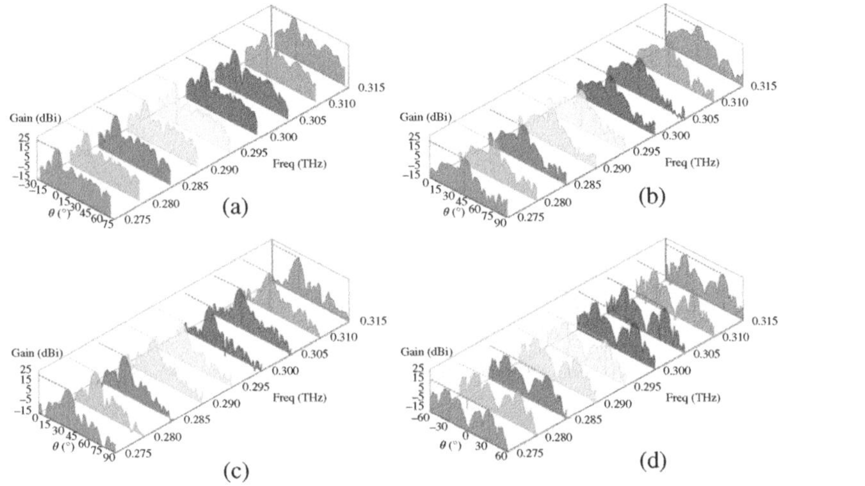

Figure 7.49 Experimental radiation gain results of the reflected EM waves for different coded function configurations as function of EM radiation frequencies ranging from 0.275 to 0.315 THz. (a) Reflection beam at $\varphi = 0°$ when fed at $\theta_{rad} = 30°$ (code function 1), (b) reflected beam at $\varphi = 40°$ when fed at $\theta_{rad} = 30°$ (code function 2), (c) reflected beam at $\varphi = 30°$ when fed at $\theta_{rad} = 0°$ (code function 3), (d) dual reflected beam at $\varphi = \pm 30°$ when fed at $\theta_{rad} = 0°$ (code function 4). Source: Reproduced with permission from [103]/John Wiley & Sons.

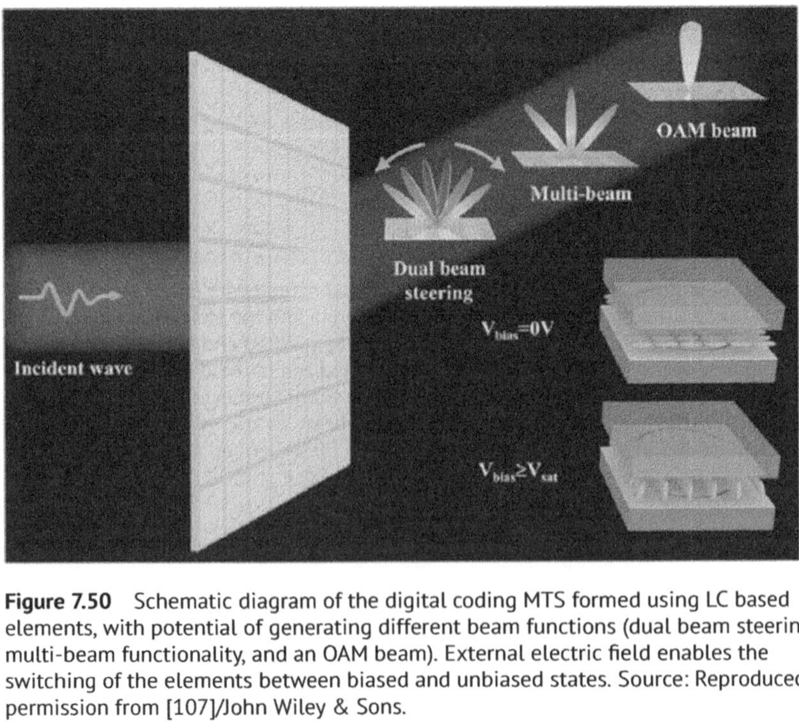

Figure 7.50 Schematic diagram of the digital coding MTS formed using LC based elements, with potential of generating different beam functions (dual beam steering, multi-beam functionality, and an OAM beam). External electric field enables the switching of the elements between biased and unbiased states. Source: Reproduced with permission from [107]/John Wiley & Sons.

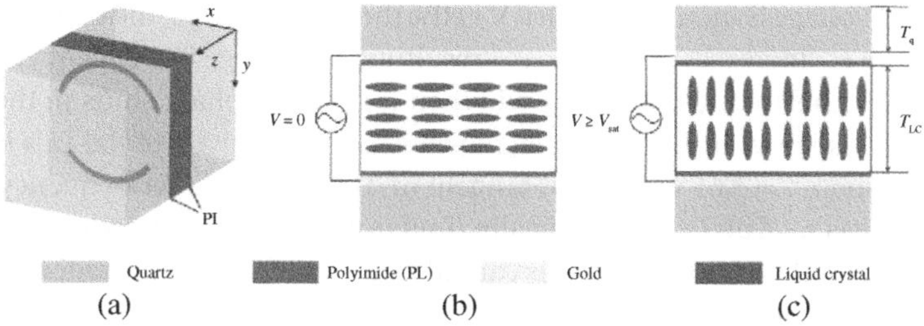

Figure 7.51 (a) 3D schematic diagram of the unit cell. (b) The side view of the unit cell in the unbiased case and (c) the biased case. ($T_q = 200\,\mu$m, $T_{LC} = 60\,\mu$m). Source: Reproduced with permission from [107]/John Wiley & Sons.

importantly, the ability to control their steering are valuable features for MIMO communications. Additional MIMO channels can be created and controlled precisely (provided there is sufficient gain generated from each beam to reach users) depending on the location and needs of the user terminals in the downlink. The advantages of OAM in the context of MIMO communications will be explained in the next subsection.

The unit cell in this work is formed by a metal-insulator metal (MIM) structure, as shown in Figure 7.51. A complementary asymmetrically split ring (CASR) metal pattern is printed onto a layer of quartz (200-μm-thick). Spin coating process is used to apply a polyimide (PI) alignment layer on the metallic layer, prior to filling of the space between the two PI layers using LCs (60-μm-thick). Under unbiased conditions, the orientation of the LC molecules in the elements is parallel to the gold layers. When biased by an external electrical field between the two metallic layers, the LC molecules' orientation then changes. When the bias voltage is higher than the threshold voltage, the LC molecules are oriented perpendicular to the metallic layers. The rotation of the LC molecules will change its relative permittivity. Its unbiased tensor is described as follows:

$$\varepsilon = \begin{pmatrix} \varepsilon_{\parallel} & 0 & 0 \\ 0 & \varepsilon_{\perp} & 0 \\ 0 & 0 & \varepsilon_{\perp} \end{pmatrix} \tag{7.29}$$

where $\varepsilon_{\perp}$ and $\varepsilon_{\parallel}$ is the LC permittivity in the minor and major axis direction of the LC molecules, respectively. On the contrary, when LCs biased (saturated case), the LC molecules are oriented perpendicular to the metallic layers, with a relative permittivity tensor of:

$$\varepsilon = \begin{pmatrix} \varepsilon_{\perp} & 0 & 0 \\ 0 & \varepsilon_{\perp} & 0 \\ 0 & 0 & \varepsilon_{\parallel} \end{pmatrix} \tag{7.30}$$

When the bias voltage is increased from 0 V to the threshold voltage, the effective permittivity $\varepsilon_{eff} = \varepsilon_{zz}$ increases from $\varepsilon_{\perp}$ to $\varepsilon_{//}$, resulting in a change in frequency in the EM response. This frequency change will then cause large phase shifts in the elements' phase response. In this example, the effective permittivity of the LCs is set to $\varepsilon_{\perp} = 2.55$ and $\varepsilon_{//} = 3.65$ with a loss tangent of 0.02. At the target frequency, the transmittance is ≈ -6 dB ($\approx 50\%$), typically similar with the reflectance of 1-bit reflective LC MTS in the THz band. This is caused by the reduction of the radiative losses by the Fano resonance [108].

Digital coding MTS samples are then fabricated to evaluate their dual-beam steering ability, as shown in Figure 7.52a, b. The prototype consists of 48 × 48

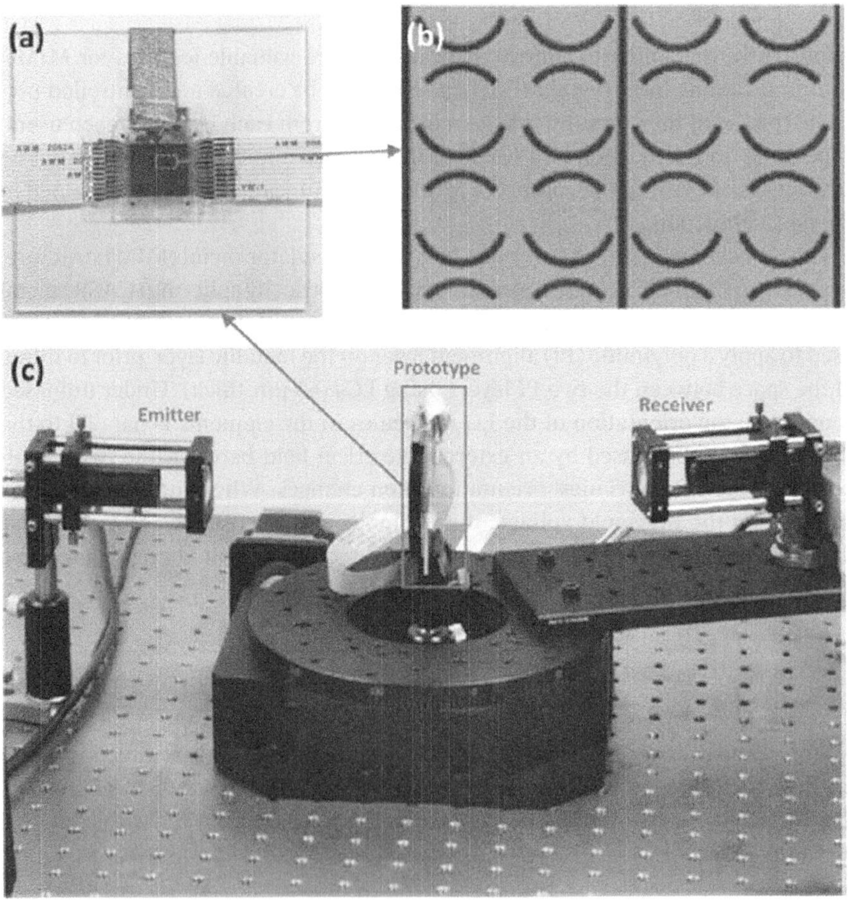

Figure 7.52 (a) Fabricated prototype. (b) Microscopic image of a group of unit cells. (c) Experimental setup for the characterization of the digital MTS performance for dual beam steering. Source: Reproduced with permission from [107]/John Wiley & Sons.

CASR elements in the form of a linear array. Gaps which are 30 μm wide are used to isolate adjacent sub-arrays. Fabrication of a metal layer on a 22 × 36 mm quartz plate is performed using the photolithography process (vacuum plating, photoresist spin coating, exposure, development, and wet etching). Upon the spin-coating of the PI alignment layer, ultraviolet glue was used to bond a quartz layer containing a spacer with a diameter of 60 μm. Finally, the space between the quartz plates is filled with the nematic LC by injection.

To assess the performance of the prototype, it was fixed at the center of a rotation stage shown in Figure 7.52c. An emitter and receiver were positioned at both ends of the MTS before the application of the voltage representative of different coding sequences. A *y*-polarized THz beam was first emitted by the photoconductor antenna to be converted into plane wave by the first polymethyl pentene (TPX) lens. Upon the deflection by the MTS and additional focusing by the second TPX lens, it is then received at the photoconductor antenna receiver at different angles. Fourier transform was then performed on the received signal to acquire their spectrum. To reduce the possibility of dynamic effects that may vary the permittivity, a 1 kHz stable square signal is selected with no offset, with the encoding for the "0" and "1" biased voltages set at 0 and 10 V, respectively.

Figure 7.53a shows the measured transmitted wave at 0.426 THz when different coding sequences were applied (/00…/, /11…/, /10…/, /1100…/, and /111000…/, with each number representing a subarray consisting of 2 × 48 sub-wavelength unit cells). The MTS is designed to split the incident wave into two beams. However, three transmitted beams were noticeable when three coding sequences (/10…/, /1100…/, and /111000…/) were measured. The additional beam at 0° may be due to the larger THz beam spot compared to the MTS. Besides that, simulations in Figure 7.53b also indicated additional beams at 0° which are significantly smaller. This may be due to the phase mismatches caused by inter-sub-array coupling, resulting in degraded side lobe levels. Another factor which may cause frequency shifts in the measurements includes differences in fabrication of the quartz thickness.

7.5.3.3 Microfluidic-Based RIS Prototypes

Finally, another newer method for tuning RIS prototypes is the microfluidic-based technique applied in the MTS proposed in [109]. The microfluidic-based chiral MTS (MSCM) is designed to be able to control and tune THz waves, as illustrated in Figure 7.54. Three bonded PDMS layers forms this MTS, as seen from Figure 7.54a, with an overall area of 30 mm × 30 mm. The top PDMS layer consists of an array of "⫟" shaped microchannels injected with liquid metal Galinstan to form the MTS structure, as depicted in the inset of Figure 7.54b. Each unit cell is sized at 1 mm × 1 mm in the *xy* direction, with the arm of the "⫟" is designed with $b = 350\,\mu m$ and $L = 300\,\mu m$. A horizontal channel connects adjacent unit cells

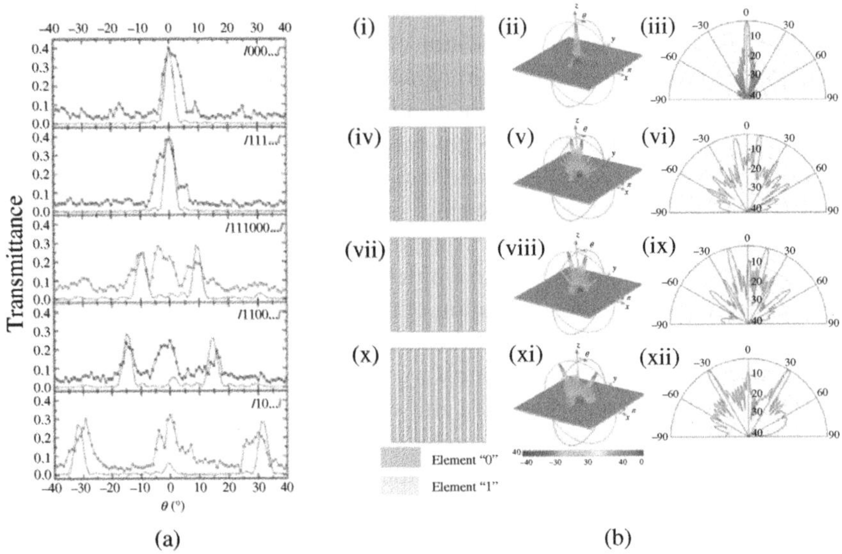

Figure 7.53 (a) Simulated and measured scattering patterns for five different coding sequence. Each number in the coding represents a 2 × 48 sub-array of unit cells. (b) Coding sequence and simulated 3D multibeam scattering patterns. (a, d, g) are coding patterns. Each bar in the figure represents a 2 × 48 sub-array of unit cells. (b, e, h) are corresponding 3D scattering patterns. (c, f, i) Are the 2D scattering patterns at $\phi = 0°$. Source: Reproduced with permission from [107]/John Wiley & Sons.

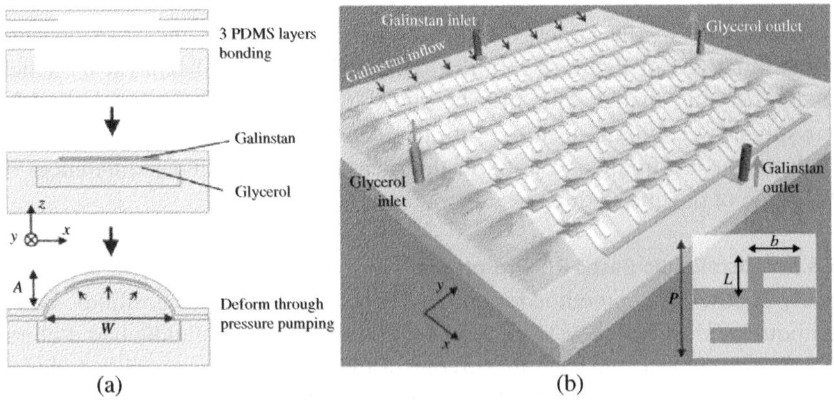

Figure 7.54 (a) Cross section view of the MSCM unit cell and (b) deformation of the MTS. Source: [109]/Optical Society of America/Public Domain.

on the same row, whereas a vertical channel connects the two asymmetrically located arm channels on two sides of the horizontal channel. Both channels are 20 μm in depth and 100 μm in width. Unit cells from all rows are connected to the Galinstan inlet to enable injection of the Galinstan along the x-direction, as show in Figure 7.54b (denoted by black arrows). This allows the filling of all microchannels in the unit cells with Galinstan and exit via the outlet on the opposite side of the inlet.

A straight microchannel is designed below each unit cell on the bottom PDMS layer along the y-direction ($W = 850$ μm). Similar to the top layer, these microchannels are connected with a Glycerol inlet and outlet. The flow of glycerol in the bottom channels is indicated by the gray, y-directed arrows in Figure 7.54b. The overall MTS structure is flat when both the inlet and outlets for glycerol and Galinstan are opened, as shown in Figure 7.54a (middle). When the outlet of the glycerol is closed while injection continues, the additional pressure in the bottom channel deforms the top PDMS layer upward. A 3D chiral structure is then formed, with a bending height of A, as seen in Figure 7.54a (bottom).

In its initial stage (without pressure pumping), the MTS is a flat 2D chiral structure without intrinsic chirality. However, interaction between the electrical dipole and the magnetic dipole in the unit cell is generated in the MTS when it is defined into a bent 3D structure, as shown in Figure 7.55. The incidence of a linearly polarized wave with electric field, E_i will excite an electric dipole $\boldsymbol{p}$ in the unit cell to generate a secondary wave with electric field, E_p. Due to its two arms being located asymmetrically, the response of the unit cell to the magnetic field is irregular. The induced magnetic dipole m has a nonzero component either parallel or antiparallel with the electric dipole p. Besides that, the magnetic dipole also radiates EM waves with electric field E_m perpendicular to E_p. Therefore, the total scattering field E_s, which is a combined E_p and E_m is not aligned to the incident E_i due to the non-zero E_m. This results in a rotated transmitted field $E_\mathrm{t} = E_\mathrm{i} + E_s$ with respect to the original E_i direction.

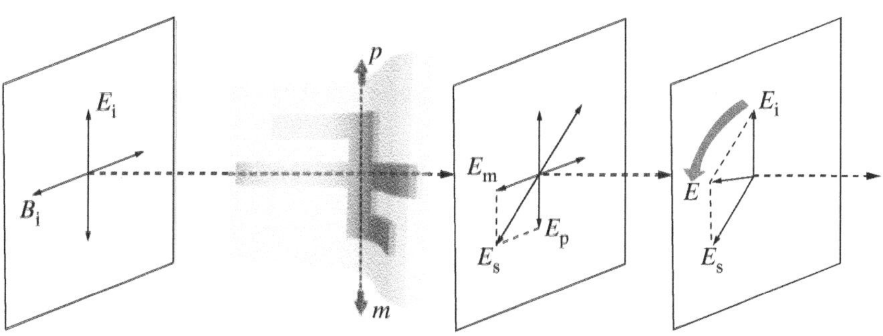

Figure 7.55 Induced by the interaction between electric and magnetic dipole. Source: [109]/Optical Society of America/Public Domain.

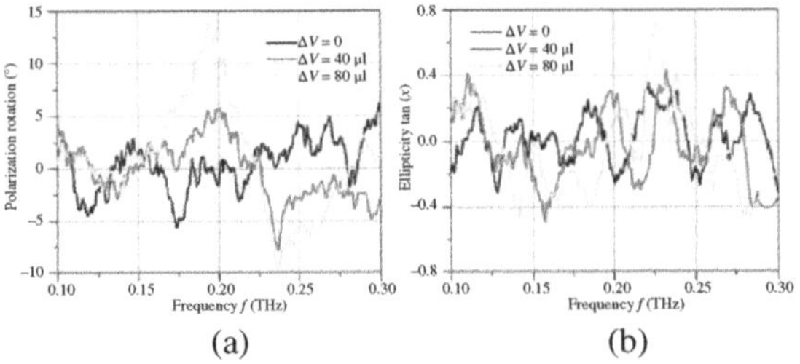

Figure 7.56 (a) The measured polarization rotation; (b) the ellipticity of the soft MTS. Source: [109]/Optical Society of America/Public Domain.

For experimental validation, two linear polarization surfaces were used, each operating as the polarizer and the analyzer. The polarizer is aligned with the direction of polarization of the incident THz wave. The MTS can be aligned with the polarizer direction (denoted as x-polarized) or its orthogonal direction (y-polarized). By rotating the analyzer, the co-polarized transmittances $\sim t_{xx}$ and $\sim t_{yy}$, and the cross-polarized transmittances $\sim t_{xy}$ and $\sim t_{yx}$ can be obtained using equations specified in the paper. The polarization rotation and ellipticity of the MTS plotted in Figure 7.56 indicated polarization rotation is around 0 between 0.1 and 0.3 THz when $\Delta V = 0$. However, several small resonances can be seen for both the polarization rotation angle and the ellipticity. This may be due to the interaction between the Fabry Perot resonance in the 2 mm thick bottom PDMS layer and the resonance in the Galinstan MTS.

The weak THz signal is also suspected as the cause of the relatively high noise observed in the measurements. This is especially evident in the cross-polarized transmittance measurements. A polarization rotation of 5° (at 0.19 THz) and −7° (at 0.24 THz) is seen as ΔV is increased to 40 μl. Then, the deformed cross section is assumed to be a half ellipse with the long axis of W and short axis of $2A$ to establish the deformation amplitude A at this state. This resulted in the estimation of the cross-section area of the half ellipse, $S_1 = \pi WA/4$. Considering that each unit cell is 1 mm in side length, with a total of 30 × 30 unit cells, the deformed cross section area is calculate as $S_2 = \Delta V/(30 \times 30 \times 1\,\text{mm})$. Therefore, if $S_1 = S_2$, it can be estimated that $A = 67\,\mu\text{m}$ when $\Delta V = 40\,\mu\text{l}$. As ΔV increased to 80 μL, A is then 133 μm. The polarization rotation is significant at 0.19 THz, which reaches 14°. A high ellipticity value of 0.7 is also observed at 0.23 THz, demonstrating the optical activity when the chiral MTS is bent.

7.6 MIMO OAM

Spatial multiplexing is no doubt the most important feature of a MIMO system [24]. Various MIMO systems have been proposed to exploit this feature. The resulting multiplexing gain depends on both the array configuration and the propagation environment [110]. For a given channel, spatial multiplexing is essentially limited by the array aperture. Recently, some researchers claim that OAM-based multiplexing (i.e. simultaneously transmitting multiple signals using different OAM modes) can enhance the capacity [111–114]. This has raised certain controversies. Moreover, OAM communications has been largely confined to line-of-sight (LoS) environments. The conventional OAM transmission/detection is extremely vulnerable to the multipath effect, but it is a very active area of research where several interesting developments have appeared lately.

The majority of the OAM works in the open literature focus on OAM generation using, e.g. spiral phase plate (SPP) [115], uniform circular array (UCA) [116], and MTS [117]. Despite the different OAM generation methods, OAM waves are essentially generated via the discrete Fourier transform in either analogue domain [115, 117] or digital domain [116]. Analogously, the conventional OAM detection is done via the inverse discrete Fourier transform using either SPP, MTS, or UCA. In this respect, the OAM generation (detection) can be regarded as a special MIMO precoding (equalization). Thus, OAM waves do not offer an additional degree-of-freedom (DoF) [118, 119]. Although OAM modes are all orthogonal, generating high order OAM modes requires larger aperture and its spatial DoF is limited by the source aperture as well, just like the MIMO array.

The UCA serves as a classic OAM wave generator in the radio frequency band [120]. A typical UCA consists of a phased antenna array, as illustrated in Figure 7.57a. In this configuration, all antenna elements are positioned around a ring and excited with arranged phases at intervals of $2\pi m/N$ (where N represents the total component number, and m is the OAM mode index). It is evident that the performance of OAM wave generation in the UCA primarily depends on the phase configuration. The antenna components can include horn antennas, dipole antennas, or patch antennas. In line with the miniaturization trend, patch antenna UCAs have garnered more attention. Figure 7.57b shows a compact patch antenna UCA design tailored for full-duplex applications [120]. This design comprises two antenna arrays and two feeding networks. When implementing UCA methods, the design of feeding networks is of utmost importance. The phase distributions demonstrate its capability to emit high-purity multi-mode OAM waves.

MTS designs for generating OAM waves have garnered significant attention due to their low profile, cost-effectiveness, and ease of fabrication using printed circuit technology [121]. MTS can be broadly categorized into three types: transmit-arrays, reflect-arrays, and leaky wave structures. These MTS achieve wave-front phase modulation through the arrangement of unit cells, while maintaining a compact, thin, and planar or conformal aperture structure. Figure 7.58 illustrates a transmit-array design for OAM wave generation, which includes a circularly-polarized feed antenna and two MTS. The focal-to-diameter ratio of

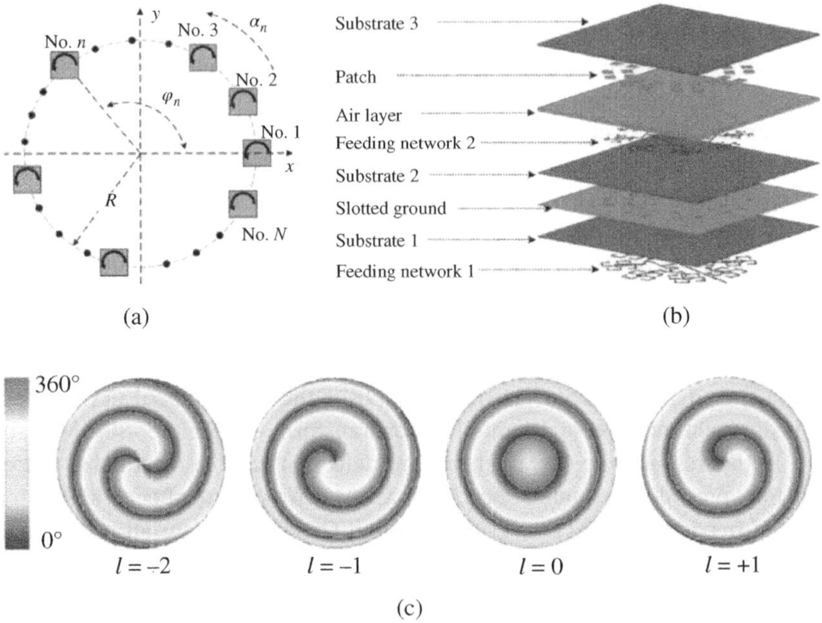

Figure 7.57 Design of a UCA OAM launcher: (a) Configuration of the UCA. (b) 3-D view of the UCA. (c) OAM waves phase distributions. Source: Zhang and Li [120]. © 2018 IEEE.

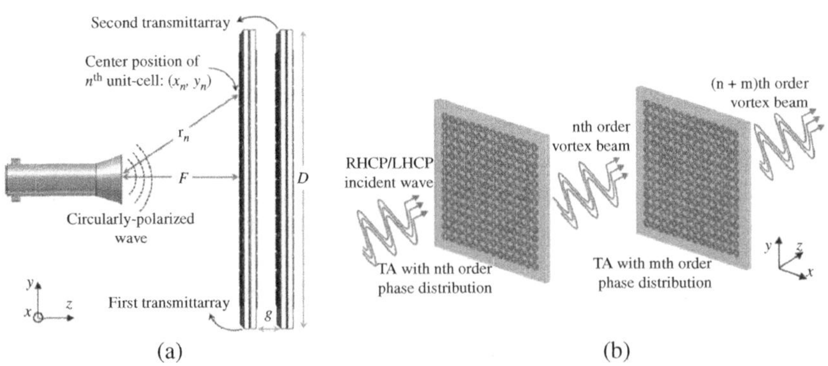

Figure 7.58 MTS transmit array for OAM generation. (a) Schematic diagram. (b) Exploded view demonstrating the performance of two cascaded MTS. Source: Zhang and Li [121], © 2018 IEEE.

the entire structure, denoted as F/D, is remarkably low at 0.5, where F represents the focal length, and D denotes the aperture diameter. To determine the required phase shift for each unit cell, one can refer to the feed antenna's phase distribution within the aperture. In Figure 7.58b, two transmit-arrays are shown, each carrying out different phase modulation. Notably, each transmit-array can convert an incident wave into an OAM wave with a specific mode index. By cascading these transmit-arrays, the mode index can be linearly stacked. Furthermore, MTS offer the flexibility to modify the number, polarization, and direction of radiated OAM beams. Therefore, MTS designs, considering their various advantages, find practical applications in OAM wave generation in near future research.

As aforementioned, OAM communications have been largely considered in LOS environments and multipath is regarded as detrimental for OAM transmission. Nevertheless, it was shown recently that OAM transmission/reception is possible even in a rich multipath environment [122]. This is possible if a MIMO equalizer (e.g. a zero-forcing receiver) [24] is employed at the receiver [122]. To demonstrate the feasibility, experiments have been conducted in a reverberation chamber (RC). The RC is a large metal cavity equipped with metallic mode stirrers and a turn-table platform. A detailed description of the RC can be found in Chapters 2 and 9 of these books. A 2×2 MIMO system based on OAM multiplexing was constructed in the RC as shown in Figure 7.59. The transmitter has two horn antennas loaded with two OAM MTS in $l = 1$ and 2 modes, respectively [124]. The receiver was a handset which includes a MIMO antenna system with eight slot elements [123, 125]. Two arbitrary slot elements of the handset MIMO antenna system were used as the receiving antennas. A vector network analyzer (VNA) was used to measure the channel transfer function (i.e. the S-parameters) at 5.1 GHz with a bandwidth of 200 MHz. Note that the constructed MIMO OAM system here is different from the conventional OAM-based MIMO system, where OAM antennas are employed at both transmitting and receiving sides and facing each other in the LoS environment. In the reverberation chamber with strong multipath components, the OAM modes' orthogonality is destroyed in the analogue domain. Hence, it is not necessary to use OAM antennas at the receiver. Nevertheless, the multiplexed signals can still be successfully separated using a MIMO equalizer in the digital domain using the two slot elements in a handset MIMO used as the receiver.

To further compare the OAM multiplexing with a conventional MIMO system, another measurement setup was constructed, where the two OAM MTS are removed, i.e. two horn antennas were used at the transmitter while maintaining the receiver setup. Figure 7.60a shows the measured channel capacities of the MIMO systems at 15-dB SNR with and without OAM multiplexing. As can be seen, the two channel capacities have almost the same average value over the operating bandwidth. Figure 7.60b shows the bit error rate (BER) performances of the two MIMO systems with and without OAM multiplexing. Note that the

Figure 7.59 Test setup of the MIMO OAM system in the RC. Source: Reproduced with permission from [123]/IEEE.

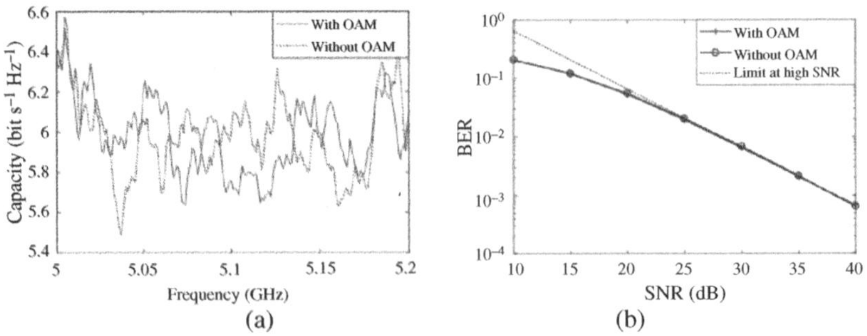

Figure 7.60 Comparison of the MIMO systems with and without OAM multiplexing: (a) Channel capacities at 15-dB SNR and (b) BERs as a function of SNR.

OFDM technique has been assumed for the BER simulations to deal with the frequency selectivity of the measured channel in the RC, whereas the zero-forcing equalizer is used at the receiver. As can be seen, the two systems have identical BER performance. From the two sets of experiments, it is safe to conclude that OAM multiplexing is possible in rich multipath environments provided that the MIMO equalizer is used at the receiver. It should be noted also that identical performance is achieved in the RC with isotropic scattering. When the angular distribution of the incoming waves becomes non-uniform, the performance of the OAM multiplexing (enabled by the MIMO equalizer) depends on many factors, such as the divergence angles of the OAM waves, the mean angle of the incident waves, and the angular spread of the propagation channel. The interested reader can refer to [123] for a more detailed discussion on this matter.

7.7 MIMO Antennas for Space Applications

The rapid growth of satellite communications capacity and advances in multi-target area communications technology have led to the development of multi-beam antennas. Compared to the traditional single-shaped wide beam method [126], multi-beam antennas emit electromagnetic waves into space via multiple spot beams, each with its own input/output channel. The single-shaped wide beam method used in conventional communications satellites uses an antenna profile to match the coverage area [127], typically a continent or country. However, the coverage area often has an irregular shape, making it difficult to efficiently reduce power wastage by excluding additional areas beyond the coverage zone. In addition, such a large antenna coverage area results in a wide beamwidth, which leads to gain degradation due to the allocation of spectrum that can only be used once. Multi-beam antenna technology significantly addresses the above issues with traditional communications satellites. Multi-beam antenna technology provides a more precise area division compared to the single-line wide beam method. In addition, the area covered by each beam is significantly reduced, resulting in a narrower beamwidth. In addition, the scarcity of spectrum resources has forced the adoption of higher frequency bands such as Ka-band, and even Q and W-bands [128]. Several operators around the world are actively researching the Ka-band [129]. Over the last two decades, communication satellites have evolved into high-capacity and high-efficiency systems such as high-throughput satellites (HTS) [126] and very high-throughput satellites (VHTS) [130]. Currently, multi-beam antennas for communication satellites can be categorized into three types mainly: multi-beam reflector antennas, multi-beam lens antennas, and multi-beam phased array antennas (MBPAA) [131].

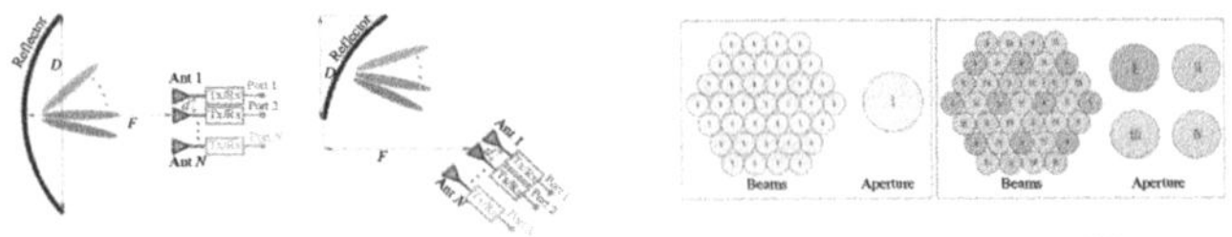

(a) (b) (c) (d)

Figure 7.61 (a) Center-fed configurations producing *N* beams. (b) Offset-fed configurations producing *N* beams. (c) Single-aperture reflector schemes. (d) Four-aperture schemes reflector schemes. Source: (a, b) Lei et al. [126], © 2022 IEEE, (c, d) Hong et al. [131], @ 2017 IEEE.

With a simple structure, well-established technology and lightweight, multi-beam reflector antennas are the most commonly used antennas in satellite communications. Figure 7.61a, b illustrates the structure of a multi-beam reflector antenna. A passive multi-beam reflector antenna typically consists of several feed antennas with a pitch of d and a reflector with a focal length of F and a diameter of D [132]. However, a single reflector may not be sufficient when a large number of tightly focused beams are required. Two configurations are therefore used, depending on the number of feeds for each beam. The first configuration, known as the single feed per beam (SFPB) scheme, forms the spot beam radiated from the antenna aperture by using a specific feed source irradiating a reflector. In contrast, the multi-feed per beam (MFPB) scheme uses a beamforming network consisting of array element units with different excitation amplitudes and phases to form different multi-beams, achieving beamforming and beam isolation using a single reflector [128].

Within the SFPB schemes, there are both single-aperture reflectors and multi-aperture reflectors. Aperture reuses in the multi-aperture reflection scheme are usually denoted by different (gray) colors. For example, a "four-color" aperture reuse scheme (Figure 7.61d) is demonstrated in [133]. Conversely, the single-aperture reflection scheme (Figure 7.61c) presents challenges in achieving high reflective illumination efficiency and sufficient isolation due to the requirements related to feed spacing, feed diameter and increased reflector volume, thereby compromising antenna coverage flexibility [134]. An alternative architecture uses a single oversized reflector, eliminating the need for beamforming networks. Although the multi-aperture reflection scheme offers advantages such as flexible control over the number and shape of beams, requiring only two reflectors for signal transmission and reception while occupying less space on the satellite, it has several drawbacks. For instance, this scheme requires many times more feeders than SFPB, and consequently, the structure of its beamforming network is typically much more complex.

With the increasing frequency of satellite communications, reflector multibeam antenna technology is proven to be suitable for high-frequency bands such as

Ka-band, regardless of the SFPB or MFPB scheme. In the lower L- and S-bands, where the wavelength is relatively long, the volume of the feed array remains large. For example, the Alphasat-I-XL satellite uses a spiral antenna as an array element in GEO orbit, resulting in an L-band feed array of more than 2 m in diameter [135]. However, as the frequency increases to Ka-band with shorter wavelengths, the feed becomes smaller, making the size of the reflector antenna more practical, regardless of the SFPB or MFPB scheme. SFPB is particularly advantageous for high-volume satellites. Telesat's Anik-F2 satellite, the first to use multi-beam technology, uses the SFPB scheme [136]. It has a 0.8° beamwidth and 45 Ka-band user beams, covering a significant portion of the United States and Canada. IPSTAR is the first true HTS combining a reflective antenna and a receiving antenna. It contains 84 1° spot beams in the Ku-band, achieving a total system capacity of 45 Gbit/s, nearly equivalent to the combined capacity of all satellites serving Asia at that time [137].

The lens antenna [138] is an application of the principles of geometric optics to electromagnetism, as shown in Figure 7.62. Lens antennas have excellent rotational symmetry. However, they also have similar disadvantages as reflector antennas, such as heavy weight, especially when used in low-frequency bands. The reduction in wavelength, however, opens up new opportunities for the development of such antennas. For example, the discrete lens antenna proposed in [134], although not considered a promising solution due to volume, weight, deployment and flexibility issues, has led to the exploration of potential remedies as discussed in [139]. These solutions include embedding an amplifier in the lens, minimizing the number of radiators in the front and rear arrays while maximizing the efficiency of the PA, and reducing the weight and depth of the radiating elements. These advances aim to overcome the limitations associated with lens antennas and improve their performance in practical applications.

MBPAAs are a further development of traditional phased array antennas, enabling the generation of multiple simultaneous beams through beamforming networks (BFN) [140]. MBPAAs can be classified into three categories based on

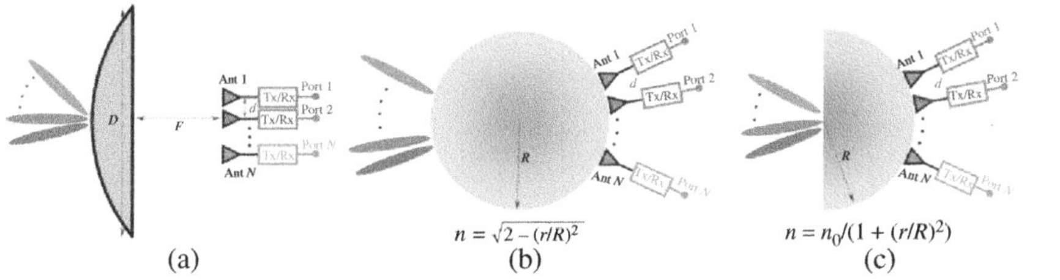

Figure 7.62 (a) Conventional homogeneous lens configurations. (b) Gradient-index Luneburg lens configurations. (c) Gradient-index half Maxwell fish-eye lens configurations. Source: Hong et al. [131], © 2016 IEEE.

the type of BFN used: i.e. analog beamforming, digital beamforming (DBF), HBF, and optical beamforming (OBF). Initially, analog beamforming was widely used thanks to its advantages in bandwidth and low power consumption. However, it faces challenges when used in large-scale multi-beam scenarios. The phased array multi-beam antenna can be further divided into circuit-based BFN and phase-shifting-based BFN. The system structure diagram in Figure 7.63a illustrates a circuit-based MBPAA using a 4 × 4 Butler matrix configuration [141]. It consists of four 90° couplers and two 45° phase shifters, allowing the generation of multiple beams with a 45° phase spacing. Another approach is to use the Blass matrix as shown in Figure 7.63b. This structure includes travelling wave feedlines, transmission lines, directional couplers, lossless phase shifters and matching loads. While the former is limited in the number of beams it can generate, the latter can generate any number of beams. However, the Blass matrix is less efficient due to the presence of matching loads.

Many early low-orbit spaceborne multibeam systems used beamforming techniques. For example, the Iridium satellite system used the Butler matrix scheme [142]. It used three active phased array plates, each consisting of 106 array elements, T/R components and a beamforming network. The array element consists of lightweight patch radiators, with each of them powered by a T/R

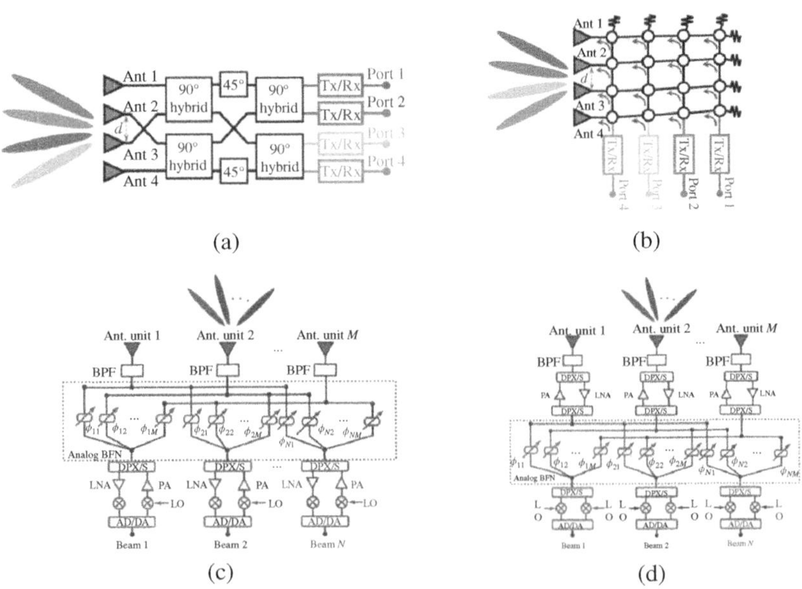

Figure 7.63 (a) 4 × 4 Butler-matrix-based MBPAA system architecture. (b) 4 × 4 Blass-matrix-based MBPAA system architecture. (c) Passive MBPAA system architecture. (d) Active MBPAA system architecture. Source: Yu et al. [140], © 2023 IEEE.

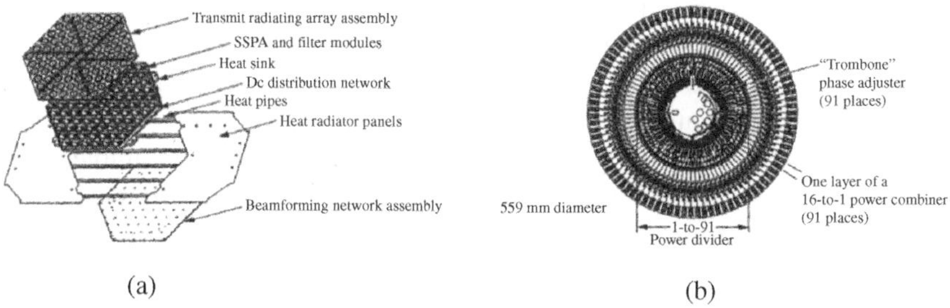

Figure 7.64 (a) S-band multibeam active transmit array antenna view. (b) S-band transmit antenna beamforming network layer. Source: Metzen [143], © 2000 IEEE.

module and coexcited by an optimized BFN. The beamforming strategy relied on a two-dimensional intersecting Butler matrix and power splitters to form 16 spot beams. Figure 7.63c, d shows the passive and active structures of the MBPAA based on phase shifters. The active module included additional transceiver modules between the BFN and the antenna unit, using PA and LNA to control the feed amplitude. Globalstar-1 (GB1) used MBPAA based on phase shifters [143]. The structure of the antenna consists of two parts, the radiating array at the top and the sealed module at the bottom. The top radiating array is a hexagonal planar array. The lower sealed module contains the solid-state PA and bandpass filter and is located above the beamformer on the heat sink, as shown in Figure 7.64.

The proposal for the Iridium Next initiative emerged in 2007 [140]. The main mission antenna (MMA) of the satellite was an L-band flat-phased array antenna consisting of 120 units. The antenna unit as shown in Figure 7.65 is a patch radiator manufactured as a separate component and attached to the MMA board during array assembly. The MMA was capable of generating 48 beams for both transmission and reception, facilitating communication with user terminals. In addition, each satellite was equipped with two Ka-band feeder link antennas operating at 20/30 GHz. These antennas enabled the establishment of two

Figure 7.65 (a) The front and (b) the back of the patch radiator. Source: Reproduced with permission from [142], © 2010 IEEE.

beams for connection to the ground gateway. In addition, the system facilitated inter-satellite communications between four satellites within the same orbit and adjacent orbital planes. This communication was made possible through the use of fixed and mobile antennas.

The Globalstar-2 (GB2) generation [144], launched in 2010, implemented semi-active phased array antennas. As shown in Figure 7.66, two pairs of passive multi-sided dome antennas (TX10, TX6) were used to generate 16 beams. Signal amplification was achieved using a multiport PA. The central beam, referred to as Beam 1, used the TX10 horn, while the remaining nine TX10 fronts formed the outer nine beams. The center circle consisted of six beams formed by the six TX6 fronts, each with eight elements per front. The multi-port PA allowed power adjustment across the beams. Similar to Globalstar-1, the L-band reception relied on an active phased array antenna with 52 array elements. The array element is a very compact 0.6λ patch in cavity. This approach is aimed at optimizing costs by using passive phase shifters and attenuators to achieve beamforming.

The most notable recent satellite endeavor is SpaceX's ambitious Starlink project, which aims to provide high-speed, low-latency broadband access services on a global scale [145]. SpaceX successfully launched 47 V1.5 Starlink satellites into orbit from Vandenberg Space Force Base using a Falcon 9 rocket on June 22, 2023. This mission brings the total number of deployed Starlink satellites to 4642, of which approximately 4312 remain operational in orbit. The Starlink satellites use Ku-band phased array antennas with beam-hopping coverage technology [146, 147]. The involved antenna structure consists of four layers: the antenna array layer, the mapping layer, the multiplex feed layer, and the beamforming

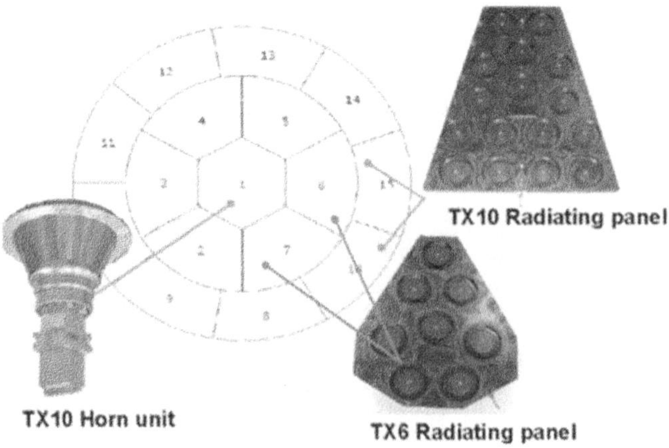

Figure 7.66 Transmitting antenna beam coverage scheme. Source: Reproduced with permission from [144], © 2023 IEEE.

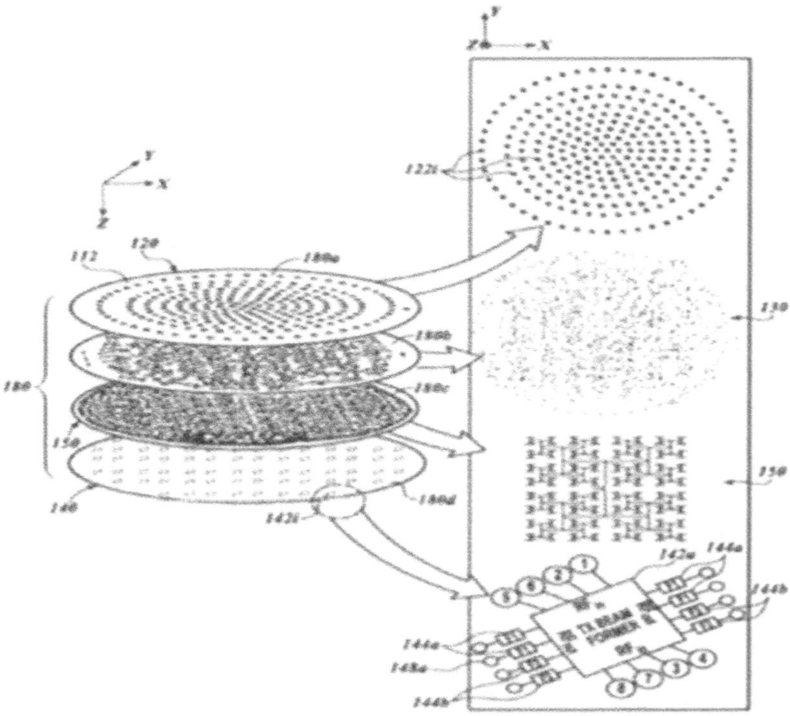

Figure 7.67 Phased array antenna overall structure. Source: Yu et al. [140].

layer as shown in Figure 7.67 [140]. Within the beamforming section is an 8-channel, 8-beam multifunction chip that uses only phase shifters. This design choice reduces the cost, size and power consumption of the chip.

Another important constellation project is OneWeb [145], which aims to establish a new generation of low-earth orbit (LEO) satellite communications systems to provide affordable and seamless broadband Internet access services worldwide. Each satellite in the constellation is capable of generating 16 elliptical Ku-band user beams and two Ka-band feed beams. OneWeb's user link antenna uses a Ku-band passive phased array antenna with 16 linear feed elements to generate fixed beams in an elongated elliptical shape. The integration of array antenna technology with integrated circuits or programmable gate arrays has given rise to DBF [148]. DBF offers significant advantages, such as the ability to generate a large number of beams independent of weight and power consumption, which are determined by signal bandwidth and radiation components. However, the key challenge is to achieve the required beam weights while using minimal computational resources and efficient algorithms. Currently, due to the limited sampling rate of digital chips, weight coefficients are typically stored in advance and retrieved using table look-up methods. Future developments aim to achieve real-time adaptive computation for updating the weighting coefficients. Another beamforming technology, known as optically controlled beamforming [149], uses optical fiber or waveguide transmission lines and achieves beam scanning through ordered photonic true delay control. The development of this technology is hindered by challenges related to high integration, miniaturization, and temperature sensitivity. Nevertheless, these challenges have been partially addressed in [150].

The growing demand for HTSs, satellite mobile communications and LEO satellite constellations has spurred the development of multi-beam antennas to provide high-band, broadband, multi-functional, and cost-effective solutions. In the case of reflector antennas, there has been a shift from the use of single fixed reflectors to the use of expandable mesh reflectors, especially in Ka-band. In addition, the combination of fixed reflectors and mesh structures has been explored as a means of constructing large Ka-band reflectors [151]. With respect to phased arrays, there is a trend to move from analog beamforming to DBF. In addition, the integration of microwave and photonic technologies has attracted considerable interest. Photonic technology offers advantages such as large bandwidth, low loss, and no crosstalk. This has led to the exploration of scanning plane reflection arrays for future communications satellite phased arrays, taking advantage of photonic technology in addition to microwave integration technology [152]. To reduce the complexity of satellite payload systems, the development of ground-based beamforming (GBBF) technology [153, 154] is emerging as a solution for the next generation of multi-beam antennas for communications satellites. GBBF technology allows beamforming to be performed on the ground, reducing the complexity and weight of the satellite payload while increasing flexibility and adaptability, a new technology that is worth following in the years to come.

In addition to multi-beam antenna technology, MIMO has many other applications in satellite communications. These include increasing data rates in satellite links through spatial multiplexing, mitigating atmospheric interference, and environmental signal fading to improve signal quality through diversity reception and adaptive modulation and coding, among others. On April 25, AST SpaceMobile announced the successful establishment of the world's first satellite-based cellular voice call. The test satellite, named Blue Walker 3, was launched at 9:20 p.m. Eastern Daylight Time on September 10, 2022. It serves as a precursor to the commercial Blue Bird satellite. AST SpaceMobile's goal is to connect universally available, non-customized mobile phones directly to satellites, enabling global ubiquitous mobile communications. In June, AST SpaceMobile conducted 4G LTE download speed tests on today's popular smartphones, achieving initial speeds of up to 10.3 Mbps. The next phase of testing will enable 5G cellular broadband. The test results for voice calls and 4G download speeds demonstrate the ability of LEO satellites to deliver communications services to existing mobile devices. MIMO technology is undoubtedly the key technology for LEO satellites to communicate with mobile devices.

References

1 Cheema, H.M., Shamim, A., and Khalid, F. (2021). *Antenna-on-Chip: Design, Challenges, and Opportunities.* Artech.

2 Lin, J.J., Gao, L., Sugavanam, A. et al. (2004). Integrated antennas on silicon substrates for communication over free space. *IEEE Electron Device Letters* 25 (4): 196–198. https://doi.org/10.1109/LED.2004.824837.

3 Zhang, Y.P., Sun, M., and Guo, L.H. (2005). On-chip antennas for 60-GHz radios in silicon technology. *IEEE Transactions on Electron Devices* 52 (7): 1664–1668. https://doi.org/10.1109/TED.2005.850628.

4 Cheema, H.M. and Shamim, A. (2013). The last barrier: on-chip antennas. *IEEE Microwave Magazine* 14 (1): 79–91. https://doi.org/10.1109/MMM.2012.2226542.

5 Chan, K.T., Chin, A., Chen, Y.B. et al. (2001). Integrated antennas on Si, proton-implanted Si and Si-on-quartz. In: *International Electron Devices Meeting. Technical Digest (Cat. No.01CH37224), 02–05 December 2001,* 40.6.1–40.6.4. Washington, DC: https://doi.org/10.1109/IEDM.2001.979659.

6 Ojefors, E., Kratz, H., Grenier, K. et al. (2006). Micromachined loop antennas on low resistivity silicon substrates. *IEEE Transactions on Antennas and Propagation* 54 (12): 3593–3601. https://doi.org/10.1109/TAP.2006.886532.

7 Kang, K. et al. (2010). A 60-GHz OOK receiver with an on-chip antenna in 90 nm CMOS. *IEEE Journal of Solid-State Circuits* 45 (9): 1720–1731. https://doi.org/10.1109/JSSC.2010.2053095.

8 Ou, Y.C. and Rebeiz, G.M. (2012). Differential microstrip and slot-ring antennas for millimeter-wave silicon systems. *IEEE Transactions on Antennas and Propagation* 60 (6): 2611–2619. https://doi.org/10.1109/TAP.2012.2194651.

9 Chen, Z.N., Liu, D., Nakano, H. et al. (2016). *Handbook of Antenna Technologies.* Springer Publishing Company Incorporated.

10 Ou, Y.C. and Rebeiz, G.M. (2012). Differential microstrip and slot-ring antennas for millimeter-wave silicon systems. *IEEE Transactions on Antennas and Propagation* 60 (6): 2611–2619. https://doi.org/10.1109/TAP.2012.2194651.

11 Song, Y. et al. (2017). A compact Ka-band active integrated antenna with a GaAs amplifier in a ceramic package. *IEEE Antennas and Wireless Propagation Letters* 16: 2416–2419. https://doi.org/10.1109/LAWP.2017.2721956.

12 Arsalan, M., Shamim, A., Roy, L., and Shams, M. (2009). A fully differential monolithic LNA with on-chip antenna for a short range wireless receiver. *IEEE Microwave and Wireless Components Letters* 19 (10): 674–676. https://doi.org/10.1109/LMWC.2009.2029760.

13 Sharawi, M.S., Dhar, S.K., Hammi, O., and Ghannouchi, F.M. (2016). Miniaturised active integrated antennas: a co-design approach. *IET Microwaves, Antennas and Propagation* 10 (8): 871–879. https://doi.org/10.1049/iet-map.2015.0588.

14 Dhar, S.K., Sharawi, M.S., Hammi, O., and Ghannouchi, F.M. (2016). An active integrated ultra-wideband MIMO antenna. *IEEE Transactions on Antennas and Propagation* 64 (4): 1573–1578. https://doi.org/10.1109/TAP.2016.2537379.

15 Gonzalez, G. (1997). *Microwave Transistor Amplifiers: Analysis and Design*, vol. 2. New Jersey: Prentice Hall.

16 Rasilainen, K., Phan, T.D., Berg, M. et al. (2023). Hardware aspects of sub-THz antennas and reconfigurable intelligent surfaces for 6G communications. *IEEE Journal on Selected Areas in Communications* 41 (8): 2530–2546. https://doi.org/10.1109/JSAC.2023.3288250.

17 Mostajeran, A., Cathelin, A., and Afshari, E. (2017). A 170-GHz fully integrated single-chip FMCW imaging radar with 3-D imaging capability. *IEEE Journal of Solid-State Circuits* 52 (10): 2721–2734. https://doi.org/10.1109/JSSC.2017.2725963.

18 Rasilainen, K. et al. (2024). Over-the-air characterization of a steerable sub-THz Si lens and on-chip antenna system. *IEEE Transactions on Microwave Theory and Techniques* 72 (5): 2883–2898. https://doi.org/10.1109/TMTT.2023.3329724.

19 Oshiro, M. et al. (2016). A 3.2 mA-RX 3.5 mA-TX fully integrated SoC for bluetooth low energy. *2016 IEEE Asian Solid-State Circuits Conference (A-SSCC)*, 07–09 November 2016, Toyama, Japan, pp. 1–4. http://doi.org/10.1109/ASSCC.2016.7844120.

20 Popplewell, P., Karam, V., Shamim, A. et al. (2008). A 5.2-GHz BFSK transceiver using injection-locking and an on-chip antenna. *IEEE Journal of Solid-State Circuits* 43 (4): 981–990. https://doi.org/10.1109/JSSC.2008.917516.

21 Huang, K.K. and Wentzloff, D.D. 60 GHz on-chip patch antenna integrated in a 0.13μm CMOS technology. *2010 IEEE Int. Conf. Ultra-Wideband, ICUWB2010 – Proc.*, vol. 1, pp. 122–125. http://doi.org/10.1109/ICUWB.2010.5615699.

22 Jing, L., Rowell, C.R., Raju, S. et al. (2016). Fabrication and measurement of millimeter-wave on-chip MIMO antenna for CMOS RFIC's. *2016 IEEE MTT-S*

Int. Wirel. Symp. IWS 2016, pp. 1–4. http://doi.org/10.1109/IEEE-IWS.2016.7585434.

23 Chang, L., Li, Y., Zhang, Z. et al. (2018). Planar air-filled terahertz antenna array based on channelized coplanar waveguide using hierarchical silicon bulk micromachining. *IEEE Transactions on Antennas and Propagation* 66 (10): 5318–5325. https://doi.org/10.1109/TAP.2018.2862360.

24 Paulraj, A., Nabar, R., and Gore, D. (2003). *Introduction to Space-Time Wireless Communications*. Cambridge University Press.

25 Kimura, Y. et al. (2005). A low-cost and very compact wireless terminal integrated on the back of a waveguide planar array for 26 GHz band Fixed Wireless Access (FWA) systems. *IEEE Transactions on Antennas and Propagation* 53 (8I): 2456–2463. https://doi.org/10.1109/TAP.2005.852320.

26 Miura, Y., Hirokawa, J., Ando, M. et al. (2011). Double-layer full-corporate-feed hollow-waveguide slot array antenna in the 60-GHz band. *IEEE Transactions on Antennas and Propagation* 59 (8): 2844–2851. https://doi.org/10.1109/TAP.2011.2158784.

27 Kim, D., Hirokawa, J., Ando, M. et al. (2014). 64 × 64-element and 32 × 32-element slot array antennas using double-layer hollow-waveguide corporate-feed in the 120 GHz band. *IEEE Transactions on Antennas and Propagation* 62 (3): 1507–1512. https://doi.org/10.1109/TAP.2013.2296318.

28 Guo, Q.Y. and Wong, H. (2021). 155 GHz dual-polarized Fabry-Perot cavity antenna using LTCC-based feeding source and phase-shifting surface. *IEEE Transactions on Antennas and Propagation* 69 (4): 2347–2352. https://doi.org/10.1109/TAP.2020.3019528.

29 Mopidevi, H. et al. (2014). Three-dimensional microfabricated broadband patch antenna for wigig applications. *IEEE Antennas and Wireless Propagation Letters* 13: 828–831. https://doi.org/10.1109/LAWP.2014.2319242.

30 Del Campo, A. and Greiner, C. (2007). SU-8: A photoresist for high-aspect-ratio and 3D submicron lithography. *Journal of Micromechanics and Microengineering* 17 (6): https://doi.org/10.1088/0960-1317/17/6/R01.

31 Wang, Y., Ke, M., Lancaster, M.J., and Chen, J. (2011). Micromachined 300-GHz SU-8-based slotted waveguide antenna. *IEEE Antennas and Wireless Propagation Letters* 10: 573–576. https://doi.org/10.1109/LAWP.2011.2158285.

32 Liu, W., Wang, H., Zhang, Z., and Feng, Z.Micromachined 40 GHz post-supported patch array using thick SU8 photoresist. *IEEE Antennas Propag. Soc. AP-S Int. Symp.*, 06–11 July 2014, Memphis, TN, USA, pp. 1083–1084. http://doi.org/10.1109/APS.2014.6904868.

33 Kildal, P.S. (2009, March). Three metamaterial-based gap waveguides between parallel metal plates for mm/submm waves. *2009 3rd European Conference* *on Antennas and Propagation*, 23–27 March 2009, Berlin, Germany, IEEE, pp. 28–32.

34 Kildal, P.S., Alfonso, E., Valero-Nogueira, A., and Rajo-Iglesias, E. (2009). Local metamaterial-based waveguides in gaps between parallel metal plates. *IEEE Antennas and Wireless Propagation Letters* 8: 84–87. https://doi.org/10.1109/LAWP.2008.2011147.

35 Kildal, P.S. (1990). Artificially soft and hard surfaces in electromagnetics. *IEEE Transactions on Antennas and Propagation* 38 (10): 1537–1544.

36 Silveirinha, M.G., Fernandes, C.A., and Costa, J.R. (2008). Electromagnetic characterization of textured surfaces formed by metallic pins. *IEEE Transactions on Antennas and Propagation* 56 (2): 405–415. https://doi.org/10.1109/TAP.2007.915442.

37 Zaman, A.U. and Kildal, P.S. (2014). Wide-band slot antenna arrays with single-layer corporate-feed network in ridge gap waveguide technology. *IEEE Transactions on Antennas and Propagation* 62 (6): 2992–3001. https://doi.org/10.1109/TAP.2014.2309970.

38 Zarifi, D., Farahbakhsh, A., Zaman, A.U., and Kildal, P.S. (2016). Design and fabrication of a high-gain 60-GHz corrugated slot antenna array with ridge gap waveguide distribution layer. *IEEE Transactions on Antennas and Propagation* 64 (7): 2905–2913. https://doi.org/10.1109/TAP.2016.2565682.

39 Vosoogh, A. and Kildal, P.S. (2016). Corporate-fed planar 60-GHz slot array made of three unconnected metal layers using AMC pin surface for the gap waveguide. *IEEE Antennas and Wireless Propagation Letters* 15: 1935–1938. https://doi.org/10.1109/LAWP.2015.2510296.

40 Zarifi, D., Farahbakhsh, A., and Zaman, A.U. (2017). A gap waveguide-fed wideband patch antenna array for 60-GHz applications. *IEEE Transactions on Antennas and Propagation* 65 (9): 4875–4879. https://doi.org/10.1109/TAP.2017.2722866.

41 Sadri-Moshkenani, P., Rashed-Mohassel, J., and Shahabadi, M. (2018). Microstrip antenna array fed by a low-loss gap-waveguide feed network. *IEEE Transactions on Antennas and Propagation* 66 (8): 4359–4363. https://doi.org/10.1109/TAP.2018.2835722.

42 Farahbakhsh, A., Zarifi, D., and Zaman, A.U. (2018). A mmWave wideband slot array antenna based on ridge gap waveguide with 30% bandwidth. *IEEE Transactions on Antennas and Propagation* 66 (2): 1008–1013. https://doi.org/10.1109/TAP.2017.2782263.

43 Vosoogh, A., Kildal, P.S., and Vassilev, V. (2017). Wideband and high-gain corporate-fed gap waveguide slot array antenna with ETSI Class II radiation pattern in V-band. *IEEE Transactions on Antennas and Propagation* 65 (4): 1823–1831. https://doi.org/10.1109/TAP.2016.2634282.

44 Liu, J., Vosoogh, A., Zaman, A.U., and Yang, J. (2019). A slot array antenna with single-layered corporate-feed based on ridge gap waveguide in the 60 GHz band. *IEEE Transactions on Antennas and Propagation* 67 (3): 1650–1658. https://doi.org/10.1109/TAP.2018.2888730.

45 Farahbakhsh, A., Zarifi, D., and Zaman, A.U. (2017). 60-GHz groove gap waveguide based wideband h-plane power dividers and transitions: for use in high-gain slot array antenna. *IEEE Transactions on Microwave Theory and Techniques* 65 (11): 4111–4121. https://doi.org/10.1109/TMTT.2017.2699680.

46 Liu, J., Vosoogh, A., Zaman, A.U., and Yang, J. (2017). Design and fabrication of a high-gain 60-GHz cavity-backed slot antenna array fed by inverted microstrip gap waveguide. *IEEE Transactions on Antennas and Propagation* 65 (4): 2117–2122. https://doi.org/10.1109/TAP.2017.2670509.

47 Razavi, S.A., Kildal, P.S., Xiang, L. et al. (2014). 2 × 2-Slot element for 60-GHz planar array antenna realized on two doubled-sided PCBs using SIW cavity and EBG-type soft surface fed by microstrip-ridge gap waveguide. *IEEE Transactions on Antennas and Propagation* 62 (9): 4564–4573. https://doi.org/10.1109/TAP.2014.2331993.

48 Pucci, E., Rajo-Iglesias, E., Vázquez-Roy, J.L., and Kildal, P.S. (2014). Planar dual-mode horn array with corporate-feed network in inverted microstrip gap waveguide. *IEEE Transactions on Antennas and Propagation* 62 (7): 3534–3542. https://doi.org/10.1109/TAP.2014.2317496.

49 Wu, S., Li, J., Chen, X. et al. (2023). A Ka-band SLM printed filtering divider-fed magnetoelectric dipole antenna array using embedded gap waveguide. *IEEE Antennas and Wireless Propagation Letters* 22 (4): 774–778. https://doi.org/10.1109/LAWP.2022.3224867.

50 Dadgarpour, A., Bayat-Makou, N., Antoniades, M.A. et al. (2020). A dual-polarized magnetoelectric dipole array based on printed ridge gap waveguide with dual-polarized split-ring resonator lens. *IEEE Transactions on Antennas and Propagation* 68 (5): 3578–3585. https://doi.org/10.1109/TAP.2020.2964940.

51 Liu, Y., Yue, Z., Jia, Y. et al. (2021). Dual-band dual-circularly polarized antenna array with printed ridge gap waveguide. *IEEE Transactions on Antennas and Propagation* 69 (8): 5118–5123. https://doi.org/10.1109/TAP.2020.3048504.

52 Hassan, A.T. and Kishk, A.A. (2020). Efficient procedure to design large finite array and its feeding network with examples of ME-dipole array and microstrip ridge gap waveguide feed. *IEEE Transactions on Antennas and Propagation* 68 (6): 4560–4570. https://doi.org/10.1109/TAP.2020.2972407.

53 Sharifi Sorkherizi, M., Dadgarpour, A., and Kishk, A.A. (2017). Planar high-efficiency antenna array using new printed ridge gap waveguide technology. *IEEE Transactions on Antennas and Propagation* 65 (7): 3772–3776. https://doi.org/10.1109/TAP.2017.2700885.

54 Ferrando-Rocher, M., Herranz-Herruzo, J.I., Valero-Nogueira, A. et al. (2019). 8 × 8 Ka-band dual-polarized array antenna based on gap waveguide technology. *IEEE Transactions on Antennas and Propagation* 67 (7): 4579–4588. https://doi.org/10.1109/TAP.2019.2908109.

55 Akbari, M., Farahbakhsh, A., and Sebak, A.R. (2019). Ridge gap waveguide multilevel sequential feeding network for high-gain circularly polarized array antenna. *IEEE Transactions on Antennas and Propagation* 67 (1): 251–259. https://doi.org/10.1109/TAP.2018.2878281.

56 Ferrando-Rocher, M., Herranz-Herruzo, J.I., Valero-Nogueira, A., and Vila-Jimenez, A. (2018). Single-layer circularly-polarized Ka-band antenna using gap waveguide technology. *IEEE Transactions on Antennas and Propagation* 66 (8): 3837–3845. https://doi.org/10.1109/TAP.2018.2835639.

57 Zang, Z., Zaman, A.U., and Yang, J. (2022). Single layer dual circularly polarized antenna array based on ridge gap waveguide for 77 GHz automotive radar. *IEEE Transactions on Antennas and Propagation* 70 (7): 5977–5982. https://doi.org/10.1109/TAP.2022.3161283.

58 Wu, K., Yao, Y., Cheng, X. et al. (2022). Wideband circularly polarized septum antenna array with ridge gap waveguide feeding network for wireless application. *IEEE Transactions on Antennas and Propagation* 70 (5): 3882–3887. https://doi.org/10.1109/TAP.2021.3137508.

59 Liu, J., Yang, F., Fan, K., and Jin, C. (2021). Unequal power divider based on inverted microstrip gap waveguide and its application for low sidelobe slot array antenna at 39 GHz. *IEEE Transactions on Antennas and Propagation* 69 (12): 8415–8425. https://doi.org/10.1109/TAP.2021.3096981.

60 Ran, J., Jin, C., Zhang, P. et al. (2022). High-gain and low-loss dual-polarized antenna array with reduced sidelobe level based on gap waveguide at 28 GHz. *IEEE Antennas and Wireless Propagation Letters* 21 (5): 1022–1026. https://doi.org/10.1109/LAWP.2022.3155566.

61 Cao, B., Wang, H., Huang, Y., and Zheng, J. (2015). High-gain L-probe excited substrate integrated cavity antenna array with LTCC-based gap waveguide feeding network for W-band application. *IEEE Transactions on Antennas and Propagation* 63 (12): 5465–5474. https://doi.org/10.1109/TAP.2015.2481483.

62 Cao, B., Wang, H., and Huang, Y. (2016). W-band high-gain TE220-mode slot antenna array with gap waveguide feeding network. *IEEE Antennas and Wireless Propagation Letters* 15: 988–991. https://doi.org/10.1109/LAWP.2015.2489721.

63 Algaba Brazalez, A., Pucci, E., Kildal, P.S. et al. (2013). Evaluation of losses of the ridge gap waveguide at 100 GHz. *IEEE Antennas Propag. Soc. AP-S Int. Symp.* 1456–1457. https://doi.org/10.1109/APS.2013.6711387.

64 Shi, Y., Zhang, J., Zeng, S., and Zhou, M. (2018). Novel W-band millimeter-wave transition from microstrip line to groove gap waveguide for MMIC integration and antenna application. *IEEE Transactions on Antennas and Propagation* 66 (6): 3172–3176. https://doi.org/10.1109/TAP.2018.2819902.

65 Vosoogh, A., Haddadi, A., Zaman, A.U. et al. (2018). W-band low-profile monopulse slot array antenna based on gap waveguide corporate-feed network. *IEEE Transactions on Antennas and Propagation* 66 (12): 6997–7009. https://doi.org/10.1109/TAP.2018.2874427.

66 Tamayo-Domínguez, A., Fernández-González, J.M., and Sierra-Castañer, M. (2021). Monopulse radial line slot array antenna fed by a 3-D-printed cavity-ended modified butler matrix based on gap waveguide at 94 GHz. *IEEE Transactions on Antennas and Propagation* 69 (8): 4558–4568. https://doi.org/10.1109/TAP.2021.3060045.

67 Moradi, M., Sharawi, M.S., and Wu, K. (2024). Exploring low-loss wideband substrate-integrated image guides (SIIG) for terahertz applications. *IEEE Transactions on Terahertz Science and Technology* 14 (1): 1–12. https://doi.org/10.1109/TTHZ.2023.3332303.

68 Moradi, M., Sharawi, M.S., and Wu, K. (2023). Analytical model of guided waves in periodically perforated dielectric structure and its applications to terahertz substrate-integrated image guide (SIIG). *IEEE Transactions on Microwave Theory and Techniques* 71 (10): 4236–4246. https://doi.org/10.1109/TMTT.2023.3260445.

69 Rahmati, E., Burasa, P., Baladi, E., and Sharawi, M.S. Fully embedded dual-element dielectric-based antenna for sub-and terahertz applications. *IEEE Open Journal of Antennas and Propagation*. https://doi.org/10.1109/OJAP.2024.3392628.

70 Gupta, A., Aggarwal, A., Sharma, A., and Kumar, V.Design and performance analysis of an aperture coupled MIMO antenna for THz applications. *2022 2nd Int. Conf. Comput. Sci. Eng. Appl. ICCSEA 2022*, 08-08 September, Gunupur, India, pp. 1–4. http://doi.org/10.1109/ICCSEA54677.2022.9936337.

71 Pant, R. and Malviya, L.(2021). A 2 × 2 MIMO-antenna array based on photonic band gap for terahertz applications. *2021 IEEE Indian Conf. Antennas Propagation, InCAP 2021*, 13–16 December 2021, Jaipur, Rajasthan, India, India, pp. 68–71. 10.1109/InCAP52216.2021.9726488.

72 Tripathi, U., Solanki, D., Malviya, P. et al. (2023). MIMO antenna design with PBG structure for THz communication. *Int. Conf. Microwave, Antenna Commun. MAC 2023*, 24–26 March 2023, Prayagraj, India, no. Ml, pp. 1–6, http://doi.org/10.1109/MAC58191.2023.10177063.

73 Das, S., Mitra, D., and Chaudhuri, S.R.B.Fractal loaded planar super wide band MIMO antenna in THz frequency range. *Asia-Pacific Microw. Conf.*

Proceedings, APMC, 08–11 December 2020, Hong Kong, China, vol. 2020, December, pp. 703–705. http//doi.org/10.1109/APMC47863.2020.9331409.

74 Muthukrishnan, K., Kamruzzaman, M.M., Lavadiya, S., and Sorathiya, V. (2023). Superlative split ring resonator shaped ultrawideband and high gain 1×2 MIMO antenna for Terahertz communication. *Nano Communication Networks* 36: 100437. https://doi.org/10.1016/j.nancom.2023.100437.

75 Mencagli, M., Martini, E., and Maci, S. (2015). Surface wave dispersion for anisotropic metasurfaces constituted by elliptical patches. *IEEE Transactions on Antennas and Propagation* 63 (7): 2992–3003. https://doi.org/10.1109/TAP.2015.2422352.

76 Aydin, K., Cubukcu, E., Ozbay, E., and Bayindir, M. (2003). Transmission and reflection properties of composite metamaterials in free space. *OSA Trends Opt. Photonics Ser.* 88 (10): 195–197.

77 Faenzi, M., Minatti, G., González-Ovejero, D. et al. (2019). Metasurface antennas: new models, applications and realizations. *Scientific Reports* 9 (1): 10178.

78 Zhang, N. et al. (2020, 2020). Programmable coding metasurface for dual-band independent real-time beam control. *IEEE Journal on Emerging and Selected Topics in Circuits and Systems* 10 (1): 20–28. https://doi.org/10.1109/JETCAS.2020.2973310.

79 Zhang, N. et al. (2022). A dual-polarized reconfigurable reflectarray antenna based on dual-channel programmable metasurface. *IEEE Transactions on Antennas and Propagation* 70 (9): 7403–7412. https://doi.org/10.1109/TAP.2022.3165872.

80 Yang, H. et al. (2016). A 1-bit 10 × 10 reconfigurable reflectarray antenna: design, optimization, and experiment. *IEEE Transactions on Antennas and Propagation* 64 (6): 2246–2254. https://doi.org/10.1109/TAP.2016.2550178.

81 Liu, S. et al. (2016). Frequency-dependent dual-functional coding metasurfaces at terahertz frequencies. *Advanced Optical Materials* 4 (12): 1965–1973. https://doi.org/10.1002/adom.201600471.

82 Gutierrez, C. E., Pallucchini, L., & Stachura, E. (2017). General refraction problems with phase discontinuity. arXiv preprint arXiv:1703.06014.

83 Yang, H. et al. (2017). A 1600-element dual-frequency electronically reconfigurable reflectarray at X/Ku-band. *IEEE Transactions on Antennas and Propagation* 65 (6): 3024–3032. https://doi.org/10.1109/TAP.2017.2694703.

84 Dai, J.Y. et al. (2020). Realization of multi-modulation schemes for wireless communication by time-domain digital coding metasurface. *IEEE Transactions on Antennas and Propagation* 68 (3): 1618–1627. https://doi.org/10.1109/TAP.2019.29524.

85 Zhao, J. et al. (2019). Programmable time-domain digital-coding metasurface for non-linear harmonic manipulation and new wireless communication systems. *National Science Review* 6 (2): 231–238. https://doi.org/10.1093/nsr/nwy135.

86 Minkov, M., Shi, Y., and Fan, S. (2017). Exact solution to the steady-state dynamics of a periodically modulated resonator. *APL Photonics* 2 (7): https://doi.org/10.1063/1.4985381.

87 Dai, J.Y., Zhao, J., Cheng, Q., and Cui, T.J. (2018). Independent control of harmonic amplitudes and phases via a time-domain digital coding metasurface. *Light: Science & Applications* 7 (1): https://doi.org/10.1038/s41377-018-0092-z.

88 Zhang, L. et al. (2019). Breaking reciprocity with space-time-coding digital metasurfaces. *Advanced Materials* 31 (41): 1–10. https://doi.org/10.1002/adma.201904069.

89 Shaltout, A., Kildishev, A., and Shalaev, V. (2015). Time-varying metasurfaces and Lorentz non-reciprocity. *Optical Materials Express* 5 (11): 2459. https://doi.org/10.1364/ome.5.002459.

90 Zhang, L. et al. (2018). Space-time-coding digital metasurfaces. *Nature Communications* 9 (1): 1–11. https://doi.org/10.1038/s41467-018-06802-0.

91 Luo, Z. et al. (2019). Digital nonlinear metasurface with customizable nonreciprocity. *Advanced Functional Materials* 29 (49): 1–10. https://doi.org/10.1002/adfm.201906635.

92 Shtaiwi, E., Zhang, H., Vishwanath, S. et al. (2021). Channel estimation approach for RIS assisted MIMO systems. *IEEE Transactions on Cognitive Communications and Networking* 7 (2): 452–465. https://doi.org/10.1109/TCCN.2021.3075413.

93 Elmossallamy, M.A., Zhang, H., Song, L. et al. (2020). Reconfigurable intelligent surfaces for wireless communications: principles, challenges, and opportunities. *IEEE Transactions on Cognitive Communications and Networking* 6 (3): 990–1002. https://doi.org/10.1109/TCCN.2020.2992604.

94 Zhang, S. and Zhang, R. (2020). Capacity characterization for intelligent reflecting surface aided MIMO communication. *IEEE Journal on Selected Areas in Communications* 38 (8): 1823–1838. https://doi.org/10.1109/JSAC.2020.3000814.

95 Heath, R.W., Gonzalez-Prelcic, N., Rangan, S. et al. (2016). An overview of signal processing techniques for millimeter wave MIMO systems. *IEEE Journal of Selected Topics in Signal Processing* 10 (3): 436–453. https://doi.org/10.1109/JSTSP.2016.2523924.

96 Wu, Q. and Zhang, R. (2019). Intelligent reflecting surface enhanced wireless network via joint active and passive beamforming. *IEEE Transactions on Wireless Communications* 18 (11): 5394–5409. https://doi.org/10.1109/TWC.2019.2936025.

97 Zhang, X., Liang, T., An, K. et al. (2022). Secure transmission in RIS-assisted cell-free massive MIMO system with low resolution ADCs/DACs*. *IEEE Wireless Communications and Networking Conference (WCNC)* 2022: 339–344. https://doi.org/10.1109/WCNC51071.2022.9771561.

98 Ning, B., Chen, Z., Chen, W. et al. (2021). Terahertz multi-user massive MIMO with intelligent reflecting surface: beam training and hybrid beamforming. *IEEE Transactions on Vehicular Technology* 70 (2): 1376–1393. https://doi.org/10.1109/TVT.2021.3052074.

99 Lin, Q.W., Wong, H., Huitema, L., and Crunteanu, A. (2022). Coding metasurfaces with reconfiguration capabilities based on optical activation of phase-change materials for terahertz beam manipulations. *Advanced Optical Materials* 10 (1): 2101699.

100 Meng, X., Nekovee, M., and Wu, D. (2021). The design and analysis of electronically reconfigurable liquid crystal-based reflectarray metasurface for 6g beamforming, beamsteering, and beamsplitting. *IEEE Access* 9: 155564–155575. https://doi.org/10.1109/ACCESS.2021.3125837.

101 Venkatesh, S., Lu, X., Saeidi, H., and Sengupta, K. (2020). A high-speed programmable and scalable terahertz holographic metasurface based on tiled CMOS chips. *Nature Electronics* 3 (12): 785–793.

102 Dai, L. et al. (2020). Reconfigurable intelligent surface-based wireless communications: antenna design, prototyping, and experimental results. *IEEE Access* 8: 45913–45923. https://doi.org/10.1109/ACCESS.2020.2977772.

103 Lin, Q.W., Wong, H., Huitema, L., and Crunteanu, A. (2022). Coding metasurfaces with reconfiguration capabilities based on optical activation of phase-change materials for terahertz beam manipulations. *Advanced Optical Materials* 10 (1): https://doi.org/10.1002/adom.202101699.

104 Kodama, C.H. and Coutu, R.A. (2016). Tunable split-ring resonators using germanium telluride. *Applied Physics Letters* 108 (23): https://doi.org/10.1063/1.4953228.

105 Pitchappa, P., Kumar, A., Prakash, S. et al. (2019). Chalcogenide phase change material for active terahertz photonics. *Advanced Materials* 31 (12): 1–7. https://doi.org/10.1002/adma.201808157.

106 Balanis, C.A. (2016). *Antenna Theory: Analysis and Design*. Wiley.

107 Liu, C.X. et al. (2021). Programmable manipulations of terahertz beams by transmissive digital coding metasurfaces based on liquid crystals. *Advanced Optical Materials* 9 (22): 1–9. https://doi.org/10.1002/adom.202100932.

108 Wu, J. et al. (2020). Liquid crystal programmable metasurface for terahertz beam steering. *Applied Physics Letters* 116 (13): https://doi.org/10.1063/1.5144858.

109 Zhang, W. et al. (2021). Microfluid-based soft metasurface for tunable optical activity in THz wave. *Optics Express* 29 (6): 8786. https://doi.org/10.1364/oe.420660.

110 Li, M., Chen, X., Zhang, A. et al. (2021). Reducing correlation in compact arrays by adjusting near-field phase distribution for MIMO applications. *IEEE Transactions on Vehicular Technology* 70 (8): 7885–7896. https://doi.org/10.1109/TVT.2021.3094314.

111 Yan, Y. et al. (2014). High-capacity millimetre-wave communications with orbital angular momentum multiplexing. *Nature Communications* 5: 1–9. https://doi.org/10.1038/ncomms5876.

112 Zhang, Z., Zheng, S., Zhang, W. et al. (2017). Experimental demonstration of the capacity gain of plane spiral OAM-based MIMO system. *IEEE Microwave and Wireless Components Letters* 27 (8): 757–759. https://doi.org/10.1109/LMWC.2017.2723719.

113 Zhang, Z., Zheng, S., Chen, Y. et al. (2016). The capacity gain of orbital angular momentum based system. Nat. Publ. Gr., no. November 2015, pp. 1–11. http://doi.org/10.1038/srep25418.

114 Ren, Y. et al. (2017). Line-of-sight millimeter-wave communications using orbital angular momentum multiplexing combined with conventional spatial multiplexing. *IEEE Transactions on Wireless Communications* 16 (5): 3151–3161. https://doi.org/10.1109/TWC.2017.2675885.

115 Tamburini, F., Mari, E., and Thide, B. (2011). Experimental verification of photon angular momentum and vorticity with radio techniques. *Applied Physics Letters* 204102: 2011–2013. https://doi.org/10.1063/1.3659466.

116 Li, X., Zhou, N., Zhu, S. et al. (2023). Communication resolution analysis of coincidence imaging based on OAM beams with equal divergence angle. *IEEE Transactions on Antennas and Propagation* 71 (3): 2891–2896. https://doi.org/10.1109/TAP.2022.3233718.

117 Meng, X., Chen, X., Chen, R. et al. (2021). Generation of multiple high-order Bessel beams carrying different orbital-angular-momentum modes through an anisotropic holographic impedance metasurface. *Physical Review Applied* 16 (4): 044063.

118 Andersson, M., Berglind, E., and Björk, G. (2015). Orbital angular momentum modes do not increase the channel capacity in communication links. *New Journal of Physics* 17 (4): 043040.

119 Edfors, O. and Johansson, A.J. (2012). Is orbital angular momentum (OAM) based radio communication an unexploited area? *IEEE Transactions on Antennas and Propagation* 60 (2): 1126–1131. https://doi.org/10.1109/TAP.2011.2173142.

120 Y. M. Zhang and J. -L. Li, “Analyses and full-duplex applications of circularly polarized OAM arrays using sequentially rotated configuration,” in *IEEE Transactions on Antennas and Propagation*, vol. 66, no. 12, pp. 7010-7020, Dec. 2018, https://doi.org/10.1109/TAP.2018.2872169.

121 Zhang, Y., Member, S., and Li, J. (2018). Analyses and full-duplex applications of circularly polarized OAM arrays using sequentially rotated configuration. *IEEE Transactions on Antennas and Propagation* 66 (12): 1. https://doi.org/10.1109/TAP.2018.2872169.

122 Chen, X. et al. (2020). Orbital angular momentum multiplexing in highly reverberant environments. *IEEE Microwave and Wireless Components Letters* 30 (1): 112–115. https://doi.org/10.1109/LMWC.2019.2952975.

123 Xue, W., Chen, X., Liu, X. et al. (2020). A revisit of orbital angular momentum multiplexing in multipath environment. *Journal of Communications and Information Networks* 5 (4): 438–446. https://doi.org/10.23919/JCIN.2020.9306017.

124 Shi, H., Wang, L., Peng, G. et al. (2019). Generation of multiple modes microwave vortex beams using active metasurface. *IEEE Antennas and Wireless Propagation Letters* 18 (1): 59–63. https://doi.org/10.1109/LAWP.2018.2880732.

125 Li, J. et al. (2019). Dual-band eight-antenna array design for MIMO applications in 5G mobile terminals. *IEEE Access* 7: 71636–71644. https://doi.org/10.1109/ACCESS.2019.2908969.

126 Lei, L., Wan, J., and Gong, Q. (2022). Multibeam antennas with reflector for high throughput satellite applications. *IEEE Aerospace and Electronic Systems Magazine* 37 (2): 34–46. https://doi.org/10.1109/MAES.2021.3121654.

127 Cherrette, A.R., Lee, S.-W., and Acosta, R.J. (1989). A method for producing a shaped contour radiation pattern using a single shaped reflector and a single feed. *IEEE Transactions on Antennas and Propagation* 37 (6): 698–706. https://doi.org/10.1109/8.29356.

128 Palacin, B. et al. (2017). Multibeam antennas for very high throughput satellites in Europe: Technologies and trends. *2017 11th Eur. Conf. Antennas Propagation, EUCAP 2017*, pp. 2413–2417. http://doi.org/10.23919/EuCAP.2017.7928493.

129 Lepeltier, P., Bosshard, P., Maurel, J. et al. (2012). Recent achievements and future trends for multiple beam telecommunication antennas. *2012 15th Int. Symp. Antenna Technol. Appl. Electromagn. ANTEM 2012*, 25–28 June 2012, Toulouse, France, *2012*. http://doi.org/10.1109/ANTEM.2012.6262426.

130 Fenech, H., Amos, S., Hirsch, A., and Soumpholphakdy, V.VHTS systems: requirements and evolution. *2017 11th Eur. Conf. Antennas Propagation, EUCAP 2017*, pp. 2409–2412. http://doi.org/10.23919/EuCAP.2017.7928175.

131 Hong, W. et al. (2017). Multibeam antenna technologies for 5G wireless communications. *IEEE Transactions on Antennas and Propagation* 65 (12): 6231–6249. https://doi.org/10.1109/TAP.2017.2712819.

132 Balanis, C.A. (2016). *Antenna Theory: Analysis and Design*. Wiley.

133 Rao, S.K. (1999). Design and analysis of multiple-beam reflector antennas. *IEEE Antennas and Propagation Magazine* 41 (4): 53–69. https://doi.org/10.1109/74.789737.

134 Toso, G., Mangenot, C. and Angeletti, P.Recent advances on space multi-beam antennas based on a single aperture. *2013 7th European Conference on Antennas and Propagation (EuCAP)*, 08–12 April 2013, Gothenburg, Sweden, pp. 454–458.

135 Elser, D et al. (2015). Satellite quantum communication via the alphasat laser communication terminal – quantum signals from 36 thousand kilometers above earth. *2015 IEEE International Conference on Space Optical Systems and Applications (ICSOS)*, New Orleans, LA, USA, pp. 1–4. http://doi.org/10.1109/ICSOS.2015.7425077.

136 Doan, D.L., Amyotte, E., Mok, C. and Uher, J.Anik-F2 Ka-band transmit multibeam antenna. *2004 10th International Symposium on Antenna Technology and Applied Electromagnetics and URSI Conference*, 20–23 July 2004, Ottawa, ON, Canada, pp. 1–4, http://doi.org/10.1109/ANTEM.2004.7860708.

137 Sawekpun, T. The iPSTAR broadband satellite. *21st AIAA International Communications Satellite Systems Conference and Exhibit*, 15 April 2003–19 April 2003, Yokohama, Japan.

138 Volakis, J.L. (2007). *Antenna Engineering Handbook*. McGraw-Hill Education.

139 Ruggerini, G., Nicolaci, P., Toso, G., and Angeletti, P.An active discrete lens antenna for Ka-band multibeam applications. *IEEE Antennas Propag. Soc. AP-S Int. Symp.*, 08–14 July 2012, Chicago, IL, USA, pp. 1–2. http://doi.org/10.1109/APS.2012.6348516.

140 Yu, L., Wan, J., Zhang, K. et al. (2023). Spaceborne multibeam phased array antennas for satellite communications. *IEEE Aerospace and Electronic Systems Magazine* 38 (3): 28–47. https://doi.org/10.1109/MAES.2022.3231580.

141 Vallappil, A.K., Rahim, M.K.A., and Member, S. (2021). Butler matrix based beamforming networks for phased array antenna systems: a comprehensive review and future directions for 5G applications. *IEEE Access* 9: https://doi.org/10.1109/ACCESS.2020.3047696.

142 Rohwer, A.B., Desrosiers, D.H., Bach, W. et al. (2010). Iridium main mission antennas – a phased array success story and mission update. *IEEE Int. Symp. Phased Array Syst. Technol.*, 12–15 October 2010, Waltham, MA, USA, pp. 504–511. http://doi.org/10.1109/ARRAY.2010.5613319.

143 Metzen, P.L. 2000. Globalstar satellite phased array antennas. *IEEE Int. Symp. Phased Array Syst. Technol.*, pp. 207–210. http://doi.org/10.1109/past.2000.858941.

144 Croq, F. et al. The GLOBALSTAR 2 antenna sub-system. *2009 3rd European Conference on Antennas and Propagation*, 21–25 May 2000, Dana Point, CA, USA, Berlin, Germany, pp. 598–602.

145 Kempton, B. and Riedl, A.Network simulator for large low earth orbit satellite networks. *IEEE Int. Conf. Commun.*, 14–23 June 2021, Montreal, QC, Canada, pp. 5–7. 10.1109/ICC42927.2021.9500439.

146 Li, Y., Luo, Z., Zhou, W., and Zhu, J. (2021). Benefits analysis of beam hopping in satellite mobile system with unevenly distributed traffic. *China Communications* 18 (9): 11–23. https://doi.org/10.23919/JCC.2021.09.002.

147 Li, Y., Fan, Y., Liu, S. et al. (2021). Overview of beam hopping algorithms in large scale LEO satellite constellation. *Proc. – 2021 IEEE 20th Int. Conf. Trust. Secur. Priv. Comput. Commun. Trust.*, 20–22 October 2021, Shenyang, China, pp. 1345–1351, http://doi.org/10.1109/TrustCom53373.2021.00187.

148 Bailleul, P.K. (2016). A new era in elemental digital beamforming for spaceborne communications phased arrays. *Proceedings of the IEEE* 104 (3): 623–632. https://doi.org/10.1109/JPROC.2015.2511661.

149 Tsokos, C., Mylonas, E., Groumas, P. et al. (2019). Optical beamforming network for multi-beam operation with continuous angle selection. *IEEE Photonics Technology Letters* 31 (2): 177–180. https://doi.org/10.1109/LPT.2018.2889411.

150 Roeloffzen, C.G.H. et al. (2018). Low-loss si3n4 triplex optical waveguides: technology and applications overview. *IEEE Journal of Selected Topics in Quantum Electronics* 24 (4): https://doi.org/10.1109/JSTQE.2018.2793945.

151 Lafond, J.C. et al. Thales Alenia Space multiple beam antennas for telecommunication satellites. *8th Eur. Conf. Antennas Propagation, EuCAP 2014*, 06–11 April 2014, The Hague, Netherlands, no. EuCAP 2014, pp. 186–190. http://doi.org/10.1109/EuCAP.2014.6901723.

152 Fan, B., Zhang, F., Ma, C. et al. (2020). Microwave photonic bistatic radar for real-time and high-resolution imaging. *IEEE Photonics Technology Letters* 32 (21): 1397–1400. https://doi.org/10.1109/LPT.2020.3026776.

153 Lu, N., Sun, X., Huang, H., and Shao, L.Anti-interference analysis of multiplexing methods for GBBF system. *Proc. 2016 8th IEEE Int. Conf. Commun. Softw. Networks, ICCSN 2016*, 04–06 June 2016, Beijing, China, pp. 236–240. http://doi.org/10.1109/ICCSN.2016.7586656.

154 Vasavada, Y., Ravishankar, C., Corrigan, J. et al. (2017). Determination of locations of calibration earth stations for ground based beam forming in satellite systems. *Proc. – IEEE Mil. Commun. Conf. MILCOM*, vol. 2017, October, pp. 331–336. 10.1109/MILCOM.2017.8170795.

8

Decoupling Techniques for MIMO Antenna Systems

8.1 Mutual Coupling and Its Impact on MIMO

Mutual coupling in antenna arrays is a physically complex phenomenon that refers to the electromagnetic interaction between antenna elements. For example, even if all the antennas in an array are transmitting, they will simultaneously receive part of the energy transmitted from each element due to the non-ideal directional radiation characteristics. At the same time, part of the incident energy received by the antenna will be re-radiated, resulting in multiple iterative interferences between the antennas. This energy exchange is called mutual coupling.

In a multi-antenna system, when the distance between the antenna elements becomes closer, the elements in the system are no longer isolated, and their original matching and radiation characteristics will be affected by the surrounding antennas. In this case, the antenna is not only subjected to the electromagnetic effect generated by its own excitation but also receives the electromagnetic effects from other excited antennas, resulting in mutual coupling between them. The existence of mutual coupling will have a great influence on the antennas in the system, which will lead to the deterioration of the matching, pattern, and efficiency.

8.1.1 Generation and Characterization of Mutual Coupling

The detailed mechanism of mutual coupling depends on whether the antennas are in transmitting mode or receiving mode. So, the mechanism of the coupling in the two modes is introduced separately. Generally, the mutual coupling effect in antenna array becomes more complex as the number of elements increases, but the mechanism of coupling stays the same [1].

8.1.1.1 Transmitting Mode

For brevity, a two-element array (i.e. antenna 1 and antenna 2) is considered to illustrate the mutual coupling in transmitting mode, as shown in Figure 8.1a.

MIMO Antenna Systems for 5G and Beyond, First Edition.
Xiaoming Chen, Ping Jack Soh, and Mohammad S. Sharawi.

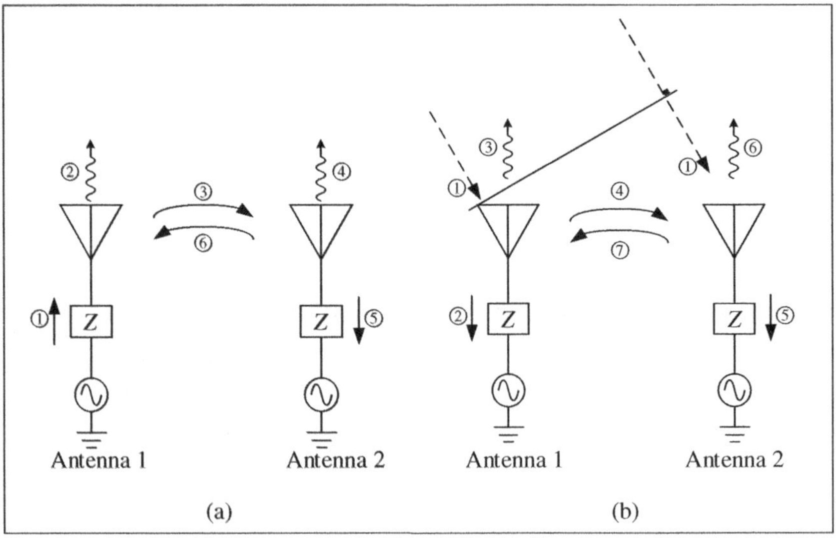

Figure 8.1 Coupling mechanism in array (a) on transmitting mode and (b) receiving mode.

Assuming that antenna 1 is excited, the energy ① from the source will be radiated to free space ② and received by antenna 2 ③. The radiator of antenna 2 will generate induced current due to the received wave ③ from antenna 1. The induced current will be re-radiated toward free space ④ and antenna 1 ⑥, the remaining energy will flow into the generator of antenna 2 ⑤. Besides, the wave ⑥ from antenna 2 will be received and re-scattered by antenna 1, and this process can iterate infinitely between antennas. However, we usually consider the first few exchanges since the scattered energy will drop sharply with the number of iterations.

The coupled wave is transmitted from antenna 1 to antenna 2 to the generator of antenna 2, where the coupled wave is superimposed vectorially on the original incident and reflected waves of antenna 2, thereby deteriorating the standing wave ratio (SWR) and changing the input impedance of antenna 2. Mutual coupling not only changes the mutual impedance of the antenna but also changes its self-impedance. Due to the existence of mutual coupling, there are many scattered fields between the antennas, which tends to a deteriorated radiation pattern.

8.1.1.2 Receiving Mode

Similarly, the array system in receiving mode consists of two antenna elements (for this illustration, but usually will have more than 2) terminated with matched loads, as shown in Figure 8.1b. Assume that the incident wave ① is received by antenna 1 at first, thus the induced current in antenna 1 will be radiated to space

③ and antenna 2 ④, and the wave ④ is superimposed with the incident wave ① entering antenna 2, and the induced current generated in antenna 2 will also be re-radiated to antenna 1 ⑦ to repeat the same process and coupling is obtained.

8.1.1.3 Coupling Path

In addition to the above-mentioned radiated waves and scattered waves in space, when an antenna *m* and an antenna *n* share the same ground, as shown in Figure 8.2, the coupling paths between them also contain the induced current on the ground and surface waves on the dielectric substrate (for printed antennas). Assume antenna *m* is excited, the induced current on the ground will flow to the adjacent antenna *n* and will be radiated by antenna *n*, affecting the radiation pattern of elements. When radiators *m* and *n* are all on the same dielectric substrate, there will also be a surface wave coupling path propagating along the dielectric substrate. Therefore, part of the energy emitted by antenna *m* will be transmitted to antenna *n* in the form of surface waves. This will not only affect the radiation pattern of antenna *n* but also the wave received by antenna *n* will generate surface waves that will end up at antenna *m*. In turn, such surface waves will reduce the radiation efficiency of the antenna system.

For example, in a microstrip patch antenna array, multiple patches are placed on the same dielectric substrate, and each unit has the same profile. The propagation of surface waves in the antenna array has a serious impact on its performance. It is necessary to analyze the surface current distribution and suppress the surface waves from reaching adjacent antenna elements. In order to improve the radiation efficiency of each antenna in the antenna array and maintain the spatial diversity and pattern diversity of the elements, ensuring proper isolation between the radiators on the same dielectric substrate is a key objective.

The coupling paths between the elements in the antenna array can be classified as space wave coupling, surface wave coupling, and induced currents. Therefore, in studying coupling suppression in antenna arrays, different decoupling

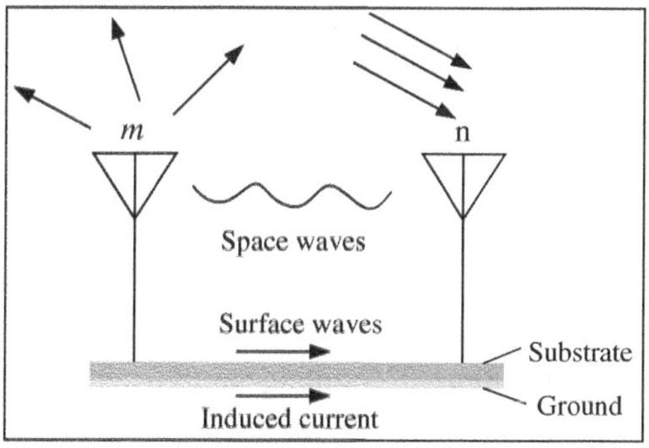

Figure 8.2 Coupling path.

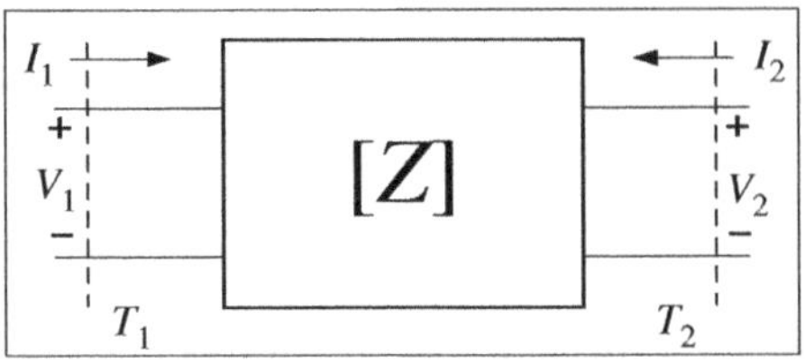

Figure 8.3 Two-port network.

structures can be designed according to the element type, its polarization mode, and array configuration, to obtain a significant decoupling effect.

In a multi-antenna system, the entire antenna system can be represented by a multi-port network. The essence of using the impedance matrix to analyze mutual coupling is to analyze the relationship between voltage and current in this multi-port network, calculate the impedance relationship between each port, and then perform mutual coupling analysis. The following will consider a passive two-port network as an example for the analysis presented.

As shown in Figure 8.3, a two-element antenna system can be equivalent to a two-port network. The voltage and current corresponding to the two ports T_1 and T_2 are V_1 and I_1 and V_2 and I_2, respectively. Then the relationship between the corresponding voltages and currents can be written as:

$$\begin{cases} V_1 = Z_{11}I_1 + Z_{12}I_2 \\ V_2 = Z_{21}I_1 + Z_{22}I_2 \end{cases} \tag{8.1}$$

Alternatively, it can be further abbreviated as:

$$[V] = [Z][I] \tag{8.2}$$

where the $[Z]$ matrix is the corresponding two-port network impedance matrix. In the actual analysis, the mutual impedances Z_{12} and Z_{21} are used to describe the mutual coupling between the two antenna elements. Similarly, the admittance matrix analysis method, which is similar to its impedance counterpart, is also commonly used to characterize mutual coupling between antenna elements. The admittance matrix describes the relationship between the current and voltage. The relationships between the currents and voltages of the two ports can be written as:

$$\begin{cases} I_1 = Y_{11}V_1 + Y_{12}V_2 \\ I_2 = Y_{21}V_1 + Y_{22}V_2 \end{cases} \tag{8.3}$$

And then written in matrix form as:

$$\begin{bmatrix} I_1 \\ I_2 \end{bmatrix} = \begin{bmatrix} Y_{11} & Y_{12} \\ Y_{21} & Y_{22} \end{bmatrix} \begin{bmatrix} V_1 \\ V_2 \end{bmatrix} \tag{8.4}$$

In the mutual coupling analysis, it is only necessary to make the mutual impedance in the impedance matrix or the mutual admittance in the admittance

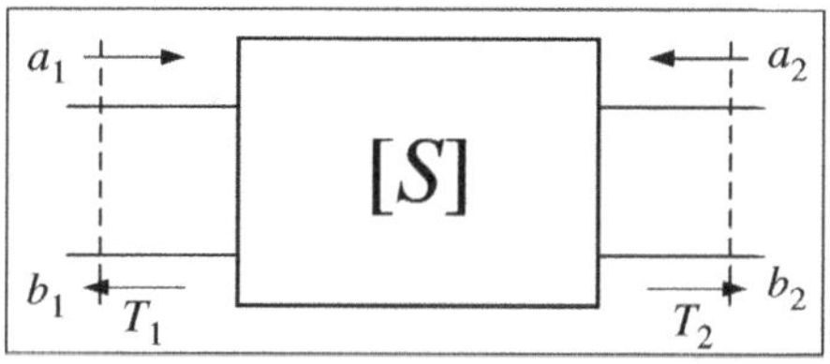

Figure 8.4 Two-port S-parameter matrix.

matrix zero. This means that the mutual influence between the antennas is eliminated, and the coupling suppression between them is accomplished.

In a multi-antenna system, the S-parameter matrix can also be used to analyze the mutual coupling. The essence is to analyze the relationship between the incident wave and the reflected wave, calculate the transmission or coupling parameters between each antenna port, and then perform mutual coupling analysis. The following discussion will demonstrate the analysis utilizing the two-port scattering matrix. Unlike the impedance and admittance matrices, the S-parameter matrix is analyzed according to the parameters of the incident and reflected waves in the network. As shown in Figure 8.4, the incident and reflected waves corresponding to ports T_1 and T_2 are a_1 and b_1, a_2 and b_2, respectively, then the linear relationship between the incident and reflected waves can be expressed as:

$$\begin{cases} b_1 = S_{11}a_1 + S_{12}a_2 \\ b_2 = S_{21}a_1 + S_{22}a_2 \end{cases} \tag{8.5}$$

And then written in matrix form as:

$$\begin{bmatrix} b_1 \\ b_2 \end{bmatrix} = \begin{bmatrix} S_{11} & S_{12} \\ S_{21} & S_{22} \end{bmatrix} \begin{bmatrix} a_1 \\ a_2 \end{bmatrix} \tag{8.6}$$

The [S] matrix is the scattering matrix of the corresponding network. In practical analysis, the S-parameters are used to describe the antenna performance. In a multi-antenna system, we use S_{ii} and S_{ji} to represent the reflection coefficient of the antenna and the coupling between the antennas, respectively. In general, isolation (as opposed to coupling) is used to describe the mutual coupling in a multi-antenna system. The higher the isolation, the better the mutual coupling suppression effect in the system (the lower the coupling between the elements). The isolation degree is based on the S-parameters, and the relationship between the isolation degree and the S-parameter between a two-element system is expressed as (assuming a passive, reciprocal system, $S_{12} = S_{21}$):

$$I = -20\log_{10}|S_{21}|(\text{dB}) \tag{8.7}$$

8.1.2 Coupling Effects on MIMO Antenna Systems

With the rapid development of communication systems, MIMO technology plays a key role in 5G communication systems and will continue to be a major technology

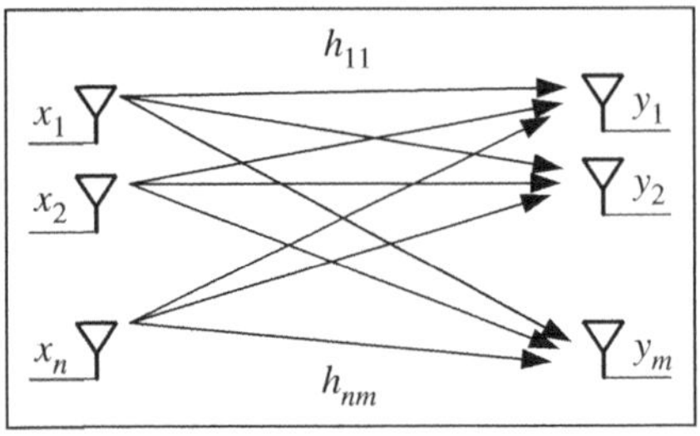

Figure 8.5 MIMO system.

for all future generations. MIMO antennas take the form of multiple transmitters and receivers, and use multi-path effects to combat fading and noise, reduce interference, and improve system data throughput and spectrum utilization. Here we will mainly introduce the key performance parameters of MIMO antenna systems, as well as the analysis and influence of mutual coupling in multi-antenna systems [2–4].

The key in MIMO technology (as well as within the antenna system) lies in multi-path transmissions, which aim to use multiple antennas to improve the transmission rate and reliability of the system in addition to improving the communication quality. As shown in Figure 8.5, the number of transmitting and receiving antennas are n and m, respectively. Assume a narrow band MIMO system, where the transmit, receive, and noise signal vectors are denoted by $\mathbf{x}$, $\mathbf{y}$, and $\mathbf{n}$, respectively, then the interrelationship between them is:

$$\mathbf{y} = \mathbf{H}\mathbf{x} + \mathbf{n} \tag{8.8}$$

where $\mathbf{x} = [x_1, x_2, \ldots, x_m]^{\mathrm{T}}, \mathbf{y} = [y_1, y_2, \ldots, y_n]^{\mathrm{T}}, \mathbf{n} = [n_1, n_2, \ldots, n_n]^{\mathrm{T}}$ and $\mathbf{H}$ represents the channel complex matrix formed by $n \times m$ channel responses.

MIMO technology is mainly divided into spatial diversity and spatial multiplexing. Spatial diversity means that we are transmitting the same signal through different transmitting antennas. Then the receiver can receive multiple signals with the same information to improve the reliability of the receiving system. Correspondingly, spatial multiplexing means that different antennas are used to transmit different signals, so that the receiving antenna can receive more signals at the same time, thereby improving the channel capacity of the system (increasing the data rates).

In theory, MIMO technology can linearly improve the channel capacity of the system when each antenna or channel is uncorrelated. However, in practical applications, MIMO systems are constrained by antenna coupling and signal correlation, which greatly affects the channel capacity of MIMO systems, making it often difficult to achieve theoretical limits. Therefore, we need to study the correlation of antennas in MIMO systems, which is also an important performance metric. In a MIMO system, the envelope correlation coefficient (ECC) is often used to evaluate the performance of a MIMO system. ECC is also a parameter to measure the mutual coupling between multiple ports of a MIMO antenna.[1] The ECC in a MIMO antenna should be less than 0.5 to indicate good field decorrelation between antennas in a MIMO array.

In practical operation, the ECC can be calculated by the S-parameter between antenna element port or the far-field pattern. In a MIMO array, The ECC calculation formula according to the S-parameter is:

$$ECC = \frac{\left|S_{ii}^{*}S_{ij} + S_{ji}^{*}S_{jj}\right|^{2}}{\left(1 - |S_{ii}|^{2} - |S_{ij}|^{2}\right)\left(1 - |S_{jj}|^{2} - |S_{ji}|^{2}\right)} \tag{8.9}$$

where S_{ii} and S_{jj} are the port reflection coefficients of the antenna elements, and S_{ij} and S_{ji} are the coupling between elements. In spite of its simplicity, the disadvantages of (8.9) have been discussed in Chapter 2.

Assuming an isotropic scattering environment, a more general formula and accurate method for calculating ECC is the one from the radiation characteristics of the antenna [5]:

$$\mathrm{ECC} = \frac{\left|\iint_{4\pi}\left[\vec{E}_i(\theta,\varphi)\cdot\vec{E}_i(\theta,\varphi)\right]\mathrm{d}\Omega\right|^{2}}{\left|\iint_{4\pi}|\vec{E}_i(\theta,\varphi)|^{2}\mathrm{d}\Omega\right|\cdot\left|\iint_{4\pi}|\vec{E}_j(\theta,\varphi)|^{2}\mathrm{d}\Omega\right|} \tag{8.10}$$

where $\vec{E}_i(\theta,\varphi)$ and $\vec{E}_j(\theta,\varphi)$ are the radiation far-field patterns of antennae i and j.

The two calculation methods above can give an estimate of the ECC between the elements in the MIMO array. Results calculated according to the antenna far-field pattern function are more accurate. However, this method requires complete data on the radiation pattern of the antenna, the amount of data is relatively large, and the calculation process is relatively complicated. Therefore, in practice, and to get a first-order approximation, the S-parameters between the antenna can be used for preliminary estimation of the ECC in advance. This method is relatively simple and only needs the S-parameters between the antenna ports. When the system is tested, a more accurate ECC value can be obtained by calculating the far-field pattern test data or from a reverberation chamber measurement [6].

8.1.2.1 Coupling Effects on Antenna Characteristics

In a MIMO system, the impedance matrix can be used to illustrate the effect of mutual coupling on the antenna impedance [7]. A multi-antenna system can be regarded as a multi-port network, and the impedance between the ports in the presence of mutual coupling is analyzed by calculating the voltages and currents

---

1 The reader is asked to check Chapter 2 for more on ECC and its calculation and measurement details.

between different ports in the network. In an N-element antenna system, the voltage at the i-th port can be calculated as:

$$V_i = \sum_{j=1}^{N} Z_{ij} I_j \tag{8.11}$$

where Z_{ij} is the mutual impedance between ports i and j ($i \neq j$). The impedance of port i is

$$Z_i = \frac{V_i}{I_i} = \sum_{j=1}^{N} Z_{ij} \frac{I_j}{I_i} \tag{8.12}$$

In the presence of mutual coupling, the input impedance of the antenna port is affected by the coupling with other elements. Even though an isolated antenna is designed to be well-matched, when it is extended to an array, the impedance of the antenna itself changes due to the coupling effect between adjacent antennas. This will lead to impedance mismatch, resulting in the deterioration of the SWR and the reduction of efficiency.

A two-element half-wavelength dipole antenna array with a center operating frequency of 2.7 GHz is designed to illustrate the effect of coupling on the antenna characteristics. The antenna structure and equivalent circuit dual-port network are shown in Figure 8.6. The two dipoles are situated along the y-axis with center-to-center separation of d. Suppose dipole 2 is absent, the theta component of the far-field function of dipole 1 can be expressed as

$$g_1(\theta, \phi) = -\frac{2C_k \eta \cos(\pi/2 \cos\theta)}{k \sin\theta} \exp\left(-jk\frac{d}{2} \sin\theta \sin\phi\right) \tag{8.13}$$

where $i = 1, 2$, $d_1 = d$, $d_2 = -d$, $k = 2\pi/\lambda$, $C_k = -jk/4\pi$, η is the wave impedance in free space. When dipole 1 is excited while dipole 2 is terminated with a load Z_L, the far-field function of dipole 1 can be calculated as [8]

$$g_{\text{emb},1} = g_1(\theta, \phi) I_1 + g_2(\theta, \phi) I_2 \tag{8.14}$$

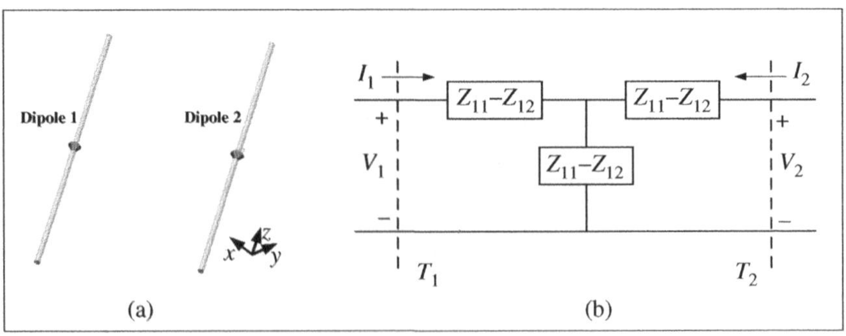

Figure 8.6 Two-element dipole array. (a) 3-D view and (b) equivalent circuit.

According to the circuit model in Figure 8.6b, the induced current at port 2 is $I_2 = Z_{12}/(Z_{22} + Z_L)$ when the excitation current at port 1 is unity ($I_1 = 1$). The term "embedded far-field function" is used to describe equation (8.14). Note that (8.14) is a good approximation for electrically small antennas. The embedded pattern of dipole 2 can be calculated in a similar way.

In order to show the effect of mutual coupling on the antenna pattern, the antenna pattern of the isolated dipole and the embedded antenna pattern of parallel dipoles with quarter wavelength spacing is illustrated in Figure 8.7. It can be seen that the antenna pattern of the isolated dipole (Figure 8.7a) has

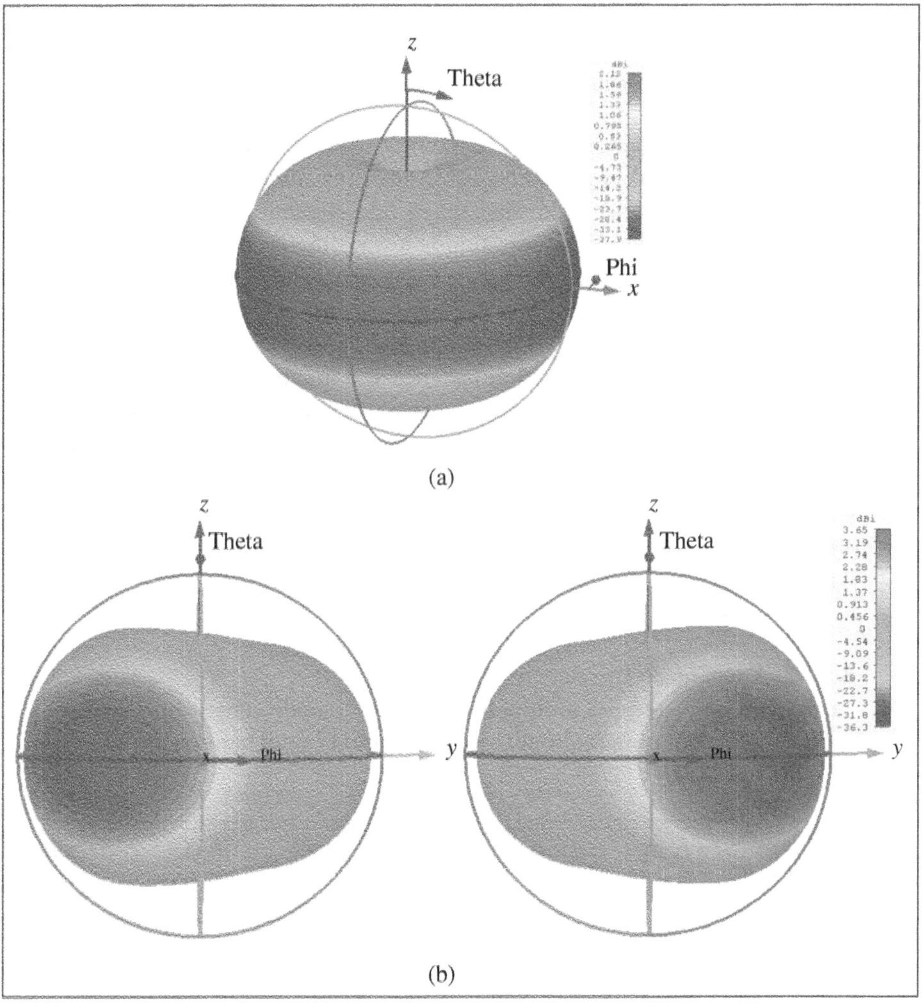

Figure 8.7 Radiation pattern of (a) isolated dipole and (b) dipole in array.

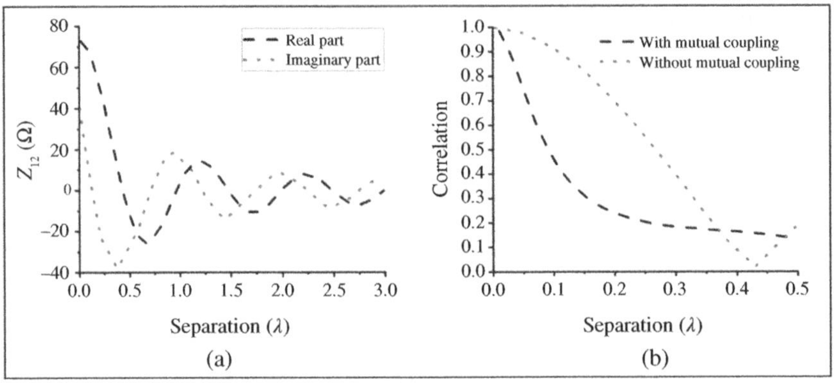

Figure 8.8 (a) Mutual impedance and (b) Correlation varies with dipole separation.

omnidirectional radiation in the horizontal plane, while the dipoles placed in parallel (in an array) radiate outward in the opposite directions due to mutual coupling[2] (Figure 8.7b). Note that when one element is excited at a time in Figure 8.7b, the pattern concentration is toward the excited antenna, i.e. left or right. The mutual coupling changes the antenna pattern, and the change of the antenna pattern also depends on the antenna type and relative position of the antenna in the array.

Figure 8.8a shows the mutual impedance as a function of the dipole antenna's spacing (normalized by wavelength). It can be seen that when the antenna spacing is small, the mutual impedance is not negligible, but as the dipole antenna spacing increases, the mutual impedance tends to zero. Note that the plot is only for parallel dipole antennas. In general, the mutual impedance tends to reduce as the antenna separation increases, but the actual variation also depends on the antenna type. The interested reader can refer to Chapter 3 for the impedances of various antenna types.

8.1.2.2 Coupling on Correlation

Taking the non-isotropic multipath propagation environment into account, the formula for calculating multi-antenna correlation can be written as [9]:

$$\rho = \frac{\iint_{4\pi} g^{H}_{\text{emb},1}(\Omega) P_{\text{a}}(\Omega) g_{\text{emb},2}(\Omega) d\Omega}{\sqrt{\iint_{4\pi} g^{H}_{\text{emb},1}(\Omega) P_{\text{a}}(\Omega) g_{\text{emb},2}(\Omega) d\Omega \cdot \iint_{4\pi} g^{H}_{\text{emb},2}(\Omega) P_{\text{a}}(\Omega) g_{\text{emb},2}(\Omega) d\Omega}} \tag{8.15}$$

2 In this literature, the pattern distortion is often attributed to the mutual coupling. Nevertheless, in essence, this kind of distortion is due to the fact that when dipole 1 is excited, dipole 2 acts as a reflector, and vice versa.

where the superscript ^{H} stands for Hermitian, P_a is the dyadic power angular spectrum of the incident waves, and Ω is the solid angle of arrival. The isotropic scattering condition, i.e. $P_a(\Omega) = I$, is frequently taken into account when the angular spectrum is unknown.

Utilizing the self and mutual impedances explicitly is another method of calculating the correlation. When there is mutual coupling, the received voltages of a system of two dipole elements can be represented as

$$\mathbf{v} = \mathbf{Z}_L[\mathbf{Z} + \mathbf{Z}_L]\mathbf{v}_{oc}$$
$$\mathbf{v} = [v_1 \ v_2]^T, \mathbf{v}_{oc} = \left[v_1^{oc} \ v_2^{oc}\right]^T \tag{8.16}$$

where $\mathbf{v}$ is the voltage matrix with coupling and $\mathbf{v}_{OC}$ is the voltage in open circuit, $\mathbf{Z}$ is the impedance matrix for the parallel dipoles, and $\mathbf{Z}_L$ is a diagonal matrix whose diagonal elements represent the same load impedance Z_L. Equation (8.16) can be expressed as

$$\begin{bmatrix} v_1 \\ v_2 \end{bmatrix} = \begin{bmatrix} \alpha & \beta \\ \beta & \alpha \end{bmatrix} \begin{bmatrix} g_1(\theta, \phi) \\ g_2(\theta, \phi) \end{bmatrix} \tag{8.17}$$

where α and β are the corresponding elements of the coupling matrix $\mathbf{Z}_L[\mathbf{Z} + \mathbf{Z}_L]^{-1}$, $g_1(\theta, \phi)$ and $g_2(\theta, \phi)$ are the far-field functions, and $g_2(\theta, \phi) = g_1(\theta, \phi)\exp(jkd\cos\phi)$ because we have identical dipoles with similar patterns and only one is shifted with respect to the other. Then the correlation in the presence of mutual coupling is

$$\rho = \frac{E\left[v_1 v_2^*\right]}{\sqrt{E\left[|v_1|^2\right] E\left[|v_2|^2\right]}}$$
$$= \frac{(|\alpha|^2 + |\beta|^2)\mathrm{Im}\{\rho_{oc}\} + 2\mathrm{Re}\{\alpha\beta^*\} + j(|\alpha|^2 - |\beta|^2)\mathrm{Im}\{\rho_{oc}\}}{\sqrt{E\left[|v_1|^2\right] E\left[|v_2|^2\right]}}$$
$$E\left[|v_1|^2\right] = |\alpha|^2 + |\beta|^2 + 2\mathrm{Re}\{\rho_{oc}\}\mathrm{Re}\{\alpha\beta^*\} + 2\mathrm{Im}\{\rho_{oc}\}\mathrm{Im}\{\alpha\beta^*\}$$
$$E\left[|v_2|^2\right] = |\alpha|^2 + |\beta|^2 + 2\mathrm{Re}\{\rho_{oc}\}\mathrm{Re}\{\alpha\beta^*\} - 2\mathrm{Im}\{\rho_{oc}\}\mathrm{Im}\{\alpha\beta^*\} \tag{8.18}$$

where the superscript* stands for complex conjugate, E stands for the expectation, and ρ_{oc} is the open-circuit correlation (correlation without mutual coupling) that can be calculated by swapping out the embedded patterns for their isolated patterns.

Figure 8.8b shows the correlation as a function of the dipole separation distance (d). As can be observed, the correlation with mutual coupling is less than that without coupling. This is because the two dipoles tend to radiate in opposite directions due to the mutual coupling (see Figure 8.8a). However, it should be highlighted that in reality, mutual coupling is a common occurrence with compact MIMO antennas and that "without mutual coupling" is simply a theoretical reference case. It can be seen that the correlation tends to grow as the dipoles get closer, as well as the mutual coupling (mutual impedance). Consequently, one may draw the conclusion that mutual coupling tends to increase correlation. Yet Figure 8.8b seems to be able to draw such a conclusion that the mutual coupling tends to lower the correlation. This does not contradict the first conclusion. The use of different references accounts for two interpretations of the same phenomenon. In case of a two-port antenna, the mutual coupling has the tendency to make the antenna pattern orthogonal to each other which affects the correlations. Nevertheless, when the array size grows larger, the mutual coupling has a negligible effect on the overall correlations. Coupling effects on the embedded element pattern from surrounding elements tend to cancel each other, resulting in almost identical embedded element pattern [10, 11].

Figures 8.9 and 8.10 show correlations of 2×2 MIMO systems with two parallel dipole antennas at the receiving side for 0.3λ and 0.5λ separation,[3] respectively. As can be seen, the correlation decreases as the azimuth spread increases. This is because the correlation distance is inversely proportional to the angular spread. It can be seen that mutual coupling reduces the correlation between the two dipole antennas with 0.3λ spacing. This is because the strong mutual coupling at 0.3λ spacing makes the two parallel dipoles radiate outward (away from each other). At a more considerable distance, this phenomenon disappears. Although the mutual coupling at 0.3λ spacing reduces the correlation, it incurs a strong absorption loss, which eventually reduces the MIMO capacity, as shown in Figure 8.9b. At 0.5λ spacing, the absorption loss due to mutual coupling is insignificant; since the correlation with mutual coupling increases, the MIMO capacity with mutual coupling is slightly lower [12–14].

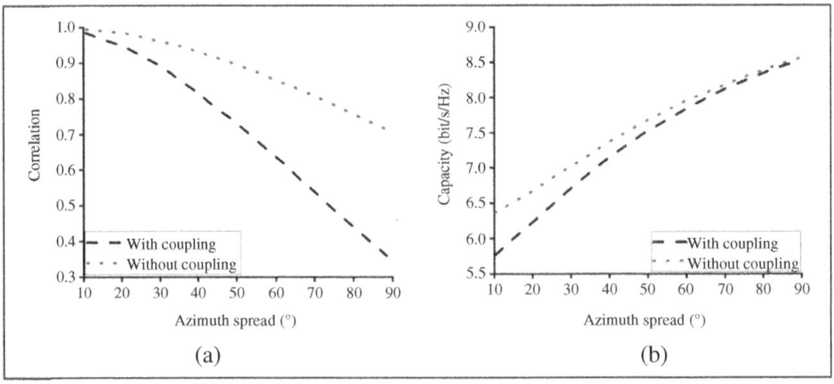

Figure 8.9 2×2 MIMO system with two 0.3λ separate with and without mutual coupling: (a) correlation, (b) capacity.

3 Here, we assume the two transmitting antennas are in the far field.

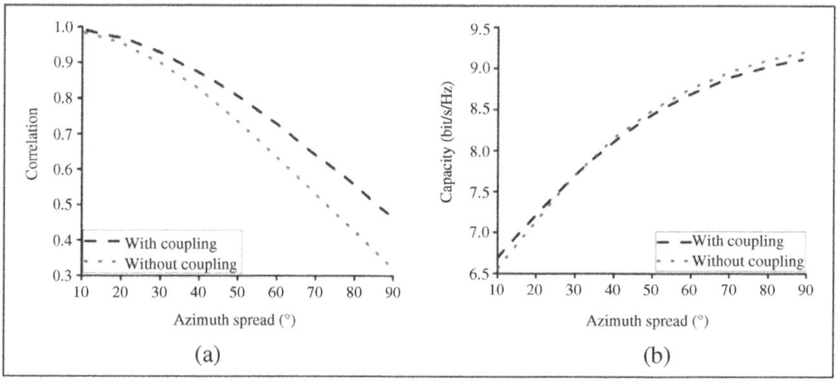

Figure 8.10 2 × 2 MIMO system with two 0.5λ separate with and without mutual coupling: (a) correlation, (b) capacity.

8.1.2.3 Coupling on Axial Ratio

We have seen that the radiation characteristics of the elements in a multi-antenna system can significantly change from those when the antennas work alone. Moreover, for antenna elements at different positions in the system (even if the antenna forms or types are the same) the respective radiation patterns will be different due to mutual coupling. Axial ratio (AR) is an important performance index of circularly polarized (CP) antennas, as it represents the purity of circular polarization, and also corresponds to the quality of the radiation pattern of a CP antenna. A 1 × 2 CP antenna array with center-to-center distance of $0.5\lambda_0$ (λ_0 is the free-space wavelength at 3.5 GHz) is used to illustrate the mutual coupling effect on the AR. As shown in Figure 8.11, compared with an isolated antenna, the AR in the array is significantly deteriorated due to the existence of mutual coupling, where the AR is above 3 dB in the operating frequency band. For CP arrays used in satellite communications, larger arrays with stringent pattern requirements are required. Thus an effective decoupling scheme should be applied to improve the performance of CP arrays, which will be introduced in the coming sections.

8.1.2.4 Coupling Effects on the Nonlinearity of Power Amplifiers

Power amplifiers (PAs) are important RF devices in MIMO transmitters, where each antenna or subarray is connected to an independent power amplifier with moderate power capacity. For compact large-scale arrays, there is not enough space to place isolators between the antenna and the PA pathways. Therefore, the mutual coupling between the elements in the MIMO array will change the active port impedance, resulting in a mismatch between the antenna and the PA. In the presence of mutual coupling, the outputs of the PAs will interfere with each other, increasing the nonlinearity of the PAs.

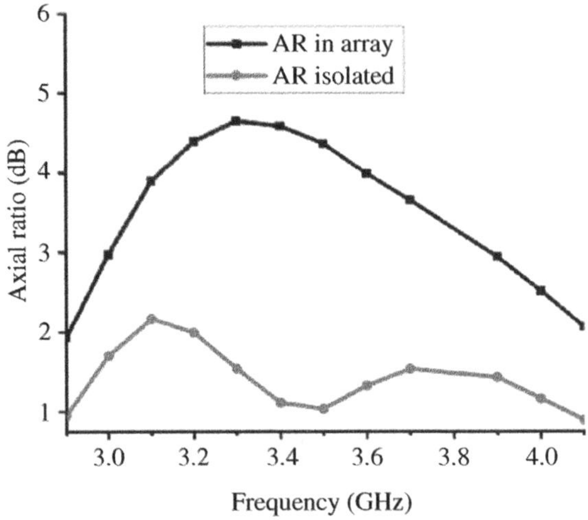

Figure 8.11 AR in isolated antenna and in array.

Researchers have proposed various PA behavior models to optimize linearity. F. M. Barradas [15] extends the DC consumption behavior of power amplifiers to the PA model to predict the linearization and efficiency of PA in the case of mutual coupling. A 4×1 array with 4 PAs was designed in [16], and the spectral results of PA outputs were presented, showing the spectral regeneration with/without mutual coupling respectively. The nonlinear distortion of phased array transmitter was studied in [17]. The simulation results show that, in the presence of mutual coupling, the distortion in the main beam band changes strongly with the change of scanning angle. A four-branched transmitter prototype was used in [18] to compare spectrum distortion. Conventional linearization techniques cannot effectively compensate for the distortion caused by the coupling of arrays. Instead of linearization techniques with high design complexity, it is a more cost-effective solution to decouple the array from the antenna perspective. Thus, a prototype antenna array with a removable decoupling structure is connected to the PA in [19] and the output spectrum of PA is measured before and after the coupling. The experimental results show that mutual coupling not only worsens the linearity, but also produces strong interference. More importantly, it shows that in the case of effective array decoupling, both interference and spurious signals are greatly suppressed, while the desired signal is enhanced.

8.2 Decoupling Techniques

Mutual coupling in multi-antenna systems mainly occur in several ways; space wave coupling, surface wave coupling and induced current in the ground. Space wave coupling means that when an antenna element is excited, electromagnetic waves will be coupled to other elements through free space. Surface wave and induced currents will be coupled to other elements through the dielectric substrate and the common ground, respectively. In practice, there is often a situation where a certain coupling mechanism is dominant, and many coupling mechanisms can appear at the same time. Therefore, the coupling phenomenon will be more complicated and difficult to solve. Usually, the distance between elements is often directly increased to improve the isolation. However, in compact and miniaturized systems (i.e. mobile terminals), this approach is no longer applicable. In order to study and suppress the electromagnetic coupling effect and improve the efficiency of the antenna systems, many techniques and methods for coupling suppression have been proposed and implemented.

Different from the previous work which directly introduces various decoupling techniques, we summarize the decoupling techniques according to antenna types, array configuration and antenna polarizations as follows.

8.2.1 Linearly Polarized Arrays

8.2.1.1 Decoupling Networks

A decoupling network refers to the coupling elimination by connecting active or passive microwave networks between antenna elements [20–22], which most applicable to the situation where the ports of the antenna are in close proximity. In general, this method considers a single antenna as a single-port network. When there is a large coupling effect between multi-antenna elements, the S-parameter matrix of the network can be extracted, and then the corresponding decoupling network can be introduced to eliminate the corresponding coupling elements in the matrix. This is to achieve the suppression of the corresponding coupling effect. Therefore, the design of the decoupling network is generally related to the S-parameter matrix of the antenna, and basically independent of the antenna form (note that the S-parameter matrix depends on the impedance matrix which is a function of the antenna type). When a decoupling network is used to suppress mutual coupling, the antenna and decoupling network can be designed separately and independently (a potential advantage of this method). The decoupling network is composed of lumped components (such as inductors and capacitors, or printed elements representing such elements or their variants) is simple in design and easy to construct. Moreover, the network is easy to connect with the antenna in the system in a compact space, which yields high integration and sometimes miniaturization of the design. General design procedures and methods for passive decoupling networks are given in [20], as shown in Figure 8.12. In the design process, the S-parameter matrix $[S^A]$ of the two antenna ports is first extracted. Next, a microstrip line with a specific

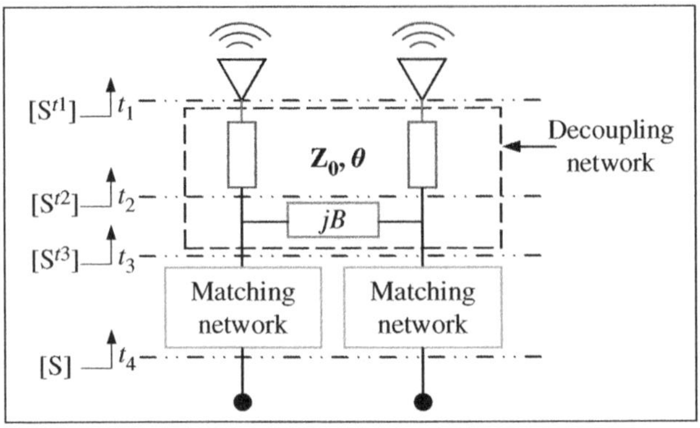

Figure 8.12 Decoupling network structure.

characteristic impedance and length is connected to the two ports respectively, so that the impedance of the two ports at the t_2 plane exhibits purely reactive characteristics. Then, a lumped reactive element (inductance, capacitance or a transmission-line section) is connected in parallel between the two ports, so that the impedances of the ports cancel each other. S_{21} in the S-parameter matrix $[S^B]$ at the t_3 plane can be made equal to 0, that is, the coupling cancellation between the two ports is realized. Then, one can load the appropriate matching network on the port to achieve good matching characteristics of the antenna. So far, through the loading of the decoupling network and the matching network, the coupling suppression of the two-element antenna system can be realized on the premise of ensuring good matching characteristics.

8.2.1.2 Defected Ground Structure

A defected ground structure (DGS) refers to the gap or slot integrated in the plane circuit and ground plane [23–26]. DGSs are usually notches or slits of different geometries etched in the ground plane. These notches will change the current distribution on the ground, which is equivalent to a resonant circuit with capacitors and inductors in parallel. The etched gaps are equivalent to creating band-stop filters between the antennas to suppress the transmission of coupled waves, increasing port isolation between array antenna elements.

DGS is widely used in antenna arrays due to its simple structure, low cost, and no increase in antenna profile. A surrogate-assisted DGS design is proposed to suppress E- and H-plane couplings simultaneously in a 2×2 microstrip antenna array in [24]. A DGS with slot splitting meandered lines was proposed to reduce the mutual coupling between two microstrip antennas with a center-to-center spacing of $0.38\lambda_0$ (λ_0 is the wavelength in the free-space), as shown in Figure 8.13. The slot splitting meandered lines are symmetrically etched on the ground between the

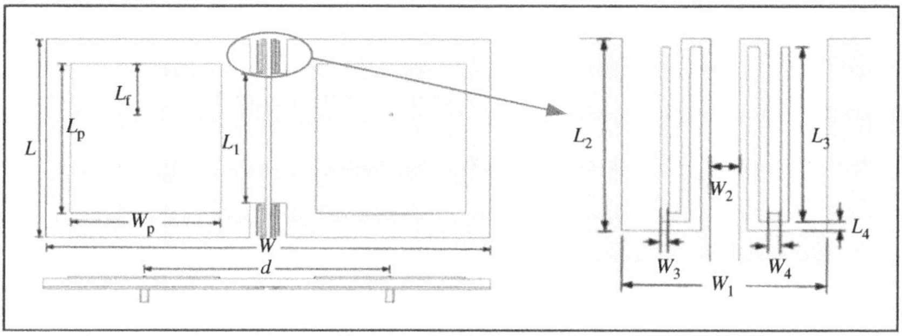

Figure 8.13 DGS with slot splitting meandered lines between two patch elements. Source: Reproduced with permission from Ref. [26]/IEEE.

microstrip antennas, and in this reference the Common/Differential mode theory was applied to the DGS to explain its decoupling mechanism [26]. When the common mode impedance is equal to the differential mode impedance, the coupling between elements can be suppressed. The measured results show that the mutual coupling is reduced by more than 20 dB after the array is applied with the DGS. The DGS can significantly reduce the coupling between microstrip patch antennas, but the slots and notches on the ground also increase the back radiation of the antenna, in addition, the decoupling bandwidth of the DGS is narrow, which is mostly suitable for reducing the coupling in narrow-band antenna arrays.

8.2.1.3 Neutralization Line

The neutralization line (NL) decoupling technique refers to the introduction of one or more microstrip transmission lines between specific positions on antennas in a multi-antenna system [27–30], which can form an additional coupling path to cancel the original coupling and thus improve the isolation between antennas. Neutralization lines can be applied to different antenna forms and application scenarios, and can achieve coupling suppression in narrowband, wideband, ultra-wideband and multi-frequency antenna systems.

Neutralization lines are often used in terminal multi-antenna systems. A WLAN MIMO antenna operating at 2.45 GHz was designed in [28]. The MIMO antennas take the form of two monopoles, located on opposite corners of the PCB, separated by a small metal ground. Since the two antennas share the same ground, when one of the antennas is excited, a part of the energy will flow into the other antenna, resulting in strong mutual coupling between the two antennas. In order to suppress the coupling, a neutralization line is introduced between the two antennas. The isolation has been increased from 9 to 20 dB, effectively suppressing the mutual coupling effect between the antennas.

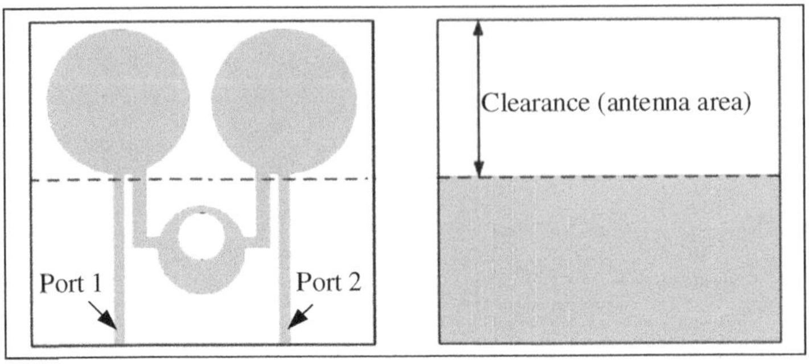

Figure 8.14 Neutralization lines in an UWB MIMO antenna system.

In multi-frequency, multi-antenna systems, the design of the neutralizing line is similar to that of a broadband multi-antenna system, which is often implemented in the form of multiple neutralization lines. For the neutralization line in an ultrawideband (UWB) multi-antenna array, it will be realized by designing some neutralization lines with broadband characteristics, as shown in Figure 8.14. Two UWB antennas are compactly arranged in [30], and the distance between the two antennas is only 2.2 mm. The isolation is only about 12 dB without the introduction of the neutralization in the operating frequency band of 3–5 GHz. In order to improve the isolation, a neutralization line with broadband characteristics was designed and proposed, which consists of two metal strip lines and a metal circular disc. Since the metal disk can create several possible decoupling current paths with different lengths to eliminate coupling currents on the ground plane, an ultra-wideband decoupling design can be achieved. The results show that within the operating bandwidth of 3–5 GHz, the broadband NL proposed can improve the in-band isolation by 12 dB, i.e. approaching 25 dB.

Different from previous decoupling technologies that focus on blocking the propagation of coupling paths, neutralization lines provide a new decoupling mechanism. Neutralization lines can introduce additional coupling paths to offset the initial coupling and achieve good decoupling effect. However, the introduction of neutralization lines will also cause some interference to the near-field radiation of the antenna. Besides that, in broadband, ultra-wideband and multi-frequency multi-antenna systems, it is often necessary to introduce multiple neutralization lines, which requires enough space to place, affecting the layout of the antenna system.

8.2.1.4 Metamaterial Structure

In recent years, metamaterials have been widely used in multi-antenna decoupling designs due to their ability to control electromagnetic waves. Metamaterials are a series of metallic periodic structures of subwavelength sizes that can achieve electromagnetic properties different from natural materials. There are many methods for metamaterial based decoupling designs, the most common of which are electromagnetic bandgap structures (EBG) [31–34] and metasurfaces [35–39].

EBG is an arrangement of specific unit structures on a dielectric substrate according to an inter-element spacing distance that forms a high impedance characteristic in a certain frequency band region, thereby suppressing the propagation of surface waves. After the electromagnetic bandgap structure is placed between or around the antennas, the near field of the antenna is rapidly attenuated due to the existence of the high impedance characteristic structure. A forbidden band (band gap) can be formed in a specific frequency band, and the coupling effect between the antenna elements can be effectively suppressed. A mushroom-like EBG was proposed for the first time in [32] to suppress surface waves between microstrip patch antennas to improve the isolation between adjacent antenna elements. After the introduction of the EBG, the isolation between the antenna elements is significantly improved from 16 dB to more than 25 dB in the operating frequency band. Due to the excellent high impedance characteristics of the EBG, many scholars have carried out more in-depth research on it and put forward a series of special structures to meet the decoupling effect in different antenna systems. For example, various types of EBG structures have been proposed, such as cross, circle, square and star in [33]. With the development of dual-frequency or multi-frequency multi-antenna systems, it is necessary to achieve good high isolation characteristics in multiple frequency bands. An EBG structure with dual-frequency characteristics is proposed to achieve coupling suppression in dual-frequency multi-antenna systems. When the structure is integrated between two dual-band antennas, the coupling between the antenna elements is significantly suppressed, and the isolation degree in the two bands is increased to 42 and 58 dB, respectively [34].

The above-mentioned decoupling methods based on the EBG can effectively suppress the propagation of surface waves and achieve high isolation between antennas. However, this structure usually requires a certain space between antenna elements, preventing the design of miniaturized arrays. Recently, a compact and high-isolation millimeter-wave array has been designed by placing the patch elements above a EGB ground (see Figure 8.15). This compactness of the array is achieved by placing the antennas and EGB in different layers and stacking them together.

Several metasurface-based decoupling methods with similar stacking scheme have emerged [35–39]. Due to the diversity of metasurfaces in the regulation of electromagnetic waves, there are also various decoupling forms within the area of antenna decoupling, such as metasurface absorbers, polarization rotators and negative permeability materials. Metasurface absorbers were proposed in [37] to achieve high isolation between two-element MIMO arrays. The absorption rate of

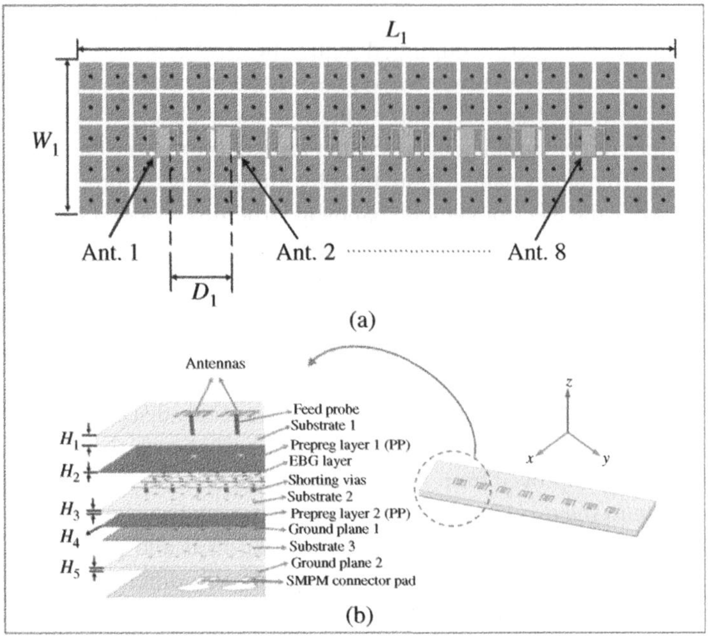

Figure 8.15 Millimeter-wave array with EBG ground: (a) top view and (b) explosive view Source: Reproduced with permission from Ref. [36]/IEEE.

the absorber at the resonant frequency is 98.7%, which achieves perfect absorption of the electromagnetic wave. A 1×4 metamaterial absorber element is linearly arranged between the two antenna elements, which effectively inhibits the propagation of surface currents, and finally achieves the isolation level of 43.7 dB. Metamaterial polarization rotators were proposed to reduce the coupling caused by spatial electromagnetic waves between dielectric resonator antennas in [38]. Another example is a polarization rotator wall that contains seven split-ring resonators (SRR) and is embedded between two antenna elements. The electric field is rotated by 90° through the polarization rotator, thus the polarization of coupled waves between the antennas is orthogonal, thereby reducing the mutual coupling. The isolation between the units is improved by more than 16 dB on average. Although metasurface decoupling superstrates has been mainly demonstrated for two-element arrays in the literature, it can actually be extended to larger arrays. For instance, a suspended metasurface composed periodic split ring resonator (SRRs) is placed above a 4×4 patch antenna array for mutual coupling mitigation as shown in Figure 8.16 [39]. This kind of metasurface is equivalent to a negative permeability (or permittivity) medium with an imaginary wave number, which turns the propagated tangential waves to evanescent ones. In

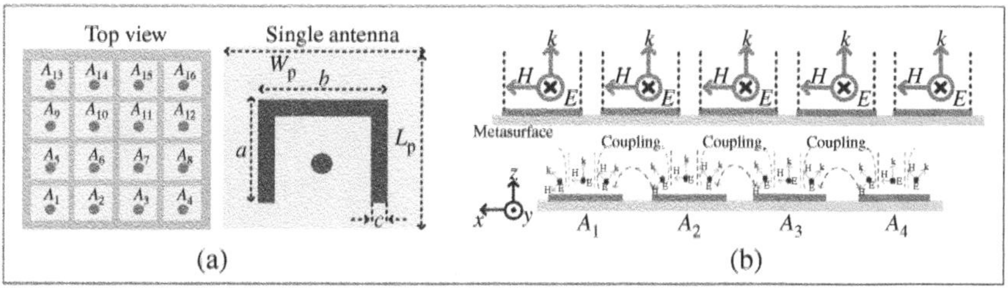

Figure 8.16 Suspended metasurface for large array decoupling: (a) top view of the 4 × 4 patch array; (b) side view of the metasurface over the patch array. Source: Reference [39]/IEEE/Public Domain.

this way, the mutual coupling can be significantly suppressed. Measured results show that the isolation is improved by 8 dB, achieving 25-dB isolation.

8.2.2 Dual Polarized Array

With the rapid deployment of 5G technology, the application scenarios of massive MIMO have been increased rapidly. Especially in base station arrays (BSA), the coupling between antennas becomes more complex with the increased number of antenna elements. Due to the space limitation of antenna systems, it is unrealistic to reduce the coupling effect by increasing the space interval between the antenna elements, not to mention the resulting grating lobes. In a large-scale dual-polarized antenna array, the antenna is subjected to co-polarization coupling and cross-polarization coupling from adjacent elements or non-adjacent elements. On the other hand, the decoupling techniques proposed in the previous section are mostly applicable to dual-element, single-polarized antennas, and cannot be easily extended to large-scale arrays. Thus, the coupling between large-scale array antennas is an urgent problem to be solved. In this section we summarize the decoupling techniques proposed in the literature for large-scale dual-polarization arrays.

8.2.2.1 Array Antenna Decoupling Surface

Array antenna decoupling surfaces (ADS) based on the neutralization mechanism were proposed in [40, 41]. The schematic of an array with an ADS is shown in Figure 8.17. An ADS consists of sets of metal patches (i.e. primary reflector patches and secondary reflector patches) printed on a dielectric substrate. When antenna 1 is excited, the ADS suspended above the array will provide new coupling paths, such as path ③ from primary patches and path ④ from secondary ones. Since the reflected signal can be controlled to be with the same magnitude and the opposite phase as that of the original coupled signal ⑤ by carefully designing the size of patches and placement height of the ADS, the mutual coupling between

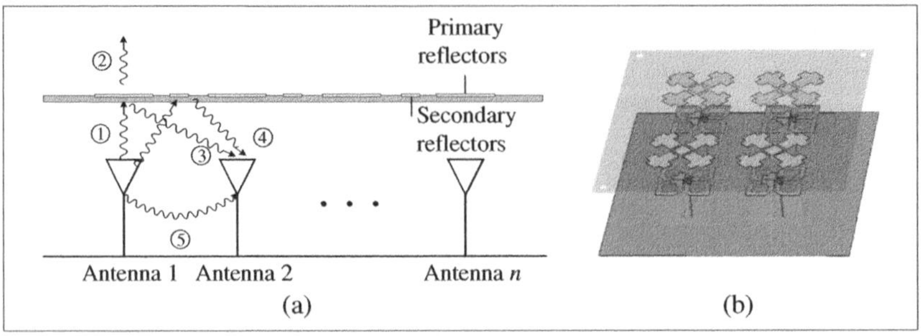

Figure 8.17 (a) Schematic of and ADS for dual-polarized antenna elements: (b) perspective view of an ADS loaded 2 × 2 dual-polarized array (with 45-mm spacing) working at 3.5 GHz.

adjacent antenna elements can be significantly reduced by the reflected waves from the ADS.

The decoupling method of the ADS is explained by the S-parameters in a two-element array with and without ADS. With an ADS, the reflected wave through the ADS between element 1 and element 2 can be expressed as

$$S_{21}^{\text{reflect}} = S_{21}^{\text{ADS}} - S_{21}^{\text{original}} \tag{8.19}$$

where S_{21}^{ADS} and S_{21}^{original} are the S-parameter in the array with or without ADS respectively. When the mutual coupling in the array is small enough, that is the $S_{21}^{\text{ADS}} = 0$, it can be concluded that

$$\begin{gathered} \left|S_{21}^{\text{reflect}}\right| = \left|S_{21}^{\text{original}}\right| \\ \text{Phase of}\left(\left|S_{21}^{\text{reflect}}\right|\right) = \text{Phase of}\left(S_{21}^{\text{original}}\right) + \pi \end{gathered} \tag{8.20}$$

A 1×8 linearly polarized patch array and 2×2 dual-polarized array were designed in [40] to demonstrate the decoupling effect using ADS. Measured results show that the mutual coupling between any two adjacent elements with ADS in patch array is significantly reduced from about −15 dB to below −30 dB. As for the decoupling in a dual-polarized array, the coupling from an adjacent element with the same polarization (co-polarization) is reduced from about −14 to −25 dB or lower. However, the coupling between adjacent dipoles with cross-polarization is difficult to control, the cross-polarized coupling is reduced from −23 to −25 dB.

An ADS consists of multiple sets of primary reflector patches and secondary reflector patches, which involves too many parameters, and the actual optimization is complicated. Compared with the cross-polarized coupling, ADS has a more obvious decoupling effect on co-polarized coupling, this is because the cross-polarization between adjacent antenna elements is mainly caused by capacitive coupling between the elements, and an ADS has a limited reduction effect on cross-polarized coupling.

8.2.2.2 Decoupling Ground

For array elements with a common ground plane (i.e. a typical BSA scenario), the mutual coupling paths contain free-space coupling and induced current flowing on the common ground plane. Decoupling ground (DG) method controls the current coupling by changing the size and shape of a small ground plane for each element to achieve opposite phase with that of the free-space coupling. A MIMO array with x-axis polarization is introduced in [42] to explain the decoupling principle of DG, as shown in Figure 8.18. Assuming that two patch antennas of x-axis polarization with small ground are placed above a large shared common ground along the x-axis, where the coupling currents between the small ground plane and the shared ground plane can be formed by displacement currents. By properly adjusting the small ground plane sizes in the x-axis direction and their distance from the shared ground (Hm), coupling paths (1) and (2) can be out of phase and cancel with each other. Metal cylinders can be added between the small ground plane and the shared ground plane as a steady support for the antenna elements. In this case, the antennas can be easily fed by coaxial cables running through the inside of the cylinders. Apparently, the coupling currents between the small ground plane and the shared ground plane are connected by the external surface of the metal cylinder. And the width of the cylinder in the x-axis direction will affect the length of the coupling current path (2). Note that Figure 8.17 illustrates the decoupling principle for E-plane coupling, while the H-plane coupling can be achieved using the same structure [42].

Two examples of dual-polarization 2 × 2 and 4 × 4 arrays were designed to verify the concept and demonstrate the effect of DG on dual-polarized coupling, and both arrays achieved minimum isolation that was greater than 25 dB with the use of a DG. Recently, the DG approach has been extended to broadband dielectric resonator antenna arrays in [43, 44].

The decoupling mechanisms of DG and ADS are similar and are based on the idea of neutralization. However, ADS provides a new spatially coupling wave path by placing a parasitic structure above the array to cancel the original spatially

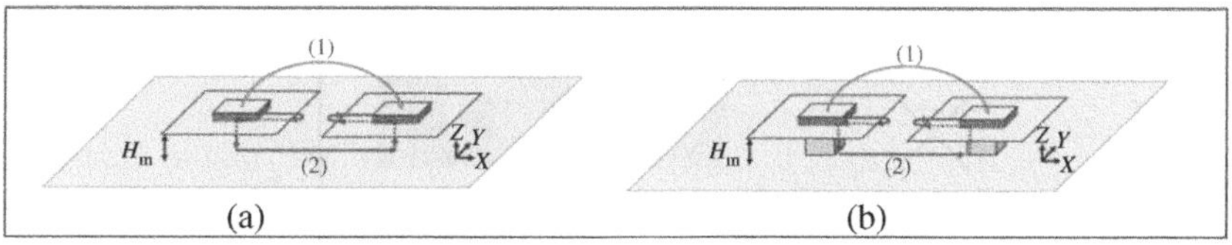

Figure 8.18 Decoupling principle of the decoupling ground without (a) and with (b) the supporting metal cylinder. Source: Reproduced with permission from Ref. [42]/IEEE.

coupled wave whereas DG adjusts the original coupling current in the ground plane to cancel the space coupling wave. Thus, the array with DG can achieve much lower profile compared ADS, but it is not suitable for loop or dipole antennas that are sensitive to the size of the ground plane.

8.2.2.3 Dielectric Superstrate

The dielectric superstrate is a novel decoupling method, it was applied to dipole array and realistic BSAs in [45–47]. Compared to complicated parameters and tedious optimization in ADS, the dielectric superstrate is placed above the antenna array, such that the magnitude and phase of the reflected waves can be controlled by adjusting the permittivity and profile of the superstrate. For example, a 2×2 base station array with dielectric superstrate was designed in [46]. The measured results showed that the dielectric superstrate could suppress the mutual coupling and provide values of the order of −25 dB in the 5G frequency band of 3.3–3.8 GHz without degrading the matching and radiation performances.

In [47], they extended the capabilities of dielectric superstrates to reduce the in-band coupling in dual-band BSAs (lower-band [LB] covering 1.9–2.5 GHz and higher-band [HB] covering 3.3–3.9 GHz), as shown in Figure 8.19. In realistic base station antennas, the LB and HB antennas are interleaved to reduce the aperture and deployment cost, however this arrangement leads to in-band and cross-band mutual coupling. As shown in Figure 8.19b, four discrete dielectric blocks are placed above the HB elements in the interleaved dual-band BSA section. Measurement results show that with the help of the dielectric block, the mutual coupling among HB elements were below −25 dB and the cross-band coupling is effectively suppressed.

Compared with the previous relevant works (such as ADS, FSS, and SRR), the dielectric superstrate could achieve superior broadband decoupling, besides the structure is simple and easy to design and optimize. In addition, radomes

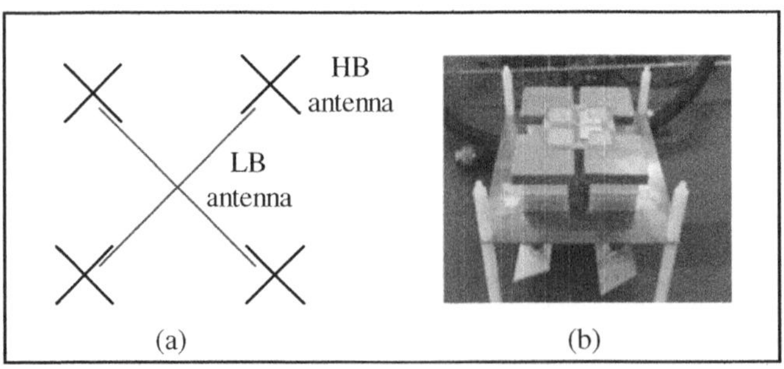

Figure 8.19 (a) Interleaved BSA and (b) antenna with dielectric superstrate
Source: Reproduced with permission from Ref. [47]/IEEE.

(the antenna covers) can be dual function, to protect the array and to provide this kind of coupling reduction if carefully designed.

8.2.2.4 SRR-Loaded Baffles

Baffles loaded by SRR have also been proposed to reduce the mutual coupling for ±45° dual-polarized MIMO antenna as shown in [48]. Each baffle consists of two rectangular SRR printed on a substrate. The SRR baffle is equivalent to a band-stop filter, as it can effectively suppress the free-space coupling of co-polarization and cross-polarization between dual-polarized array elements. In a 1×4 array with center-to-center distance of $0.5\lambda_0$ as shown in Figure 8.20, each element is surrounded by 4 baffles to enhance the isolation. The measured results shown that both the co-polarized coupling and cross-polarized coupling are reduced to levels below −25 dB across the whole operating frequency band.

When compared to ADS and DG techniques which are based on the neutralization scheme, the SRR baffle is based on the partition scheme, same as the frequency selective surfaces (FSS) [49, 50], which is applied to suppress the propagation of space waves to minimize interference between antenna elements.

8.2.2.5 Hybrid Decoupling Structures

Mutual coupling reduction methods can be divided into partition- and neutralization-based methods. Partition-based methods reduce mutual coupling by suppressing surface currents or space electromagnetic waves propagation, while the neutralization-based ones introduce an additional wave to cancel the original coupling. A hybrid decoupling structure is proposed in [51], which consists of a neutralization dielectric board, metal baffle, and feeding network to enhance the isolation in a compact large-scale base station array. Both neutralization- and partition-based schemes were utilized to reduce the mutual coupling, as illustrated in Figure 8.21. The neutralization dielectric board provides a new signal path to cancel the original mutual coupling, and the metal baffle is employed to suppress the propagation of the space electromagnetic wave. By

(a) (b)

Figure 8.20 (a) BSA element and (b) array with SRR-loaded baffle. Reproduced with permission from Ref. [48]/IEEE.

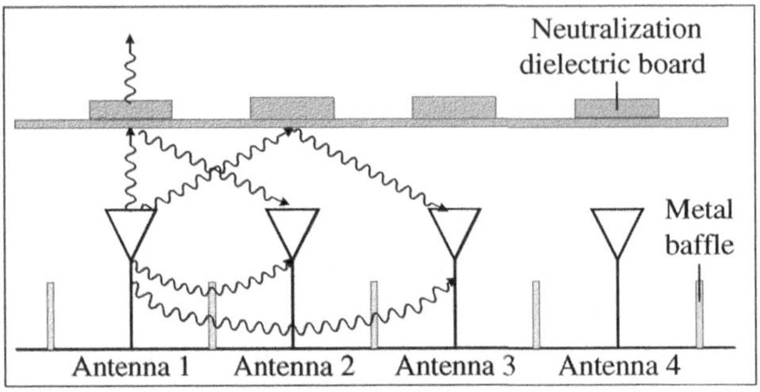

Figure 8.21 Schematic of hybrid decoupling structure. Source: Reproduced with permission from Ref. [51]/IEEE.

placing a dielectric board above the array, the electromagnetic wave emitted by Antenna 1 can be reflected through the dielectric board to produce an additional wave path. When the reflected and coupling wave have the same amplitude but opposite phase, they cancel each other, reducing the mutual couplings; the metal baffles are placed between two adjacent elements to block the propagation of spatially coupled waves between two neighboring antennas.

Figure 8.22 shows the simulated and measured S-parameters before and after decoupling. It can be seen that both the co-polarization and cross-polarization coupling between adjacent antennas (subarrays) are significantly reduced across the whole frequency band.

As shown in Table 8.1, compared with the previous decoupling methods on BSAs, the horizontal spacing is reduced by 30%, and the relative bandwidth is

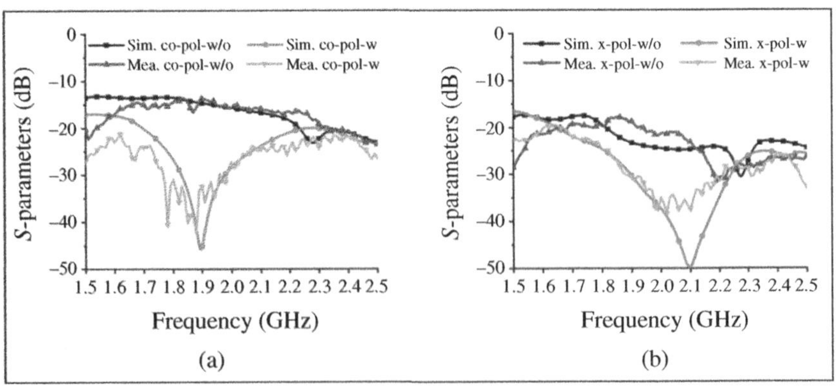

Figure 8.22 Coupling between adjacent element with and without decoupling. (a) Co-polarized and (b) cross-polarized. Source: Reproduced with permission from Ref. [51]/IEEE.

Table 8.1 Comparison of BSA decoupling techniques.

Reference	Decoupling structure	Antenna type	Array config.	Antenna distance	Isolation	Bandwidth
[40]	ADS	BS array	2×2	$0.53\lambda_0$	>25 dB	14%
[48]	SRR-baffle	BS array	1×4	$0.50\lambda_0$	>25 dB	14%
[51]	NDB + metal baffle	BS array	4×4	$0.38\lambda_0$	>20 dB	25%

nearly twice that of previous works, which achieves obvious decoupling effect in larger scale arrays. Due to the simple geometry and significant isolation enhancement, the proposed hybrid decoupling method in [51] has great potential to be applied to compact large-scale base station arrays for 5G communication.

8.2.3 Circularly Polarized Array

For linearly-polarized and dual-polarized antenna arrays, various decoupling methods were proposed to improve the isolation of the arrays. However, there are few reports in the literature to reduce the mutual coupling between circularly polarized (CP) antennas. CP antennas can reduce the "Faraday rotation" effect in ionospheric propagation and are widely used in satellite communication systems. Meanwhile, CP antennas can effectively reduce multipath interference or fading in various applications, including wireless local area network (WLAN) and radio frequency identification devices (RFID). Compared with linearly polarized antennas, CP antennas are more attractive in some scenarios and have been widely studied because of their remarkable characteristics. In CP antenna arrays, the coupling between elements not only deteriorates the port matching of the antenna elements but also affects the polarization of the antenna and deteriorates the AR of the circularly polarized antenna.

Decoupling techniques of CP antennas and linearly polarized antennas are essentially the same. Both of them improve the isolation of the array by suppressing the propagation path of coupled waves or introducing new neutralizing waves, the difference is that the polarization of the coupled wave becomes circular polarization.

In [52], they designed a slotted array DGS to reduce the coupling in microstrip circularly polarized antenna arrays, where slots are etched around each element, which resembled a band-stop filter. The measured results show that the coupling is less than −40 dB in the entire working frequency band, while the DGS structure inevitably increases the reverse radiation of the antenna. A hybrid parasitic resonator decoupling structure with an H-shaped structure and a meander line was proposed in [53] to improve the isolation of a CP antenna. The H-shaped structure and the meander line structure, respectively, reduce the coupled waves of the two linearly polarized components decomposed by the circularly polarized wave, which indirectly reduces the coupling. However, this scheme is not suitable for large arrays. A full-duplex circularly polarized antenna was designed in [54]. By introducing a W-shaped parasitic structure around the two elements, the asymmetrical W-shaped structure realizes the CP radiation of the antenna element. The introduction also suppresses the propagation of surface currents between antennas, reducing the mutual coupling between them. However, due to structural limitations, it is generally not applicable to other antennas. Metamaterial absorbers were designed to suppress the coupling between 2×2 CP antenna arrays in [55]. The isolation wall was composed of double-layer metamaterial absorbers and placed symmetrically between the CP antennas. In the frequency band, the absorber unit has a wave absorption rate of more than 80%, where the absorber absorbs the space waves between the elements to help enhance the isolation in the array, and the measured results show that the coupling between the adjacent elements and the elements on the diagonal is reduced by 11 and 8 dB. However, it is large in size and takes up a lot of space, making it unsuitable for compact arrays.

A novel decoupling structure with polarization rotation (DSPR) property was proposed recently to reduce the mutual coupling between co- and cross-polarized antennas. The DSRP element is a metal strip printed on a substrate backed with ground plane (i.e. a conventional meta-atom for polarization conversion). Unlike the conventional polarization-converting metasurface, air gaps are deliberately inserted between the DSRP element to adjust its polarization conversion rate so that it can provide neutralization waves for both orthogonal linear polarizations.[4] Its potential application for mutual coupling reduction of CP antennas has been demonstrated in [56]. It is well known that a CP wave can be decomposed into two orthogonal linear polarizations with a phase difference of 90°. Hence, when the coupling of both polarizations can be effectively reduced, the coupling between CP antennas can have a significant reduction. The proposed DSPR method can be classified as a neutralization-based decoupling principle, which introduces a new wave path with the same amplitude and opposite phase as the original coupling wave to cancel each other.

Figures 8.23a, b show the decoupling mechanism of the DSPR. Different from the previous works based on neutralization principle, the DSPR generates two reflected waves with orthogonal polarizations. As can be seen the proposed DSPR are placed above a 1×2 array, assuming that antenna 1 is u-polarized, while antenna 2 can be co-polarized (u-polarized) or cross-polarized (v-polarized). When antenna 1 is excited, the wave ① with u-polarization is radiated toward free

4 The interested reader can refer to [55] for detailed design of the DSPR.

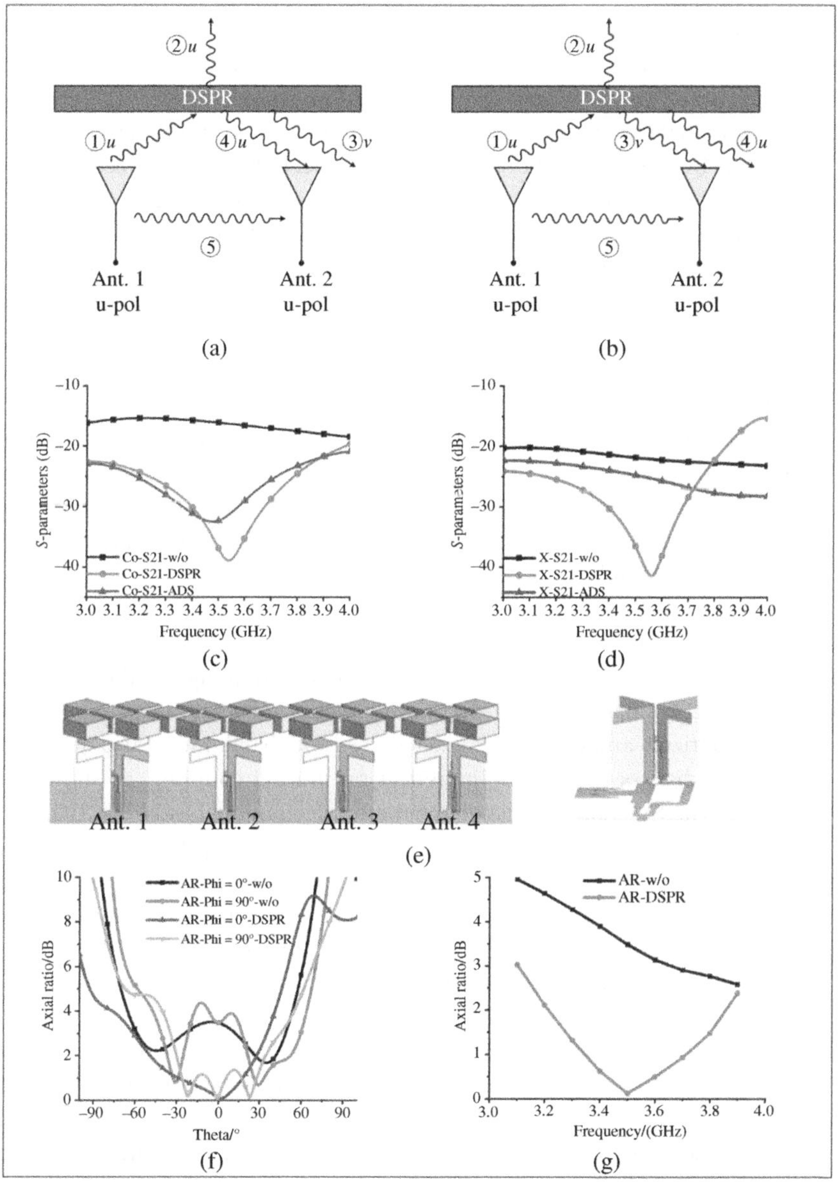

Figure 8.23 Decoupling principle of DSPR in (a) co-polarized and (b) cross-polarized. Decoupling effect of DSPR and ADS in (a) co-polarized and (b) cross-polarized. (e) DSPR in 1 × 4 CP antenna array. (f) Axial ratio in array with and without DSPR and (g) axial ratio over the frequency. Source: Reproduced with permission from Ref. [56]/IEEE.

space or the DSPR. The wave ① radiated toward DSPR will generate two reflected waves, i.e. a cross-polarized wave ③ with v-polarization and a co-polarized wave ④ with u-polarization; the transmitted wave ② through the DSPR is still a u-polarized wave. When antenna 2 is u-polarized, wave ⑤ represents the original co-polarized coupling between antenna 1 and 2, u-polarized wave ④ generated by DSPR can be controlled to cancel the co-polarized mutual coupling; similarly, when the antenna 2 is v-polarized, the wave ⑤ represents the original cross-polarized coupling, the wave ③ can be controlled to reduce the cross-polarized mutual coupling. Thus, when the coupling of both polarizations can be reduced effectively, the coupling between CP antennas achieves a significant reduction.

In order to intuitively show the superiority of the DSPR, the decoupling effects of ADS structure based on the neutralization scheme on co-polarization and cross-polarization coupling are compared, as shown in Figures 8.23c, d. For co-polarized coupling, both ADS and DSPR have obvious reduction effects, less than −30 dB in the working frequency band. However, ADS only reduces the cross-polarized coupling by about 3 dB, and DSPR significantly suppresses the cross-polarization coupling, S_{21} is reduced to below −40 dB while the DSPR structure is loaded.

According to the above introduction, the DSPR structure has the same decoupling effect on the co-polarized and cross-polarized coupling, so theoretically, the DSPR structure has the same decoupling ability for a CP antenna. A 1×4 circularly polarized antenna array excited by broadband microstrip network is designed to discuss the decoupling effect on mutual coupling, as shown in Figure 8.23e. The SWR of antenna element after decoupling is still less than 1.6 in the 3.3–3.8 GHz frequency band, for the mutual coupling between adjacent elements, i.e. S_{12} is reduced from −19 to −40 dB, and S_{23} is reduced from −18 to −30 dB, as for the coupling between the remaining elements in the 1×4 CP array, less than −25 dB in the entire operating frequency band. In addition to the obvious suppression of circularly polarized coupling, another attractive advantage of DSPR is the improvement of AR.

Figure 8.23f shows the AR of middle antenna 2 at the center frequency 3.5 GHz in the phi = 0° and phi = 90° plane. In the antenna array, the AR of the circular polarization is deteriorated due to the coupling between the elements, with an AR exceeding 3 dB. After loading the DSPR structure, the AR of the antenna is significantly improved and kept at less than 2 dB. Figure 8.23g shows the variation of the AR of antenna 2 with frequency. It can be seen that the AR of the antenna 2 is less than 3 dB in the entire working frequency band after the coupling is reduced, with stable circularly polarized radiation pattern maintained in the entire working frequency band.

8.3 Challenges and Practical Considerations

The increasing number of antennas, combined with the limited physical space in practical application scenarios, leads to a dilemma: on the one hand, the increasing number of antennas and working frequency band is inevitable; on the other hand, the trend of physical space is compactness and miniaturization. Considering the gradual evolution of specific antenna forms and requirements, as well as the coexistence of the 5G frequency bands with the original 2/3/4G frequency bands and Wi-Fi frequency bands, especially, with the large-scale commercial use of 5G system and the huge market for mobile devices, different types of multi-antenna systems face the following challenges:

For the base station antenna, when 5G antennas are co-located with 4G antennas in the common telecom tower, the width of 5G base station antennas is generally larger than 4G ones, resulting in a larger windward side and a higher wind resistance coefficient.[5] Owing to the strict requirements on the wind resistance coefficient of telecom towers, the challenges of multi-frequency multi-antenna co-aperture and compact array design need to be addressed. Thus, the array aperture and wind resistance coefficient as well as the volume and weight of the antenna need to be reduced.

For the channel and propagation between antennas, the evolution of antenna number and frequency band also brings new challenges. The propagation characteristics of sub-10 GHz and millimeter waves are completely different from the 4G frequency bands, which directly affects the shape and performance of the antenna. More specifically, super-large MIMO and new antenna morphology have new propagation characteristics and channel distribution, and ultimately affect the throughput, reliability, and delay of the system. Besides, terminal antenna needs to take into account the channel distribution and the influence of human body, besides being required to have specific performance to meet the needs of different application scenarios.

In recent years, the development momentum of the Internet of Things (IoT) continues to improve, and consumer-oriented smart terminals (such as smartphones and tablets) have great market potential. In fact, for the massive Internet of Everything (IoE), it is also expected that the problem of mutual interference with multiple devices and multiple antennas with limited space will become more severe. For example, more than forty antennas have been integrated and installed on a car, and with the further popularization of the Internet of Vehicles (IoV) application, new terminals and handsets will also encounter the

---

5 Note that the wind resistance coefficient is an important mechanical factor to be considered for BSA installations. When the BSA has a large aperture (surface), it tends to swing in strong wind, degrading the beamforming performance of the BSA.

challenge of multiple antennas (e.g. there are typically dozens of antennas in a very limited space with a clearance requirement less than 1 mm). The application of multi-antenna systems on various application scenarios and various carriers has encountered some common challenges, and diverse interference coupling suppression techniques can be extended to different devices and scenarios in the future to achieve better performance.

8.4 Conclusions

In this chapter, mutual coupling effects on MIMO antennas and decoupling techniques for various array antennas were discussed. It was shown that the coupling wave tends to deteriorate the SWR and radiation performances, especially the AR in CP antennas. In a transmitter array, the mutual coupling would change the load impedances of the PAs, and thus further increases the MIMO transmitter's nonlinearity. It has been experimentally proven that when the coupling in the array is reduced, the spurious and interfering signals in the transmitter are suppressed, and the amplitude of expected signal increases.

In order to enhance the isolation between antenna elements, various decoupling techniques were proposed in the literature. This chapter generalized the decoupling techniques according to antenna types, array configuration and antenna polarizations. Decoupling networks, DGS, EBG, FSS and metamaterial absorber could be applied to two-element linearly polarized arrays. As for realistic dual-polarized base station arrays, effective decoupling techniques include ADS, decoupling ground, dielectric superstrate, hybrid decoupling structure, among others. Moreover, decoupling techniques such as slot DGS, hybrid parasitic resonators, and DSPR were introduced to reduce the coupling in CP antenna array, in addition to effective isolation enhancement, DSPR can significantly alleviate the AR deterioration by couplings in CP arrays. These decoupling techniques were proven to be effective in reducing the coupling. However, in a practical design, it is necessary to adopt specific decoupling techniques according to the antenna form, the polarization of the coupling wave, and the transmission path to achieve effective decoupling.

References

1 Allen, J.L. and Diamond, B.L. (1966). *Mutual Coupling in Array Antennas*. Lexington, MA: MIT Lincoln Lab. Tech. Rep. EDS-66-443.

2 Sharawi, M.S. (2017). Current misuses and future prospects for printed multiple-input, multiple-output antenna systems [wireless corner]. *IEEE Antennas and Propagation Magazine* 59 (2): 162–170.

3 Larsson, E.G., Edfors, O., Tufvesson, F., and Marzetta, T.L. (2014). Massive MIMO for next generation wireless systems. *IEEE Communication Magazine* 52 (2): 186–195.

4 Zaidi, A., Athley, F., Medbo, J. et al. (2018). *5G Physical Layer: Principles, Models and Technology Components*. Academic Press.

5 Sharawi, M., Hassan, A.T., and Khan, M.U. (2017). Correlation coefficient calculations for MIMO antenna systems: a comparative study. *International Journal of Microwave and Wireless Technologies* 9: 1991–2004.

6 Chen, X., Kildal, P.-S., Carlsson, J., and Yang, J. (2013). MRC diversity and MIMO capacity evaluations of multi-port antennas using reverberation chamber and anechoic chamber. *IEEE Transactions on Antennas and Propagation* 61 (2): 917–926.

7 Chen, X., Zhang, S., and Li, Q. (2018). A review of mutual coupling in MIMO systems. *IEEE Access* 6: 24706–24719.

8 Kildal, P.S. and Rosengren, K. (2003). Electromagnetic analysis of effective and apparent diversity gain of two parallel dipoles. *IEEE Antennas Wireless Propagation Letters* 2 (1): 9–13.

9 Quist, B.T. and Jensen, M.A. (2009). Optimal antenna radiation characteristics for diversity and MIMO systems. *IEEE Transaction on Antennas and Propagation* 57 (11): 3474–3481.

10 Chen, X., Pei, H., Li, M. et al. (2020). Revisit to mutual coupling effects on multi-antenna systems. *Journal of Communications and Information Networks* 5 (4): 411–422.

11 Bhattacharyya, A.K. (2006). *Phased Array Antennas: Floquet Analysis, Synthesis, BFB, and Active Array Systems*. Hoboken, NJ: Wiley.

12 Chen, K.-H. and Kiang, J.-F. (2016). Effect of mutual coupling on the channel capacity of MIMO systems. *IEEE Transactions on Vehicular Technology* 65 (1): 398–403.

13 Masouros, C., Sellathurai, M., and Ratnarajah, T. (2013). Large-scale MIMO transmitters in fixed physical spaces: the effect of transmit correlation and mutual coupling. *IEEE Transaction on Communications* 61 (7): 2794–2804.

14 Shi, H., Gong, S., and Zheng, T. (2012). The effect of mutual coupling on the channel performance of MIMO communication system, *ISAPE2012*, 22–26 October 2012, Xi'an, China, pp. 335–339.

15 Barradas, F.M., Cunha, T.R., Cabral, P.M., and Pedro, J.C. (2017). Modeling PA linearity and efficiency in MIMO transmitters. *2017 IEEE MTT-S International Microwave Symposium (IMS)*, 04–09 June 2017, Honolulu, HI, USA, pp. 1999–2002.

16 Barradas, F.M., Tomé, P.M., Gomes, J.M. et al. (2017). Power, linearity, and efficiency prediction for MIMO arrays with antenna coupling. *IEEE Transactions on Microwave Theory and Techniques* 65 (12): 5284–5297.

17 Fager, C., Hausmair, K., Buisman, K. et al. (2017) Analysis of nonlinear distortion in phased array transmitters. *2017 Integrated Nonlinear Microwave and Millimetre-wave Circuits Workshop (INMMiC)*, 20–21 April 2017, Graz, Austria, pp. 1–4.

18 Fager, C., Eriksson, T., Barradas, F. et al. (2019). Linearity and efficiency in 5G transmitters: new techniques for analyzing efficiency, linearity, and linearization in a 5G active antenna transmitter context. *IEEE Microwave Magazine* 20 (5): 35–49.

19 Peng, F., Yang, F., Liu, B., and Chen, X. Experimental investigation of decoupling effect on the nonlinearity of power amplifiers in transmitter array. *ACES Journal* 38 (1): 7–14.

20 Chen, S.-C., Wang, Y.-S., and Chung, S.-J. (2008). A decoupling technique for increasing the port isolation between two strongly coupled antennas. *IEEE Transaction on Antennas and Propagation* 56 (12): 3650–3658.

21 Venkatasubramanian, S.N., Li, L., Lehtovuori, A. et al. (2017). Impact of using resistive elements for wideband isolation improvement. *IEEE Transactions on Antennas and Propagation* 65 (1): 52–62.

22 Meng, H. and Wu, K.-L. (2017). An LC decoupling network for two antennas working at low frequencies. *IEEE Transactions on Microwave Theory and Techniques* 65 (7): 2321–2329.

23 Chiu, C., Cheng, C., Murch, R.D., and Rowell, C.R. (2007). Reduction of mutual coupling between closely-packed antenna elements. *IEEE Transactions on Antennas and Propagation* 55 (6): 1732–1738.

24 Qian, B., Huang, X., Chen, X. et al. (2022). Surrogate-assisted defected ground structure design for reducing mutual coupling in 2×2 microstrip antenna array. *IEEE Antennas and Wireless Propagation Letters* 21 (2): 351–355.

25 Wei, K., Li, J., Ling, W. et al. (2016). S-shaped periodic defected ground structures to reduce microstrip antenna array mutual coupling. *Electronics Letters* 52 (15): 1288–1290.

26 Qian, B., Chen, X., and Kishk, A.A. (2021). Decoupling of microstrip antennas with defected ground structure using the common/differential mode theory. *IEEE Antennas and Wireless Propagation Letters* 20 (5): 828–832.

27 Ban, Y., Chen, Z., Chen, Z. et al. (2014). Decoupled hepta-band antenna array for WWAN/LTE smartphone applications. *IEEE Antennas Wireless Propagation Letters* 13: 999–1002.

28 Su, S.-W., Lee, C.-T., and Chang, F.-S. (2012). Printed MIMO-antenna system using neutralization-line technique for wireless USB-dongle applications. *IEEE Transactions on Antennas and Propagation* 60 (2): 456–463.

29 Cheng, Y.-F. and Cheng, K.-K.M. (2020). Compact wideband decoupling and matching network design for dual-antenna array. *IEEE Antennas and Wireless Propagation Letters* 19 (5): 791–795.

30 Zhang, S. and Pedersen, G.F. (2016). Mutual coupling reduction for UWB MIMO antennas with a wideband neutralization line. *IEEE Antennas and Wireless Propagation Letters* 15: 166–169.

31 Rajo-Iglesias, E., Quevedo-Teruel, Ó., and Inclan-Sanchez, L. (2008). Mutual coupling reduction in patch antenna arrays by using a planar EBG structure and a multilayer dielectric substrate. *IEEE Transactions on Antennas and Propagation* 56 (6): 1648–1655.

32 Yang, F. and Rahmat-Samii, Y. (2003). Microstrip antennas integrated with electromagnetic band-gap (EBG) structures: a low mutual coupling design for array applications. *IEEE Transactions on Antennas and Propagation* 51 (10): 2936–2946.

33 Assimonis, S.D., Yioultsis, T.V., and Antonopoulos, C.S. (2012). Design and optimization of uniplanar EBG structures for low profile antenna applications and mutual coupling reduction. *IEEE Transactions on Antennas and Propagation* 60 (10): 4944–4949.

34 Tan, X., Wang, W., Wu, Y. et al. (2019). Enhancing isolation in dual-band meander-line multiple antenna by employing split EBG structure. *IEEE Transactions on Antennas and Propagation* 67 (4): 2769–2774.

35 Dadgarpour, A., Zarghooni, B., Virdee, B.S. et al. (2017). Mutual coupling reduction in dielectric resonator antennas using metasurface shield for 60-GHz MIMO systems. *IEEE Antennas and Wireless Propagation Letters* 16: 477–480.

36 Zhao, L., He, Y., Zhao, G. et al. (2022). Scanning angle extension of a millimeter wave antenna array using electromagnetic band gap ground. *IEEE Transactions on Antennas and Propagation* 70 (8): 7264–7269.

37 Garg, P. and Jain, P. (2020). Isolation improvement of MIMO antenna using a novel flower shaped metamaterial absorber at 5.5 GHz WiMAX band. *IEEE Transactions on Circuits and Systems II: Express Briefs* 67 (4): 675–679.

38 Farahani, M., Pourahmadazar, J., Akbari, M. et al. (2017). Mutual coupling reduction in millimeter-wave MIMO antenna array using a metamaterial polarization-rotator wall. *IEEE Antennas and Wireless Propagation Letters* 16: 2324–2327.

39 Tang, J., Faraz, F., Chen, X. et al. (2020). A metasurface superstrate for mutual coupling reduction of large antenna arrays. *IEEE Access* 8: 126859–126867.

40 Wu, K.-L., Wei, C., Mei, X., and Zhang, Z.-Y. (2017). Array-antenna decoupling surface. *IEEE Transactions on Antennas and Propagation* 65 (12): 6728–6738.

41 Wei, C., Zhang, Z., and Wu, K. (2020). Phase compensation for decoupling of large-scale staggered dual-polarized dipole array antennas. *IEEE Transactions on Antennas and Propagation* 68 (4): 2822–2831.

42 Zhang, S., Chen, X., and Pedersen, G.F. (2019). Mutual coupling suppression with decoupling ground for massive MIMO antenna arrays. *IEEE Transactions on Vehicular Technology* 68 (8): 7273–7282.

43 Song, S., Chen, X., Da, Y., and Kishk, A. (2022). Broadband dielectric resonator antenna array with enhancement of isolation and front-to-back ratio for MIMO application. *IEEE Antennas and Wireless Propagation Letters* 21 (7): 1487–1491.

44 S. Song, Y. Da, B. Qian, X. Huang, X. Chen, Y. Li, and A. A. Kishk, "Dielectric resonator magnetoelectric dipole arrays with low cross polarization, backward radiation, and mutual coupling for MIMO base station applications," *China Communications*, vol. 21, no. 6, pp. 290–304, June 2024.

45 Liu, F., Guo, J., Zhao, L. et al. (2020). Ceramic superstrate-based decoupling method for two closely packed antennas with cross-polarization suppression. *IEEE Transactions on Antennas and Propagation* 69: 1751–1756.

46 Da, Y., Zhang, Z., Chen, X., and Kishk, A. (2021). Mutual coupling reduction with dielectric superstrate for base station arrays. *IEEE Antennas and Wireless Propagation Letters* 20 (5): 843–847.

47 Da, Y., Chen, X., and Kishk, A.A. (2022). In-band mutual coupling suppression in dual-band shared-aperture station arrays using dielectric block loading. *IEEE Transactions on Antennas and Propagation* 70 (10): 9270–9281.

48 Li, M., Chen, X., Zhang, A. et al. (2020). Split-ring resonator-loaded baffles for decoupling of dual-polarized base station array. *IEEE Antennas and Wireless Propagation Letters* 19 (10): 1828–1832.

49 Zhu, Y., Chen, Y., and Yang, S. (2019). Decoupling and low-profile design of dual-band dual-polarized base station antennas using frequency-selective surface. *IEEE Transactions on Antennas and Propagation* 67 (8): 5272–5281.

50 Karimian, R., Kesavan, A., Nedil, M., and Denidni, T.A. (2017). Low-mutual coupling 60-GHz MIMO antenna system with frequency selective surface wall. *IEEE Antennas and Wireless Propagation Letters* 16: 373–376.

51 Liu, B., Da, Y., Chen, X., and Kishk, A.A. (2022). Hybrid decoupling structure based on neutralization and partition schemes for compact large-scale base station arrays. *IEEE Antennas and Wireless Propagation Letters* 21 (2): 267–271.

52 Gao, D., Cao, Z., Fu, S. et al. (2020). A novel slot-array defected ground structure for decoupling microstrip antenna array. *IEEE Transactions on Antennas and Propagation* 68 (10): 7027–7038.

53 Chen, Z., Li, M., Liu, G. et al. (2019). Isolation enhancement for wideband, circularly/dual-polarized, high-density patch arrays using planar parasitic resonators. *IEEE Access* 7: 112249–112257.

54 Wei, K. and Zhu, B. (2019). The novel w parasitic strip for the circularly polarized microstrip antennas design and the mutual coupling reduction between them. *IEEE Transactions on Antennas and Propagation* 67 (2): 804–813.

55 Zhang, J., Li, J., and Chen, J. (2019). Mutual coupling reduction of a circularly polarized four-element antenna array using metamaterial absorber for unmanned vehicles. *IEEE Access* 7: 57469–57475.

56 Liu, B., Chen, X., Tang, J. et al. (2022). Co- and cross-polarization decoupling structure with polarization rotation property between linearly polarized dipole antennas with application to decoupling of circularly polarized antennas. *IEEE Transactions on Antennas and Propagation* 70 (1): 702–707.

9

Measurement Methods for MIMO Antenna Systems

9.1 Over-the-Air Tests in Reverberation Chamber

The reverberation chamber (RC) is one of the most popular approaches for over-the-air (OTA) tests. This OTA testing method is cost-effective and can provide a large testing zone. Due to this reason, it has become a standard OTA testing approach in Cellular Telecommunications Industry Association (CTIA) [1]. The RC is basically an electrically large metal cavity with various mode-stirring mechanisms. According to the resonant mode theory, the electromagnetic field in the RC can be solved by the Helmholtz equation. The result of this equation can be represented by the weighted sum of different modes. Due to the RC's complex boundary condition, this solution process can be rather complicated. As a result, researchers often resort to statistical models for evaluation of the RC measurement. According to statistical theory, the electromagnetic field in a well-stirred RC is spatially uniform and isotropic. Spatial uniformity means the mean-square values of the amplitude of the field at different positions are equal to a constant, while isotropy means the incoming/departing waves are equally distributed in the angular domain [2].

Due to the statistical property of electromagnetic fields in the RC, we need to pay attention to their probability density function (PDF). According to the central limit theorem, the real and imaginary parts of the rectangular components of the field follow Gaussian distribution. Specifically, the PDF of the rectangular field component is [2]

$$\mathrm{PDF}(E_{xr}) = \frac{1}{\sigma\sqrt{2\pi}} \exp\left(-\frac{E_{xr}^2}{2\sigma^2}\right) \tag{9.1}$$

where σ^2 is the variance of the field component. According to the probability theory, the magnitude of the rectangular field component follows the Raleigh distribution

MIMO Antenna Systems for 5G and Beyond, First Edition.
Xiaoming Chen, Ping Jack Soh, and Mohammad S. Sharawi.

$$\text{PDF}(|E_x|) = \frac{|E_x|}{\sigma^2} \exp\left(-\frac{|E_x|^2}{2\sigma^2}\right), \tag{9.2}$$

while the squared magnitude of the field component follows exponential distribution (i.e. the chi-square distribution with two degrees of freedom),

$$\text{PDF}\left(|E_x|^2\right) = \frac{1}{2\sigma^2} \exp\left(-\frac{|E_x|^2}{2\sigma^2}\right). \tag{9.3}$$

Note that the total electric field E follows the chi-square distribution with six degrees of freedom.

In order to improve the uniformity of the electromagnetic field in the RC, various stirring techniques have been introduced to superimpose as many electromagnetic modes as possible. Common stirring techniques include mechanical stirring, source stirring, and frequency stirring. Mechanical stirring is achieved by changing the boundary conditions of the electromagnetic field in the RC (through, e.g. the rotation of the built-in metallic mode stirrers). In essence, mechanical stirring can also be achieved by changing the shape or volume of the RC, but this implementation is relatively complex and therefore not common. Source stirring is usually achieved by changing the position of the probe antenna via a turn-table platform (or switching between multiple probe antennas) in the RC. Frequency stirring considers the electric field modes at the frequency points near the testing frequency of the equipment to be tested and realizes the superposition of more electromagnetic modes within the acceptable bandwidth to achieve the purpose of stirring.

In OTA tests, there are three important metrics, which are total radiated power (TRP), total isotropic sensitivity (TIS), and throughput. Specifically, TRP and TIS are used to characterize wireless devices' transmitting and receiving capabilities, respectively. These metrics of the device under test (DUT) can all be measured in the RC efficiently.

9.1.1 Total Radiated Power

The TRP is defined as the maximum power that a wireless device can radiate. The standard testing setup of TRP in the RC is shown in Figure 9.1.

The RC for OTA test is usually equipped with mode stirrers, turntable platform, measurement antenna, reference antenna, apart from the DUT. A complete TRP test in the RC contains two steps, which are the reference measurement and the DUT measurement. Besides that, due to certain requirements about the wireless channels, the loading condition in the RC needs to be adjusted for a suitable coherence bandwidth [1].

The first step of this measurement is the reference measurement, which is used to determine the average path loss in the RC under certain loading conditions.

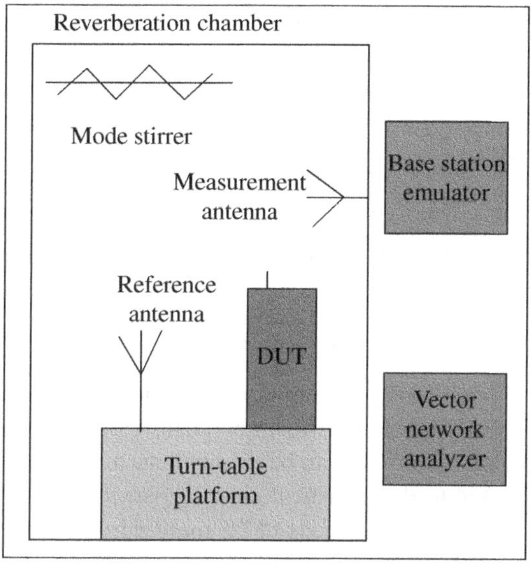

Figure 9.1 Standard setup of TRP tests in the RC.

In this step, the measurement antenna and reference antenna are connected to the vector network analyzer (VNA). The power transfer function (PTF) for this measurement (G_{REF_TRP}) can be calculated by the sample data of different stirring positions and different frequency points [1]

$$G_{REF\_TRP} = \frac{\left\langle |S_{21,REF}|^2 \right\rangle_{N_{SF}}}{\eta_R \eta_M \left(1 - |\Gamma_R|^2\right) \left(1 - |\Gamma_M|^2\right)} \tag{9.4}$$

where η_M and Γ_M are the radiation efficiency and reflection coefficient of the measurement antenna, respectively, η_R and Γ_R denote the radiation efficiency and reflection coefficient of the reference antenna, respectively, $S_{21,\,REF}$ denotes the S parameter and $\langle \cdot \rangle$ denotes the ensemble average. Note that $N_{SF} = N_S \times N_F$, where N_S denotes the number of samples obtained by mechanical stirring, and N_F denotes the number of sampling due to frequency stirring [3].

The DUT measurement is similar to the reference measurement, except that a base station emulator (BSE) is used instead of the VNA in this step. The measurement antenna is connected to the BSE, via which a link with the DUT is established. The BSE asks the DUT to work under its maximum power during the measurements. The power received by the measurement antenna at each stirring state is denoted as P_M and this procedure is repeated until all the stirring states have been sampled. The equivalent PTF in the DUT measurement can be calculated by [4]

$$G_{DUT\_TRP} = \frac{\langle P_M \rangle_{N_S} / \hat{P}_{TRP}}{\eta_M \left(1 - |\Gamma_M|^2\right)} \tag{9.5}$$

where $\hat{P}_{\mathrm{TRP}}$ represents the TRP estimate of the DUT. The true value of this metric is a constant that depends only on the DUT itself. However, the measured TRP in the RC is a random variable (RV) that abides by a specific distribution. Due to the RC's uniformity, the PTFs of the reference measurement and the DUT measurement are equal, i.e.

$$G_{\mathrm{DUT\_TRP}} = G_{\mathrm{REF\_TRP}}. \tag{9.6}$$

Combining (9.4), (9.5), and (9.6), $\hat{P}_{\mathrm{TRP}}$ can be calculated by:

$$\hat{P}_{\mathrm{TRP}} = \frac{\langle P_{\mathrm{M}} \rangle_{N_{\mathrm{S}}}}{\left\langle |S_{21,\mathrm{REF}}|^2 \right\rangle_{N_{\mathrm{SF}}}} \eta_{\mathrm{R}} \left(1 - |\Gamma_{\mathrm{R}}|^2\right). \tag{9.7}$$

The S-parameters and received powers are measured, and the properties of the reference antenna are known beforehand. Using these parameters, the estimated value of the TRP can be readily determined.

Due to the statistical characteristics of the RC, the measurement uncertainty of the TRP test needs to be analyzed. First, rewriting the form of (9.7) by using $\left\langle \left| S_{21,\mathrm{REF}}^{\mathrm{equivalent}} \right|^2 \right\rangle$ to replace $\langle P_{\mathrm{M}} \rangle / P_{\mathrm{TRP}}$ as [5]

$$\hat{P}_{\mathrm{TRP}} = P_{\mathrm{TRP}} \frac{\left\langle \left| S_{21,\mathrm{REF}}^{\mathrm{equivalent}} \right|^2 \right\rangle_{N_{\mathrm{SI}}}}{\left\langle |S_{21,\mathrm{REF}}|^2 \right\rangle_{N_{\mathrm{SF}}}} \eta_{\mathrm{R}} \left(1 - |\Gamma_{\mathrm{R}}|^2\right) \tag{9.8}$$

where P_{TRP} is the true value of the TRP of the DUT. $N_{\mathrm{SI}} = N_{\mathrm{S}} \times N_{\mathrm{I}}$, N_{I} denotes the independent sampling number due to inherent stirring caused by the signal bandwidth of the modulated LTE signal. Note that the measurement uncertainty depends on the signal bandwidth for a given RC configuration. Specifically, this influence is due to the inherent frequency stirring of the single-carrier discrete Fourier transform spread orthogonal frequency division multiplexing (DFTS-OFDM) waveform used in the LTE device. This parameter can be calculated as [6]

$$N_{\mathrm{I}} = \frac{\mathrm{BW}_{\mathrm{S}}}{\langle B_{\mathrm{C}} \rangle_{\mathrm{BW}_{\mathrm{S}}}} \tag{9.9}$$

where BW_{S} is the signal bandwidth of the DUT and $\langle B_{\mathrm{C}} \rangle_{\mathrm{BW}_{\mathrm{S}}}$ is the average coherent bandwidth over the given signal bandwidth.

In an ideal RC, RVs $X = \left| S_{21,\mathrm{REF}}^{\mathrm{equivalent}} \right|^2$ and $Y = |S_{21,\mathrm{REF}}|^2$ follow the exponential distribution [2]:

$$f(t) = \frac{C_{\mathrm{RC}}}{Q} \exp\left(-\frac{C_{\mathrm{RC}}}{Q} t\right) \tag{9.10}$$

where Q denotes the quality factor of the RC, and $C_{\mathrm{RC}} = 16\pi^2 V / \lambda^3$, with V being the volume of the RC and λ being the wavelength [7].

Assuming the mechanical stirring number N_S, the frequency stirring number N_F, and the signal bandwidth stirring number N_I are independent, RVs X_i ($i = 1, 2, \ldots, N_{SF}$) and Y_j ($j = 1, 2, \ldots, N_{SI}$) are independent and identically distributed. For RVs $R = \sum_{N_{SF}} X_i$ and $G = \sum_{N_{SI}} Y_j$, these two variables follow the chi-square distribution and the degrees of freedom for this distribution are $2N_{SF}$ and $2N_{SI}$, respectively [8]

$$f(t) = \frac{B^N}{\Gamma(N)} t^{N-1} \exp(-Bt) \tag{9.11}$$

where Γ denotes the Gamma function with $\Gamma(N) = (N-1)!$ and $B = C_{RC}/2Q$. For the RV R, $N = N_{SF}$. For the RV G, $N = N_{SI}$.

The distribution of RV $Z = G/R$ can be determined by using the multivariate transformation

$$f(z) = \frac{\Gamma(N_{SI} + N_{SF})}{\Gamma(N_{SI})\Gamma(N_{SF})} \frac{z^{N_{SI}-1}}{(1+z)^{N_{SI}+N_{SF}}}. \tag{9.12}$$

The expectation and variance of the RV $K = (N_{SF}/N_{SI}) \times Z$ can be obtained by the distribution of the RV Z

$$\begin{aligned} E(K) &= \frac{N_{SF}}{N_{SF}-1} \\ \mathrm{VAR}(K) &= \frac{N_{SF}^2(N_{SI}+N_{SF}-1)}{N_{SI}(N_{SF}-2)(N_{SF}-1)^2}. \end{aligned} \tag{9.13}$$

The measured value of TRP $\hat{P}_{TRP}$ can be expressed as

$$\hat{P}_{TRP} = \eta_R \left(1 - |\Gamma_R|^2\right) P_{TRP} \times K. \tag{9.14}$$

The expectation and variance of the measured TRP can be calculated as

$$\begin{aligned} E(\hat{P}_{TRP}) &= \frac{N_{SF}}{N_{SF}-1} P_{TRP}\eta_R \left(1 - |\Gamma_R|^2\right) \\ \mathrm{VAR}(\hat{P}_{TRP}) &= \frac{N_{SF}^2(N_{SI}+N_{SF}-1)}{N_{SI}(N_{SF}-2)(N_{SF}-1)^2} \left[P_{TRP}\eta_R \left(1 - |\Gamma_R|^2\right)\right]^2. \end{aligned} \tag{9.15}$$

As can be seen, the measurement uncertainty of the TRP tests in the RC involves the true value of the TRP, which is usually unknown in practice. Due to this reason, one usually uses the relative standard deviation to assess the uncertainty of the TRP tests. The empirical relative standard deviation can be calculated with normalized testing results. And the analytical relative standard deviation of TRP tests in RC can be calculated as

$$\sigma_{TRP} = \sqrt{\frac{N_{SI}+N_{SF}-1}{N_{SI}(N_{SF}-2)}}. \tag{9.16}$$

9.1.2 Total Isotropic Sensitivity

The TIS is another common figure of merit (FoM) that is usually used to evaluate the receiving capacity of wireless devices. This metric denotes the minimum power received by the DUT when it reaches a specific throughput or error rate threshold. The TIS measurement setup based on the RC is similar to the TRP measurement, with its testing procedure also consisting of reference measurement and DUT measurement. The reference measurement has the same purpose as that in the TRP measurement in determining the average path loss in the RC. In the DUT measurement, the output power of the BSE is gradually reduced until the selected performance indicator (throughput or error rate) reaches a specified level. The corresponding output power of the BSE at this specific threshold is then recorded, which is denoted as P_{BSE}. This procedure is repeated until all stirring states have been sampled. The equivalent PTF can be calculated as:

$$G_{\mathrm{DUT\_TIS}} = \frac{\langle \hat{P}_{\mathrm{TIS}}/P_{\mathrm{BSE}} \rangle_{N_\mathrm{S}}}{\eta_\mathrm{M}\left(1-|\Gamma_\mathrm{M}|^2\right)} = \frac{\langle 1/P_{\mathrm{BSE}} \rangle_{N_\mathrm{S}}}{\eta_\mathrm{M}\left(1-|\Gamma_\mathrm{M}|^2\right)} \hat{P}_{\mathrm{TIS}} \tag{9.17}$$

where $\hat{P}_{\mathrm{TIS}}$ is the estimate of the true value of the TIS for the DUT, and this metric is an RV as well, while the true value of the TIS P_{TIS} is constant and depends only on the DUT. Note that the measured TIS is harmonic mean of all the samples instead of the arithmetic mean of all the samples as in the TRP test.

By virtue of the field uniformity in the RC, the PTFs of reference measurement and DUT measurement are equal, i.e.

$$G_{\mathrm{REF\_TIS}} = G_{\mathrm{DUT\_TIS}}. \tag{9.18}$$

So, the estimate of the TIS can be calculated as

$$\hat{P}_{\mathrm{TIS}} = \frac{1}{\eta_\mathrm{R}\left(1-|\Gamma_\mathrm{R}|^2\right)} \frac{\left\langle |S_{12,\mathrm{REF}}|^2 \right\rangle_{N_{\mathrm{SF}}}}{\langle 1/P_{\mathrm{BSE}} \rangle_{N_\mathrm{S}}}. \tag{9.19}$$

The measurement uncertainty of the TIS measurement can be analyzed analogously as the TRP measurement. Introducing the equivalent PTF of the RC $\left(\left|S_{21,\mathrm{DUT}}^{\mathrm{equivalent}}\right|^2\right)$, (9.19) becomes

$$\hat{P}_{\mathrm{TIS}} = \frac{1}{\eta_\mathrm{R}\left(1-|\Gamma_\mathrm{R}|^2\right)} \frac{\left\langle |S_{21,\mathrm{REF}}|^2 \right\rangle_{N_{\mathrm{SF}}}}{\left\langle \left|S_{21,\mathrm{DUT}}^{\mathrm{equivalent}}\right|^2 \right\rangle_{N_{\mathrm{SI}}}} P_{\mathrm{TIS}} \tag{9.20}$$

where N_{SI} is the independent sampling number including the inherent stirring [4]. This effect is caused by the signal bandwidth of the LTE device. As a result, the expectation, variance, and relative standard deviation of the TIS measurement can be derived, respectively, as follows [9]

$$E(\widehat{P}_{\mathrm{TIS}}) = \frac{N_{\mathrm{SI}}}{N_{\mathrm{SI}}-1} P_{\mathrm{TIS}} \frac{1}{\eta_{\mathrm{R}}\left(1-|\Gamma_{\mathrm{R}}|^2\right)}$$

$$\mathrm{VAR}(\widehat{P}_{\mathrm{TIS}}) = \frac{N_{\mathrm{SI}}^2(N_{\mathrm{SI}}+N_{\mathrm{SF}}-1)}{N_{\mathrm{SF}}(N_{\mathrm{SI}}-2)(N_{\mathrm{SI}}-1)^2}\left[P_{\mathrm{TIS}}\frac{1}{\eta_{\mathrm{R}}\left(1-|\Gamma_{\mathrm{R}}|^2\right)}\right]^2$$

$$\mathrm{STD}_{\mathrm{TIS}}^{\mathrm{reletive}} = \sqrt{\frac{N_{\mathrm{SI}}+N_{\mathrm{SF}}-1}{N_{\mathrm{SF}}(N_{\mathrm{SI}}-2)}}. \tag{9.21}$$

9.1.3 Throughput Testing in RC

Throughput is the actual data transmission rate of the system. It has been shown in [10] that the throughput of IEEE 802.11n and LTE systems can be measured in RC to distinguish between good- and poor-performing MIMO devices. A theoretical throughput model based on threshold receiver has been proposed in [11] to complement the throughput measurement. According to the threshold receiver model, the block error rate (BLER) of a wireless device in a static additive white Gaussian noise (AWGN) channel can change rapidly from almost 0% to 100% when the signal to noise ratio (SNR) falls below a specific threshold level [11].

$$P_{\mathrm{e}}(\gamma) = \begin{cases} 1, & \gamma < \gamma_{\mathrm{th}} \\ 0, & \gamma > \gamma_{\mathrm{th}} \end{cases} \tag{9.22}$$

where γ is the received SNR and γ_{th} denotes the threshold SNR value. In this case, the throughput can be simply modeled as:

$$T_{\mathrm{put}}(\gamma) = T_{\mathrm{put,max}}\left[1 - P_{\mathrm{e}}(\gamma)\right] \tag{9.23}$$

where $T_{\mathrm{put,max}}$ is the maximum throughput for a given link. The results in [11] show that the throughput curves in conducted measurements exhibit obvious sharp transitions, as predicted by (9.23), which validate the effectiveness of the threshold receiver model.

For a dynamic fading channel, the instantaneous SNR and BLER will vary with time. One can obtain the average BLER based on the threshold receiver model as [11]:

$$\overline{P}_{\mathrm{e}}(\overline{\gamma}) = \int_0^{\infty} P_{\mathrm{e}}(\gamma) f(\gamma;\overline{\gamma})\mathrm{d}\gamma \tag{9.24}$$

where $\overline{\gamma}$ is the average value of γ; $f(\gamma;\overline{\gamma})$ represents the PDF of fading SNR γ for a given $\overline{\gamma}$. By substituting specific formulas, one obtains:

$$\overline{P}_{\mathrm{e}}(\overline{\gamma}) = \int_0^{\gamma_{\mathrm{th}}} f(\gamma;\overline{\gamma})\mathrm{d}\gamma = F(\gamma;\overline{\gamma}) \tag{9.25}$$

where $F(\gamma;\overline{\gamma})$ represents the cumulative distribution function (CDF) of fading SNR γ for a given $\overline{\gamma}$. Then the throughput model of a single-input single-output system with fixed modulation and coding scheme (MCS) can be expressed as:

$$T_{\text{put}}(\overline{\gamma}) = T_{\text{put,max}}\,[1 - \overline{P}_{\text{e}}(\overline{\gamma})]. \tag{9.26}$$

It can be seen that under the assumption of threshold receiver model, the throughput is related to the CDF of the instantaneous SNR. Note that frequency-flat fading channel is assumed here, while the channel's frequency-selectivity can be readily modeled coherently combining the independent subcarriers determined by (9.9) [12–14].

The simple theoretical throughput model has been further expanded in [12, 13] to predict the throughput of MIMO systems in spatial-multiplexing mode. For example, the throughput of a 2×2 MIMO system can be modeled as follows:

$$T_{\text{put}}(\overline{\gamma}) = 2T_{\text{put,max}}\,[1 - \overline{P}_{\text{e}}(\overline{\gamma})]. \tag{9.27}$$

To decouple the MIMO channel (with correlations and power imbalances) into parallel links (i.e. multiple non-interfering SISO channels), the zero-forcing (ZF) or minimum mean square error (MMSE) equalizer can be used [14]. Firstly, the MIMO channel can be modeled as:

$$\mathbf{y} = \mathbf{H}\mathbf{x} + \mathbf{n} \tag{9.28}$$

where $\mathbf{H}$ represents the flat MIMO channel; $\mathbf{y}$, $\mathbf{x}$, and $\mathbf{n}$ are the received signal vector, transmitted signal vector, and noise vector, respectively. In an open-loop MIMO system, there is no precoding. Let x_i be the ith element of $\mathbf{x}$ and $\mathbf{h}_i$ be the ith column of $\mathbf{H}$. Then the above equation can be further expressed as:

$$\mathbf{y} = \mathbf{h}_i x_i + \sum_{i\neq j} \mathbf{h}_j x_j + \mathbf{n} \tag{9.29}$$

where the first item on the right side represents the expected received signal for the ith data stream and the second item denotes the interference from the other data streams on the ith data stream. When using the ZF receiver to deal with interference and noise, in order to use the throughput model based on threshold receiver, we need to decouple the MIMO channel matrix to obtain the SNR corresponding to each stream. The effective SNR of the ith stream (after the ZF equalization) can be expressed as:

$$\gamma_i = \frac{\overline{\gamma}_0}{[(\mathbf{H}^H\mathbf{H})^{-1}]_{i,i}} \tag{9.30}$$

where $\overline{\gamma}_0 = E\left[|x_i|^2\right]/E\left[|n_i|^2\right]$; $[\mathbf{X}]_{i,i}$ represents the ith diagonal term of the matrix $\mathbf{X}$. After obtaining the instantaneous SNR γ_i of each data stream, the corresponding throughput of each data stream can be obtained based on the CDF of γ_i. The total throughput is the sum of all data stream throughput:

$$T_{\text{put}}(\overline{\gamma}) = \sum_i T_{\text{put,max},i} \cdot \{1 - \overline{P}_{\text{e}}(\overline{\gamma})\} \tag{9.31}$$

where $T_{\text{put,max},i}$ denotes the maximum throughput corresponding to the ith data stream. In addition, to study the impact of antenna effect (correlation and power imbalance) on throughput, the Kronecker channel model is used:

$$\mathbf{H} = \mathbf{R}_{\text{r}}^{1/2}\mathbf{H}_{\text{w}}\mathbf{R}_{\text{t}}^{1/2} \tag{9.32}$$

where $\mathbf{H}_{\text{w}}$ represents the independent identically distributed (i.i.d.) Rayleigh fading channel with unit variance complex Gaussian variables; $\mathbf{R}_{\text{t}}$ and $\mathbf{R}_{\text{r}}$ represent the correlation matrices at the transmit and receive antennas including possible power imbalances, respectively.

$$\mathbf{R}_{\text{r}} = \boldsymbol{\Xi}_{\text{r}} \circ \boldsymbol{\Phi}_{\text{r}} \tag{9.33}$$

where $\boldsymbol{\Phi}_{\text{r}}$ denotes a normalized correlation matrix of receive side; $\boldsymbol{\Xi}_{\text{r}} = \sqrt{\boldsymbol{e}_{\text{r}}}\sqrt{\boldsymbol{e}_{\text{r}}}^{\text{T}}$ with $\boldsymbol{e}_{\text{r}}$ denoting a column vector consisting of the antenna efficiencies at all the receive ports; $\circ$ represents the Hadamard product (i.e. element-wise product). $\mathbf{R}_{\text{t}}$ can be determined similarly.

Figure 9.2 shows the relative throughputs of the measurement and theoretical model for 2×2 open-loop MIMO system [14]. It can be seen that the throughput models with MMSE and ZF receivers effectively predict the measured throughput.

In addition, the throughput multiplexing efficiency [15] has been proposed to characterize the impact of antenna effects (correlation and power imbalance) on MIMO throughput. Its physical meaning is that for the existence of correlation or power imbalance, a greater SNR is required to achieve the same throughput level as in the ideal situation. The difference in SNR can quantitatively describe the decrease in throughput performance caused by imperfect antennas. For a 2×2 MIMO system, the closed expression for multiplexing efficiency can be derived based on the throughput model as [15]:

$$e_{\text{mux}} = 2 \cdot \left(\frac{1}{e_1} + \frac{1}{e_2}\right)^{-1} \cdot (1 - |\rho|^2) \tag{9.34}$$

where e_i is the antenna efficiency of the ith antenna port, and ρ denotes correlation coefficient. The impact of correlation on the throughput of closed-loop MIMO systems has been analogously investigated in [16], where an inversely weighted mean squared error (MSE) matrix was proposed as a precoding selection criterion for the throughput model that shows good agreement with the measurement. This implies that the information-theoretic water-filling technique is not applied in the real LTE MIMO system. Further, the results in [16] demonstrate that in the case of high correlation, the closed-loop MIMO system outperform its open-loop MIMO system in the regions of low and medium SNR.

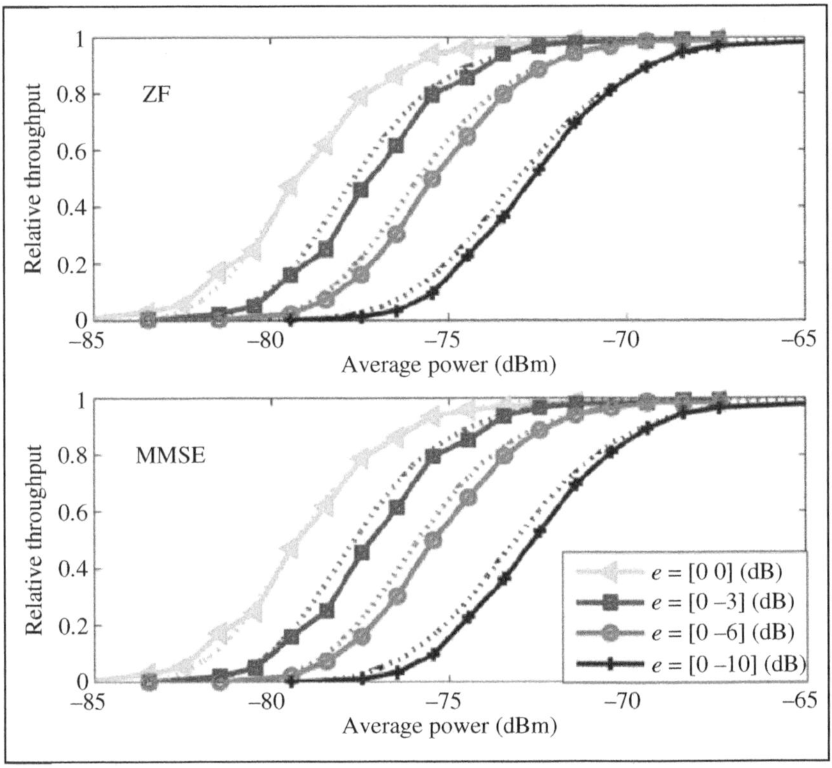

Figure 9.2 Relative throughputs of 2 × 2 open-loop MIMO system. Dashed curves denote the throughput models with the receiver (upper is ZF receiver and lower is MMSE receiver with 25-dB $\overline{\gamma}_0$); solid curves represent measurements. Source: Reproduced with permission from Ref. [14]/IEEE.

9.2 Multi-Probe Anechoic Chamber

The multi-probe anechoic chamber (MPAC) based OTA testing system mainly consists of BSE, channel emulator, power amplifier, anechoic chamber, OTA probes, and device under test (DUT). The schematic diagram of the two-dimensional (2D) MPAC setup for user equipment (UE) testing is shown in Figure 9.3. The BSE outputs the test signal to the channel emulator. Then the channel emulator simulates the real wireless communication environment to fade the input signal, and add information such as delay and Doppler. The fading signal is mapped to the OTA probes in the anechoic chamber after being amplified by power amplifier, and the signal emitted by the probes generates an environment with certain power distribution and polarization information in the test volume. Ultimately, the emulation channel experienced by DUT is almost close to the real communication

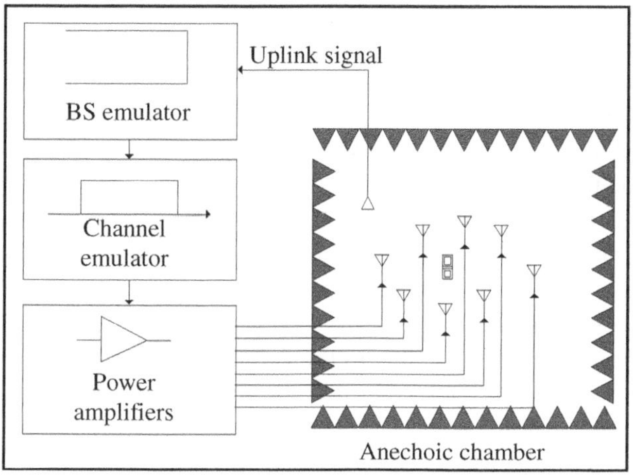

Figure 9.3 Illustration of a 2D MPAC setup for MIMO OTA testing.

environment. Furthermore, the uplink signal is directly connected to the BSE through an uplink feedback antenna. The BSE conducts statistical processing of the feedback information to obtain the performance test results of the DUT.

Spatial domain information is essential for the performance of multi-antenna devices. The MPAC method generates test signals with specific distribution through the OTA probes, thereby effectively building the spatial characteristics of the target channel. In addition, information such as delay and Doppler can be reproduced in the channel emulator, and what needs to be completed in the anechoic chamber is the reconstruction of the spatial characteristics at the DUT side.

9.2.1 Algorithms for Channel Emulation in MPAC

The main idea of channel emulation is to reproduce the target channel on the DUT side in the test volume by controlling the signals emitted from the OTA probes. Two commonly used channel emulation methods in MPAC setup are pre-fading signal synthesis (PFS) [17] and plane wave synthesis (PWS) [18]. These two methods rebuild the spatial characteristics of the DUT side in two different ways. The geometry-based stochastic channel model (GSCM) is usually adopted as the target channel model. This section mainly summarizes the target channel model, the PFS and the PWS method.

9.2.1.1 Target Channel Model

MIMO OTA testing reconstructs the actual propagation environment or standard channel model to make the simulated channel experienced by DUT close to the real propagation environment. This method provides a repeated and controllable laboratory testing environment. Only when the channel is accurately reconstructed can the DUT process the channel information effectively and ultimately obtain the correct performance test results. Therefore, channel reconstruction is the key issue to MIMO OTA testing. GSCM only focuses on the channel characteristics before the first reflection and after the last reflection of the signal (i.e. only considering the first and last scatters), ignoring the specific reflection process of the signal during the propagation and thus simplifying the complexity of the channel modeling. Commonly used GSCM includes SCM(E) [19], WINNER [20], 3GPP [21], QuaDRiGa [22], etc.

In the actual propagation environment, due to the existence of obstacles such as tall buildings and trees, the signals emitted by the base station need to undergo multiple reflections before reaching the receiver. The reflection paths are complicated and diverse, so the terminal will receive multipath signals, and the arrival direction and time of each path signal are different. Generally, the channel is divided into n clusters, and each cluster has its own departure angle, arrival angle, and delay. In addition, the signal is not discretely distributed in space, so the cluster represents the center direction of a beam. To characterize the continuous distribution of a cluster within a certain range, cluster is further subdivided into numerous subpaths. For example, in the SCME model, each cluster is subdivided into 20 subpaths symmetrically distributed around the cluster direction.

For a MIMO system, the channel transfer function of the nth cluster between the transmitting antenna s and the receiving antenna u can be represented as [21]:

$$h_{u,s,n}(f,t) = \sqrt{\frac{P_n}{M}} \sum_{m=1}^{M} F_s^{\mathrm{Tx}}\left(\Omega_{n,m}^{\mathrm{Tx}}\right) F_u^{\mathrm{Rx}}\left(\Omega_{n,m}^{\mathrm{Rx}}\right) \cdot \exp(j\phi_{n,m})$$

$$\cdot \exp\left(\frac{j2\pi\left(\bar{r}_{\mathrm{rx},n,m}^{\mathrm{T}} \cdot \bar{d}_{\mathrm{rx},u}\right)}{\lambda}\right) \exp\left(\frac{j2\pi\left(\bar{r}_{\mathrm{tx},n,m}^{\mathrm{T}} \cdot \bar{d}_{\mathrm{tx},s}\right)}{\lambda}\right) \exp(j2\pi(\nu_{n,m}t - f\tau_n)) \quad (9.35)$$

where P_n denotes the power of the nth cluster; M is the number of the subpaths; F_s^{Tx} and F_u^{Rx} are the radiation pattern of the transmitting antenna s and the receiving antenna u, respectively; $\Omega_{n,m}^{\mathrm{Tx}}$ and $\Omega_{n,m}^{\mathrm{Rx}}$ denote the angle of departure (AoD), angle of arrival (AoA) for the mth subpaths of the nth cluster, respectively; $\phi_{n,m}$ is the random initial phase; $\bar{r}_{\mathrm{rx},n,m}$ and $\bar{r}_{\mathrm{tx},n,m}$ are the spherical unit vectors for the arrival and departure angles of the mth subpath of the nth cluster, respectively; $\bar{d}_{\mathrm{rx},u}$ and $\bar{d}_{\mathrm{tx},s}$ represent the position vectors of the receiving antenna u and the transmitting antenna s, respectively; λ is wavelength; $\nu_{n,m}$ is the Doppler frequency of the mth subpath of the nth cluster; τ_n is the delay of the nth cluster. For simplicity, only vertical polarization is considered hereafter.

9.2.1.2 PFS Method

Since the weights obtained by the PFS method are real numbers and do not require strict phase calibration, it has been widely adopted in the industry. The basic idea of the PFS method is to allocate appropriate power weights to each probe through optimization algorithms, thereby reproducing the spatial characteristics of the target channel on the DUT side. The synthetic object of PFS method is cluster, and the fading signals of the same cluster are Independent and identically distributed RVs. The clusters can be mapped to probes by emitting fading signals from probes distributed in different directions in the anechoic chamber. Therefore, each cluster is independently synthesized by the discrete probes in the anechoic chamber. For multi-cluster channel models, each cluster will perform independent channel simulations. Spatial correlation is used to characterize the spatial characteristics of the DUT side. The power weights of the probes are solved by minimizing the error between the spatial correlation of the target channel and the simulated channel.

Spatial correlation is defined as the statistical similarity between signals received by two different antennas. In addition, spatial correlation and power angle spectrum (PAS) are Fourier-transformed relationships. For a better explanation, assuming that the antennas are all isotropic and only vertical polarization is considered, the results can be easily extended to dual-polarization channels. The spatial correlation of the target channel can be expressed as [23]:

$$\rho_{\mathrm{Tar}} = \int_{-\pi}^{\pi} \exp\left(\mathrm{j}\frac{2\pi}{\lambda}(\bar{d}_{\mathrm{rx},u1} - \bar{d}_{\mathrm{rx},u2})\bar{r}_{\mathrm{rx}}(\phi)\right) P(\phi)\mathrm{d}\phi \tag{9.36}$$

where ϕ is the AoA of the incoming wave; $\bar{r}_{\mathrm{rx}}(\phi)$ denotes the unit vectors at the AoA of ϕ; $P(\phi)$ is the PAS.

For the PFS method, the spatial correlation obtained from the simulated channel based on MPAC can be expressed as [23]:

$$\rho_{\mathrm{PFS}} = \sum_{k=1}^{K} g_k \cdot \exp\left(j\frac{2\pi}{\lambda}(\bar{d}_{\mathrm{rx},u1} - \bar{d}_{\mathrm{rx},u2}) \cdot \bar{r}_{\mathrm{rx}}(\phi_k)\right) \tag{9.37}$$

where K is the total number of OTA probes; g_k denotes the power weight of the kth OTA probe; ϕ_k is the spherical angle of the kth OTA probe.

At this time, the space correlations of the target channel and the simulated channel are obtained separately. By minimizing the 2-norm of the deviation between the target spatial correlation and the simulation spatial correlation, the corresponding power weights can be obtained by solving the convex optimization problem [24]:

$$\begin{aligned} &\min_{g_n} \|\rho_{\mathrm{Tar}} - \rho_{\mathrm{PFS}}(\mathbf{g})\|^2 \\ &\text{s.t} \quad \mathbf{g}_1 = 1, 0 \leq g_k \leq 1 (\forall k \in [1, K]) \end{aligned} \tag{9.38}$$

After determining the power weights of the OTA probes, the channel coefficients of the simulated channel under the PFS method can be obtained. It is worth noting that the channel coefficients ignore the path loss in free space and the patterns of the OTA probes.

$$h_{u,s,n}^{\text{PFS}}(f,t) = \sqrt{\frac{P_n}{M}} \sum_{k=1}^{K} \sum_{m=1}^{M} F_s^{\text{Tx}}\left(\Omega_{n,m}^{\text{Tx}}\right) F_u^{\text{Rx}}\left(\Omega_k^{\text{OTA}}\right) \sqrt{g_k} \exp(j\phi_{n,m,k})$$
$$\cdot \exp\left(\frac{j2\pi\left(\bar{r}_{\text{tx},n,m}^{\text{T}} \cdot \bar{d}_{\text{tx},s}\right)}{\lambda} + \frac{j2\pi\left(\bar{r}_{\text{rx},n,m}^{\text{T}} \cdot \bar{d}_{\text{rx},u}\right)}{\lambda} \right) \exp(j2\pi(v_{n,m}t - f\tau_n)) \quad (9.39)$$

where $\Omega_{n,k}^{\text{OTA}}$ is the arrival angle of the kth probe; $\phi_{n,m,k}$ denotes the random initial phase of the kth probe at the mth subpath of the nth cluster.

9.2.1.3 PWS Method

The basic idea of the PWS method is to allocate the appropriate complex weight for each probe so that the target plane wave field can be accurately simulated in the test volume. The core is the restructuring of the target field. Due to the far communication distance between the base station and the terminal, it is generally believed that the signal ultimately reaches terminal in the form of a plane wave. The arrival angle in the channel model can be considered as a plane wave in the corresponding direction. By assigning appropriate complex weight to each OTA probe, the plane waves in any direction can be synthesized on the DUT side. The expected signal environment can be considered achieved by reproducing the plane waves of the specified arrival angle in the target channel within the test volume. Unlike the PFS method, the PWS method can reproduce each subpath of the cluster.

Assuming that there are J sampling points in the test volume, and the sampling meets the space Nyquist sampling theorem. In order to achieve plane waves in the angular direction of arrival in the channel model, there are:

$$\boldsymbol{FG} = \boldsymbol{T} \quad (9.40)$$

where $\boldsymbol{F} = \{\alpha_{j,k}\} \in \mathbb{C}^{J\times K}$ is the coefficient transfer matrix between K OTA probes and J sampling points in the anechoic chamber; $\alpha_{j,k}$ represents the attenuation of power and phase change during the propagation of electromagnetic waves from the kth probe to the jth sampling point, i.e. $\alpha_{j,k} = L(d_{k,j}) \exp\left(-j\frac{2\pi}{\lambda} d_{k,j}\right)$; $d_{k,j}$ denotes the distance between the kth probe and the jth sampling point in the anechoic chamber; $\boldsymbol{G} = \{\omega_k\} \in \mathbb{C}^{K\times 1}$ represents the complex weight vector of K OTA probes; $\boldsymbol{T} = \{e_z(\bar{r}_j)\} \in \mathbb{C}^{J\times 1}$ denotes the target field vector of J sampling points; $\bar{r}_j$ is the position vector of the jth sampling point; $e_z(\bar{r}_j) = E_0 \exp(-j\bar{\beta}\bar{r}_j)$; $\bar{\beta}$ is the wave vector; E_0 is the field intensity of a plane wave. Let $E_0 \equiv 1$. It can be seen that $e_z(\bar{r}_j)$ reflects the phase difference between the sampling point and the reference point.

After obtaining $\boldsymbol{F}$ and $\boldsymbol{T}$, the complex weight vector $\boldsymbol{G}$ can be solved using the least squares method. When the number of sampling points equals the number of OTA probes ($J = K$), there are:

$$\boldsymbol{G} = \boldsymbol{F}^{-1}\boldsymbol{T} \tag{9.41}$$

When the number of sampling points is greater than the number of OTA probes ($J < K$), there are:

$$\boldsymbol{G} = (\boldsymbol{F}^H\boldsymbol{F})^{-1}\boldsymbol{F}^H\boldsymbol{T} \tag{9.42}$$

After obtaining the complex weights of K probes, the spatial correlation obtained by the PWS method based on the simulated channel of MPAC can be expressed as [25]:

$$\rho_{\mathrm{PWS}} = \frac{1}{\beta_0}\sum_{m=1}^{M}\sum_{k=1}^{K}F(\Omega_k)\omega_{m,k}\sum_{k'=1}^{K}F^H\left(\Omega_{k'}\right)\omega_{m,k'}^H \tag{9.43}$$

where $F(\Omega_k)$ is the field pattern of the DUT antenna; $\omega_{m,k}$ denotes the complex weight of the kth probe to the mth subpath; β_0 is the normalization factor, can be expressed as:

$$\beta_0 = \sqrt{\sum_{m=1}^{M}\left\|\sum_{k=1}^{K}F(\Omega_k)\omega_{m,k}\right\|^2}\cdot\sqrt{\sum_{m=1}^{M}\left\|\sum_{k'=1}^{K}F\left(\Omega_{k'}\right)\omega_{m,k'}\right\|^2} \tag{9.44}$$

According to the PWS method, the channel coefficients of the simulated channel can be obtained as:

$$h_{u,s,n}^{\mathrm{PWS}}(f,t) = \sqrt{\frac{P_n}{M}}\sum_{m=1}^{M}\sum_{k=1}^{K}F_s^{\mathrm{Tx}}\left(\Omega_{n,m}^{\mathrm{TX}}\right)F_u^{\mathrm{Rx}}\left(\Omega_{n,k}^{\mathrm{OTA}}\right)\omega_{n,m,k}\exp(j\phi_{n,m})$$
$$\exp\left(\frac{j2\pi\left(\hat{r}_{\mathrm{tx},n,m}^{\mathrm{T}}\cdot\overline{d}_{\mathrm{tx},s}\right)}{\lambda_0}+\frac{j2\pi\left(\hat{r}_{\mathrm{rx},n,m}^{\mathrm{T}}\cdot\overline{d}_{\mathrm{rx},u}\right)}{\lambda_0}\right)\exp(j2\pi(v_{n,m}t - f\tau_n)) \tag{9.45}$$

where $\omega_{n,m,k}$ denotes the complex weight of the mth subpath of the nth cluster on the kth probe; $\phi_{n,m}$ is the random initial phase of the m th subpath of the nth cluster.

Note that since the PFS method optimizes the power weights of the OTA probes, it only requires power calibration. The PWS method optimizes the complex weights of the OTA probes, so the calibration of the PWS method requires both power calibration and phase calibration.

9.2.2 Testing of UE in MPAC

Before the massive roll-out of MIMO devices, it is crucial to test the radio performance of UE in the early-stage prototyping and mid-term improvement stage. It helps researchers identify product defects during the design phase and improve them in a timely manner. Here we will mainly introduce the comparison of the two-channel emulation methods in the MPAC system, the metrics of the channel emulation accuracy, and the requirements of the MPAC system design [26–29].

9.2.2.1 Comparison of Channel Emulation Methods

Figure 9.4 shows the MPAC setups for standardized 2D UE OTA testing. The DUT is located in the center of the anechoic chamber, with OTA probes evenly distributed around it. Assuming the DUT is in the far field of the OTA probes. Generally, the PFS and PWS methods exhibit different emulation accuracy under different conditions. In order to compare the performance of these two channel emulation methods, a series of simulation verifications is conducted.

Assume that 16 OTA probes are evenly distributed around the DUT (forming a circle), and the minimum angular spacing between the OTA probes is 22.5°, as shown in Figure 9.4. In addition, the single cluster channel model is selected as the target channel. Two types of single cluster models were considered to check the emulation accuracy. In one of the clusters' incoming wave directions, there exists only an OTA probe (i.e. AoA = 0°), and the other cluster is located in the middle of the two adjacent OTA probes (i.e. AoA = 11.25°). In addition, the angular spreads of arrival (ASA) of the single cluster channel model is increased from 5° to 35°, with intervals of 10°. By comparing the differences between the emulated channel spatial correlation and the target channel spatial correlation in these two cases, the emulation performance of the PFS and PWS methods is comprehensively analyzed.

Figures 9.5 and 9.6 show the spatial correlation of target and emulated channels in different antenna spacing in two different cases, respectively. It can be seen that the PWS method fits the theoretical values better than the PFS method in

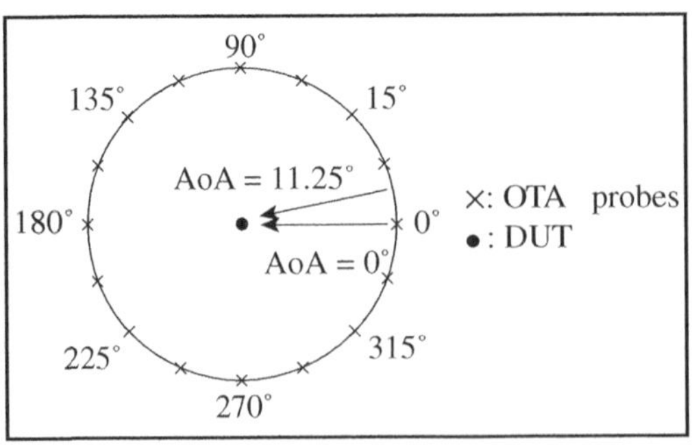

Figure 9.4 Illustration of 2D OTA probe configuration.

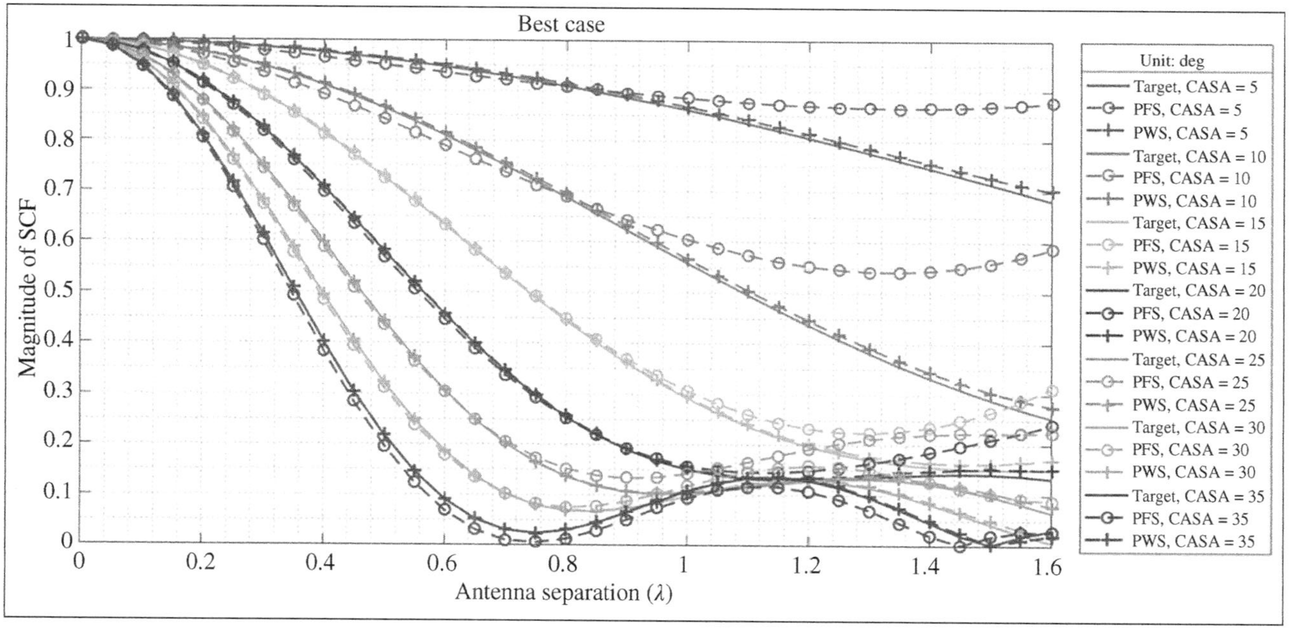

Figure 9.5 Spatial correlations of target channel, PFS method-emulated channels, and PWS method simulation channel in the case of AoA = 0°. Source: Reproduced with permission from Ref. [25]/IEEE.

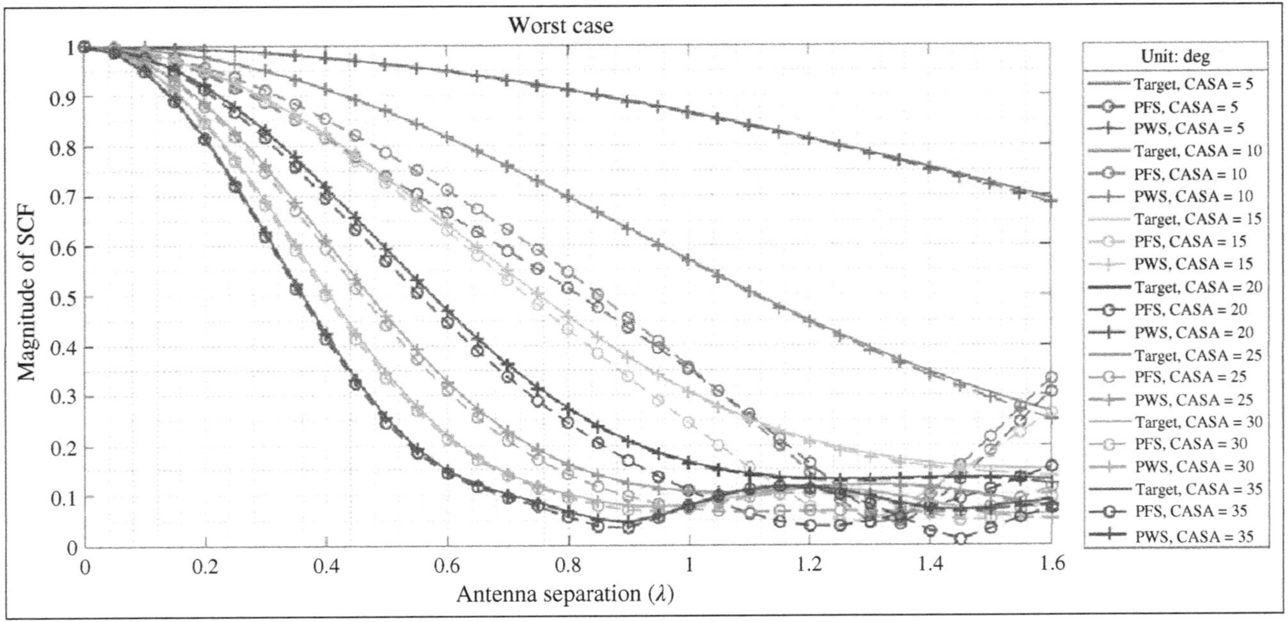

Figure 9.6 Spatial correlations of target channel, PFS method emulated channels, and PWS method simulation channel in the case of AoA = 11.25°. Source: Reproduced with permission from Ref. [25]/IEEE.

both cases. This is because when the number of OTA probes is sufficient, the PWS method can accurately reproduce the arbitrary plane wave field. In addition, the PFS method makes it difficult to accurately reproduce the target cluster with a small ASA. This is because when the ASA is smaller than the angular separation between the OTA probes, the cluster becomes very specular in angular domain. At this time, since there is no OTA probe in the direction of the specular cluster, the PFS method will not be able to reproduce it. Nonetheless, with the increase of ASA, the deviation between the PFS and PWS methods will gradually decrease, indicating that both the PFS and PWS methods can effectively reproduce clusters with larger ASA.

9.2.2.2 Channel Emulation Accuracy

To evaluate the accuracy of emulated channels, we summarize several commonly used metrics.

1. Spatial correlation error. The spatial correlation error is defined as the absolute value of the difference between the spatial correlation of the target channel and the emulated channel, i.e. $|\rho - \hat{\rho}|$. The PFS method often chooses the spatial correlation error to assess the emulation accuracy. When the spatial correlation is at a high level, $|\rho - \hat{\rho}|$ is more critical, as its tiny differences can lead to drastic changes in MIMO system performance [28].
2. Field synthesis error. The field synthesis error is defined as the absolute value of the difference between the target field and the emulated field, i.e. $|E - \hat{E}|$. The PWS method often chooses the field synthesis error to assess the emulation accuracy. The error threshold of the electric field's error vector magnitude (EVM) is typically set at −15 dB.

$$\mathrm{EVM} = 10\log_{10}\frac{\|E - \hat{E}\|_2^{\ 2}}{\|E\|_2^{\ 2}} \tag{9.46}$$

3. Capacity error. The capacity error is proposed to reflect the data rate error [29], which is defined as the absolute value of the difference between the target channel and the emulated channel capacity, i.e. $|C - \hat{C}|$. Generally, the error threshold of Channel capacity is set to 0.2 bits/s/Hz.
4. Throughput error. The channel capacity reflects the theoretical upper bound of data rate, which is not achievable in practice. The throughput reflects the real-time data rate in the actual MIMO systems. The throughput error is defined as the absolute value of the difference between the target channel and the emulated channel throughput, which reflects the end-to-end performance of the system.

9.2.2.3 Requirements of UE Test System Design

1. Number of OTA probes. The number of OTA antennas required in MPAC setups is usually determined by the size of DUT, the locations of the OTA probes, the channel model and the acceptable error levels. In theory, synthesizing a completely accurate target field in the test zone requires an infinite number of OTA probes, which is not affordable. Some studies have proposed a rule of thumb for the number of OTA probes required [18, 30]. The minimum number of OTA probes required can be expressed as:

$$K_{\text{MIN}} = 2(\lceil kr_0 + n_1 \rceil)^2 + 4(\lceil kr_0 + n_1 \rceil) \tag{9.47}$$

where $\lceil \cdot \rceil$ represents round-up operation; k is the wave number; r_0 denotes the radius of the test area; n_1 is a modest integer used to regulate the simulation accuracy. In addition, the relationship between n_1 and r_0 can be expressed as [31]:

$$n_1 = o\left(\sqrt[3]{kr_0}\right) \tag{9.48}$$

2. Test zone size. The particular size of the testing area that can be supported in MPAC settings is one of the important challenges that needs to be addressed. The throughput error metric is usually selected to represent the size of the test zone. The synthetic MPAC setup has been suggested in [32] to solve this issue, as shown in Figure 9.7. In the synthesized MPAC setup, the channel transmission coefficient is embedded into the channel emulator, and throughput measurement is then carried out via conduction. During the measurement, formula (9.45) allows for flexible settings of various MPAC configurations and antenna spacings.
3. Physical dimension of the setup. In order to save costs, a small anechoic chamber can be used. However, it also brings the corresponding problem, which is that the incident wave will generate curved phase fronts. In order to reduce the spherical effects in small MPAC setups, a plane wave compensating approach

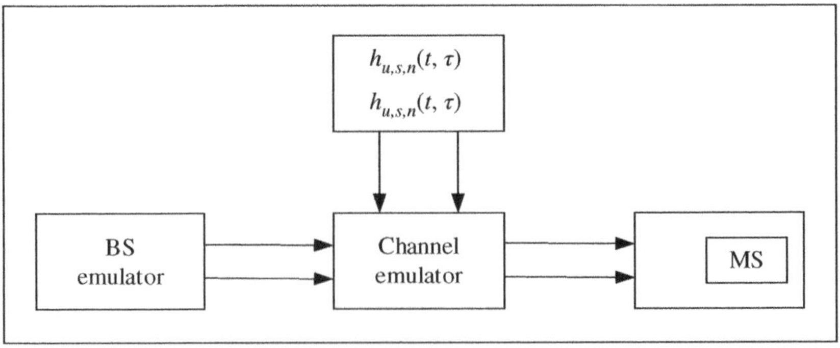

Figure 9.7 Synthetic MPAC setup.

has been presented in [33]. The planar wave compensation technique simulates a far-field OTA probe through several near-field OTA probes, and they are connected to programmable power attenuators and phase shifters, but only one output of the channel emulator. Multiple probes can improve compensation accuracy through flexible placement, and the far-field angle positions of the OTA probes can be obtained through the particle swarm optimization (PSO) algorithm. The results show that this compensation technique can improve the performance of the system in terms of spatial correlation, field distribution, and other aspects. In other words, it can reduce traditional far-field standards to construct a small MPAC setup to save costs.

9.2.2.4 3D Channel Model Emulation in MPAC

In practical scenarios, the propagation environment of signals is usually 3D. The 3D UE OTA testing environment can be formed by placing multiple probe rings with different elevations in the MPAC setups. The results in [34] demonstrate that symmetrical placement between the probe ring produces better simulation performance. The appropriate probe configuration to produce the highest simulation accuracy can be found once the target channel model is established. In the 3D UE OTA testing, the sphere-shaped test zone is assumed, but the results in [35] show that because the probes in the azimuth plane is more than the elevation plane, it is more inclined to generate an ellipsoid-shaped test zone around DUT.

For 3D multi-probe setup, three probe selection algorithms are proposed in [36], and the proposed algorithms were verified in the PFS and PWS methods, respectively. The results show that the selected probe subset based on the algorithms shows good simulation accuracy. However, the number of probe subsets selected in [36] is fixed. Two better probe selection algorithms have been proposed in [37], namely the decremental selection algorithm (DSA) and the error threshold selection algorithm based on alternating search (SAAS). These two algorithms minimize the number of required probes while meeting the specified simulation accuracy.

In order to further improve emulation accuracy, a flexible MPAC setup for 3D UE OTA testing has been proposed in [38]. The probes are installed on the semi-arc track, separately so that they can move according to the defined optimal position. The genetic algorithm and multi-shot algorithm are introduced to validate the proposed 3D MPAC setup.

9.2.3 Testing of Massive BS in MPAC

With the rapid development of communication systems, massive MIMO technology plays a key role in 5G communication systems. On the basis of MIMO technology, massive MIMO further improves system data throughput and spectrum utilization. The radio test of the massive MIMO devices under the condition

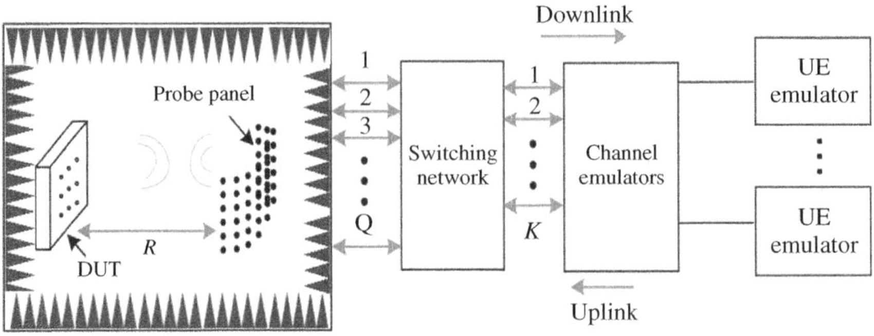

Figure 9.8 Illustration of 3D sectored MPAC setup.

of fading wireless channels is essential for the development of 5G BS operating in the millimeter wave frequency band.

BS is usually installed on walls or high places to serve a certain angle sector. In addition, due to the fact that BS is usually located far away from the scatters, the angular power distribution of BS should be limited to a certain angular region. In response to this situation, a 3D sectored MPAC setup has been proposed for testing massive BS in [39], as shown in Figure 9.8. The setup includes the anechoic chamber, the BS array to be tested, the probe panel covering a certain angle area, the switch network for selecting a specific number of active probes from the probe panel, the channel emulator, and the UE emulator.

Here we will mainly introduce the comparison of the two-channel emulation methods in the 3D sectored MPAC system, the evaluation metrics of the channel emulation accuracy, and the requirements of the 3D sectored MPAC system design.

9.2.3.1 Comparison of Channel Emulation Methods

In the 3D sectored MPAC system, we need to select a part of the probe panel through the switch network as active probes for simulation. The spatial angle mapping (SAM) method is used as the selection scheme for active probes in [39]. Firstly, all clusters in the channel model are arranged in descending power order, and then the clusters with the higher power are sequentially assigned active probes closest to the AOAs. If the number of active probes is greater than the number of clusters in the probe panel, more probes are assigned to the cluster with higher power. A forward allocation (FA) algorithm with better performance has been proposed in [40] to select the active probes.

For massive MIMO OTA testing, PAS may be more suitable for characterizing the accuracy of channel reconstruction than spatial correlation. The classic

Bartlett beamforming algorithm can be used to estimate the PAS of the target channel [41]:

$$p(\Omega) = a^H(\Omega)\boldsymbol{R}a(\Omega) \tag{9.49}$$

where $a(\Omega)$ denotes the steering vector in the direction of the spatial angle Ω; $\boldsymbol{R}$ represents the spatial correlation matrix of the target channel. The PAS estimate of the emulated channel can be represented as:

$$\hat{p}(\Omega) = a^H(\Omega)\hat{\boldsymbol{R}}a(\Omega) \tag{9.50}$$

where $\hat{\boldsymbol{R}}$ denotes the spatial correlation matrix of the emulated channel. In addition, the total variation distance of PAS is adopted as the evaluation metric to compare the performance of the two channel emulation methods. The definition of the total variation distance of PAS will be introduced later.

Assuming the DUT is an 8×8 antenna array with a unit spacing set to 0.5 λ. The probe panel contains $Q = 629$ probes, covering an azimuth range of 135° and an elevation range of 60°. In addition, the scenario where the probe panel contains $Q = 1241$ probes is also considered, increasing the coverage range of azimuth angle to analyze the impact of power spectrum leakage on testing performance. The distance between the DUT and the probe panel is set to 2 m. CDL-C is selected as the target channel model. The center frequency of the simulation is set to 28 GHz.

Figure 9.9 shows the total variation distance of PAS in two scenarios with different number of active probes. The results show that when the number of active probes is less than the number of clusters in the probe panel, the emulation accuracy of PWS is higher than the PFS method. This is because when the number of active probes is small, the corresponding relative angle spread will also be smaller, making up for the disadvantage of the number of active probes. However, as the number of active probes increases, the advantage of small angle spread will be significantly reduced. At this time, the PFS method will be superior to the PWS method. In addition, when the number of active probes continues to increase, the

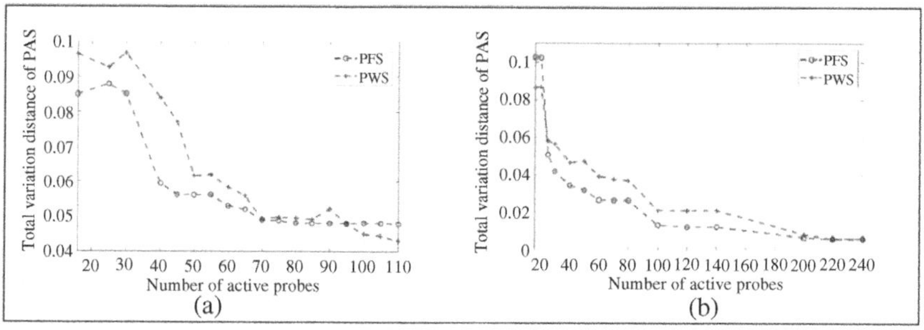

Figure 9.9 The deviation of PAS between the PFS and PWS method in (a) $Q = 629$ scenario and (b) $Q = 1241$ scenario.

deviation between the two channel emulation methods will gradually decrease, and the PWS will eventually show a better emulation accuracy.

However, the more active probes, the higher the system cost. Generally, only a small number of active probes are used to save costs while ensuring that the emulation accuracy meets the requirements. In this situation, due to the better robustness of the PFS method in undersampling, the PFS method is more suitable for 3D massive MIMO OTA testing.

9.2.3.2 Channel Emulation Accuracy

To evaluate the accuracy of massive MIMO OTA testing system, we summarize several commonly used metrics.

1. Spatial correlation error. The spatial correlation error evaluates the deviation between the emulated channel and the target channel in terms of spatial correlation. The weighted RMS spatial correlation error can be expressed as [39]:

$$e_{\mathrm{p}} = \sqrt{\frac{1}{L}\sum_{l=1}^{L} |\rho_l - \hat{\rho}_l|^2 \max(|\rho_l|, |\hat{\rho}_l|)} \tag{9.51}$$

where ρ_l and $\hat{\rho}_l$ represent the spatial correlation of target and emulated channel, respectively; L is the number of the positions. Due to the existence of weighting coefficient, the corresponding error has little impact on the final result when the spatial correlation is small.

2. Total variation distance of PAS. The total variation distance of PAS measures the ability of the emulated channel to reconstruct the target PASs. Firstly, the assumed DUT array and the Bartlett beamformer are used to estimate the PASs. This is equivalent to filtering the actual PAS of the channel through the limited aperture of the DUT array. Then normalize the estimated target PAS and emulated PAS separately, so that they can be interpreted as PDF. Finally, the total variation distance of PAS can be defined as:

$$D_{\mathrm{p}} = \frac{1}{2}\int \left| \frac{P(\Omega)}{\int P(\Omega')\mathrm{d}\Omega'} - \frac{\hat{P}(\Omega)}{\int \hat{P}(\Omega')\mathrm{d}\Omega'} \right| \mathrm{d}\Omega \tag{9.52}$$

where $P(\Omega)$ and $\hat{P}(\Omega)$ represent the PAS estimates calculated for the target channel model and the OTA-emulated channel, respectively. The value range of D_{p} is from 0 to 1, where 0 represents that the emulated channel PAS estimate is completely consistent with the target channel, while 1 represents that it is completely different.

3. Beam peak distance. The beam peak distance metric is defined as the angular distance between the probability-weighted average directions of the allocated beam. This and the next metric assume fixed beams and the code book of discrete antenna weights is used. By scanning the powers of all fixed beams at each moment, we can get the highest power beam at this time. Then, the beam peak distance can be expressed as:

$$D_b = \left\| \sum_{b=1}^{B} \Omega_b p_r(\Omega_b) - \Omega_b p_o(\Omega_b) \right\| \tag{9.53}$$

where B denotes the number of beams at a certain moment; Ω_b is the solid angle of the bth beam at this moment; $p_r(\Omega_b)$ and $p_o(\Omega_b)$ represent the probabilities of beam allocation in the target channel and OTA emulated channel, respectively.

4. Total variation distance of beam allocation distributions. This beam selection metric can be expressed as:

$$D_s = \sum_{b=1}^{B} \frac{|p_r(\Omega_b) - p_o(\Omega_b)|}{2} \tag{9.54}$$

The value range of D_s is from 0 to 1, where 0 denotes completely consistent and 1 denotes completely different.

9.2.3.3 Requirements of 3D Massive BS Test System Design

1. Flexible probe selection algorithm. In order to fully reconstruct the characteristics of the target channel, an infinite number of probes are required. However, this is not affordable. In reality, only a certain number of probes will be selected as active probes to use. On this premise, how to select active probes within the probe panel to achieve the highest emulation accuracy is the key to the issue. The multi-shot algorithm has been proposed in [34, 42] for probe selection. By removing the probes with the lower contribution after each iteration, the K probes with the highest contribution can ultimately be selected. The artificial bee colony (ABC) algorithm has been proposed in [42] to optimize probe selection by imitating the behavior of bees. This algorithm just needs to compare the benefits and drawbacks of the problem, providing a faster convergence speed. In [33], PSO algorithm has been proposed to optimize probe selection by simulating the clustering behavior of insects, herds and birds. The core of the algorithm is a fast update of speed and position.
2. Size of testing setup. The distance from the DUT to the probe panel (i.e. measurement distance R) in the massive MIMO OTA testing setup directly affects the size of the required anechoic chamber. The impact of measurement distance R on emulation performance at the frequency of 28 GHz has been evaluated in [43]. In [44], the minimum value of the required measurement distance R has been studied under the condition that a certain simulation accuracy is satisfied. The research on measurement range R is mainly used to determine whether the test needs to meet the traditional far-field criteria. The results show that the far-field requirement might be properly relaxed for the measurement distance R in the massive MIMO OTA testing setup. For example, in the case of 28 GHz, the OTA performance of the BS array with a maximum size of 0.15 m can be measured by the OTA setup with measurement distance R about 1 m [44].

9.3 Radiated Two-Stage Method

The conventional two-stage method was first proposed by Agilent in the United States for MIMO OTA testing [45]. Depending on whether to use conduction or OTA for testing in the second stage, the two-stage method can be divided into the conducted two-stage (CTS) method and the RTS method. For the CTS method, due to its inability to evaluate issues such as equipment self-interference, the initial CTS method has evolved into the RTS method in [46].

The implementation principles of RTS and MPAC are basically similar. Both of them reproduce the real propagation scenarios (i.e. multipath environments) in the anechoic chamber and then perform throughput measurements. However, the MPAC method constructs a multipath fading channel environment by transmitting test signals from multiple probes simultaneously, while the RTS method combines the antenna pattern of DUT with the channel model in a channel emulator to construct a multipath fading environment.

The implementation of the RTS method is based on two stages: the first stage is to measure the antenna pattern of the DUT in a SISO OTA testing anechoic chamber, which requires the chip of the DUT to have reporting ability. The second stage is to calculate and emulate the measured signal by combining the channel model and antenna pattern. Then, the measurement signals are fed into the terminal receiver through OTA for throughput measurement. Note that in the RTS method, it is necessary to decouple the propagation of signals in the anechoic chamber space by adding calibration equations, offsetting the signal crossover introduced between the measurement antennas and the antennas to be tested.

We summarize the implementation principles and key technologies of the RTS method as follows.

9.3.1 Principle of Radiated Two-Stage Method

9.3.1.1 Conducted Two-Stage Method

Initially, the RTS method was developed from the CTS method, and the testing process of the CTS method is shown in Figure 9.10. The CTS method is divided into two stages:

- The first stage is to measure the antenna pattern of the MIMO terminal in an anechoic chamber.
- The second stage is to combine the measured antenna pattern with the selected channel model to obtain the measurement signal, and then directly feed the measurement signal calculated in the channel emulator into the MIMO terminal through the RF cables.

Therefore, the CTS method does not require an anechoic chamber for throughput testing.

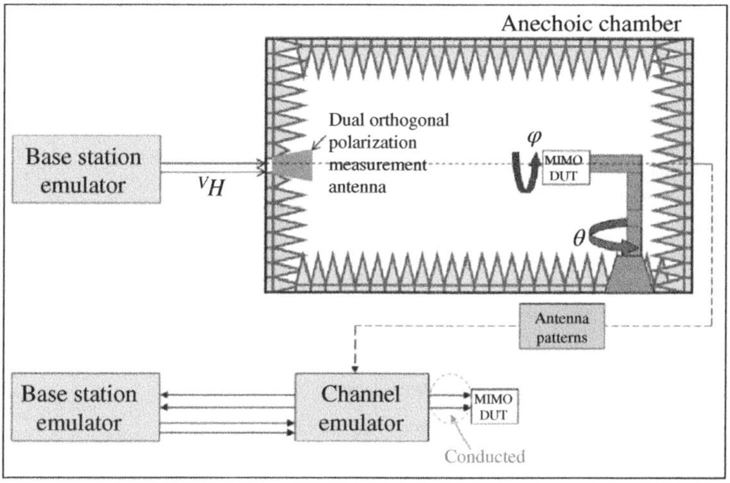

Figure 9.10 Schematic diagram of CTS method. Source: Reproduced with permission from Ref. [46]/IEEE.

However, in the second stage of the CTS method, the DUT is directly connected to the channel emulator through the RF cables, which changes the state of the DUT. The self-interference generated by the DUT will be ignored, resulting in measurement results that cannot reflect the overall performance of the DUT. As shown in Figure 9.11, when using the RF cables to directly connect the receiver feed point, the antennas of the DUT itself will be disconnected. At this time, the radiation noise generated by the DUT itself cannot be coupled into the receiver through the antennas, resulting in measurement results that outperform the actual user experience [47]. In addition, due to the fact that standard connectors are not usually used for RF ports during testing, inconsistencies in RF connections may lead to fluctuations in measurement results. As a result, the cables connected to the RF ports cannot be correctly calibrated, which will further affect the accuracy of the test results [48]. Generally, conducted measurements are usually used to evaluate the performance of individual receiver chips.

In order to solve these problems in the CTS method, the RTS method is proposed.

9.3.1.2 RTS Method for MIMO OTA Test

Unlike the CTS method, which uses RF cables to directly connect the DUT in the second stage, the RTS method uses OTA measurement signal to feed the terminal receiver. The implementation principle of the RTS method is shown in Figure 9.12. The channel emulator calculates the measurement signal by combining the channel model and pattern information (similar to the CTS method). The measurement signal output by the channel emulator is fed into the measurement antennas in

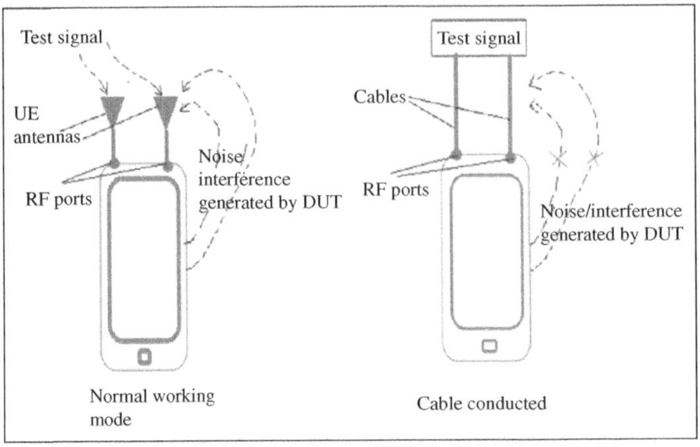

Figure 9.11 Self-interference and noise of MIMO terminals coupled to antennas. Source: Reproduced with permission from Ref. [46]/IEEE.

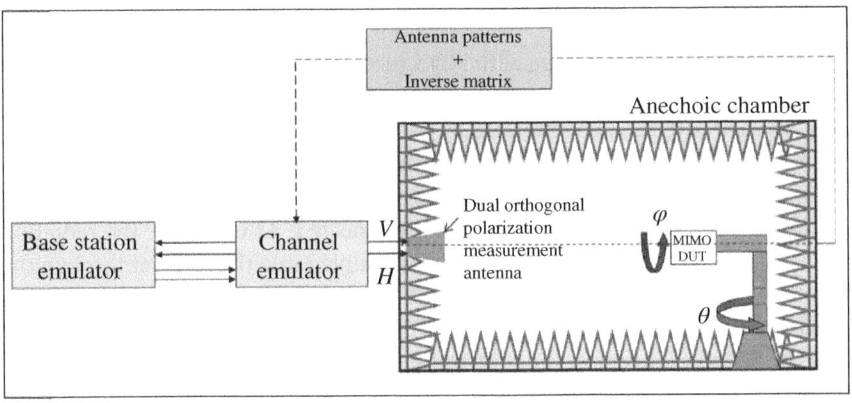

Figure 9.12 Schematic diagram of RTS method. Source: Reproduced with permission from Ref. [46]/IEEE.

the anechoic chamber through RF cables and then radiated by the measurement antennas to the position of the DUT. The final measurement signal is received by the MIMO terminal after being coupled by the DUT antenna itself.

Taking the 2×2 MIMO system as an example, in the anechoic chamber, the signal radiated by any measurement antenna will be received by two receiving antennas at the same time, as shown in Figure 9.13. The measurement antenna feed points are marked as BT_1 and BT_2, and the terminal receiving antenna feed points are marked as S_1 and S_2. Then the electromagnetic propagation between BT_1, BT_2 and S_1, S_2 can be expressed as the propagation matrix $\boldsymbol{A}$ [49]:

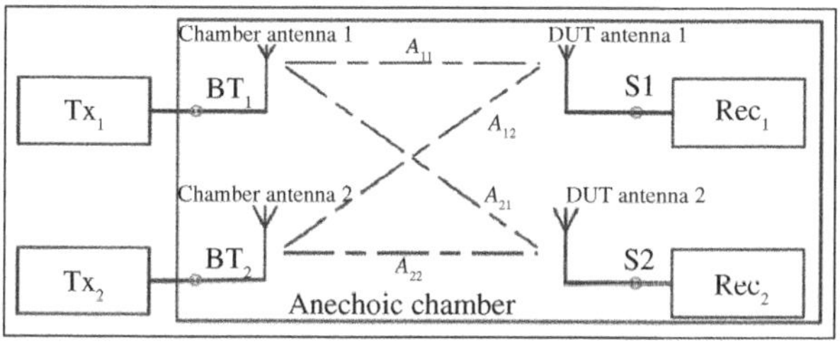

Figure 9.13 Propagation of measurement signals in an anechoic chamber. Source: Reproduced with permission from Ref. [49]/IEEE.

$$\boldsymbol{A} = \begin{bmatrix} a_{11}e^{j\varphi_{11}} & a_{12}e^{j\varphi_{12}} \\ a_{21}e^{j\varphi_{21}} & a_{22}e^{j\varphi_{22}} \end{bmatrix} \tag{9.55}$$

where a_{ij} and φ_{ij} represent the field strength and phase changes from the jth measurement antenna to the ith receiving antenna, respectively. It includes information such as the gain and phase of the measurement antenna, the complex gain of the spatial path, and the gain and phase of the receiving antenna.

The output port of the channel emulator is connected to the measurement antenna feed point by an RF module, where the RF module performs matrix transformation on the input signal, and the applied RF matrix is marked as $\boldsymbol{M}$. As shown in Figure 9.14, the output signal of the channel emulator superimposes through the $\boldsymbol{M}$ matrix and $\boldsymbol{A}$ matrix, and then enters the MIMO terminal receiver, i.e.

$$y_{\mathrm{r}}(t) = \boldsymbol{A} \cdot \boldsymbol{M} \cdot y_{\mathrm{z}}(t) \tag{9.56}$$

where $y_{\mathrm{r}}(t)$ denotes the signal received by the MIMO terminal. It is always possible to find a nonsingular $\boldsymbol{A}$ matrix, and let $\boldsymbol{M}$ matrix be the inverse of $\boldsymbol{A}$ matrix:

$$\boldsymbol{M} = \boldsymbol{A}^{-1} \tag{9.57}$$

Then the received signal of the MIMO terminal satisfies:

$$y_{\mathrm{r}}(t) = \boldsymbol{A} \cdot \boldsymbol{A}^{-1} \cdot y_{\mathrm{z}}(t) = y_{\mathrm{z}}(t) \tag{9.58}$$

This means that the system setup of the RTS method enables the signal $y_{\mathrm{z}}(t)$ to be fed into the terminal receiver, as shown in Figure 9.14. The difference from the CTS method is that the terminal in Figure 9.14 is in a complete machine state, and its noise coupling conforms to the actual terminal working state. Therefore, the results can be used to measure the real terminal performance. This signal isolation propagation method that establishes a one-to-one correspondence in the overall machine state is also known as "virtual cable" method.

In general, the testing process of the RTS method is divided into the following two stages [51–53]: the first stage is to measure the antenna pattern of the DUT in

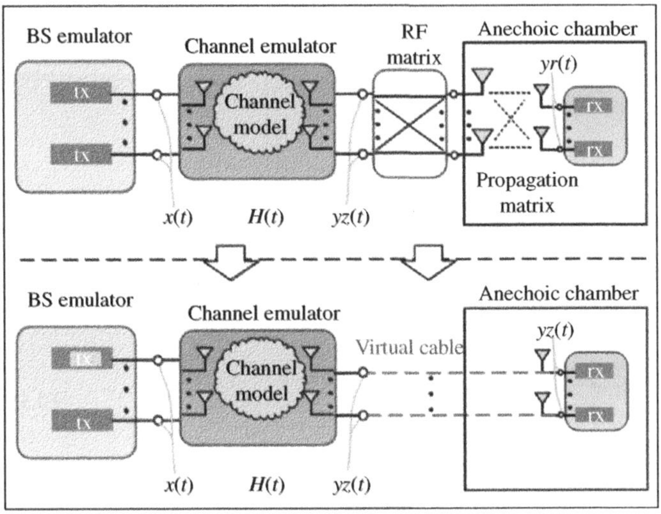

Figure 9.14 Establishment of virtual cable in the RTS method. Source: Reproduced with permission from Ref. [50]/IEEE.

the anechoic chamber and calculate the inverse matrix of the propagation matrix in the anechoic chamber. In the second stage, the measured antenna pattern and inverse matrix files are loaded into the channel emulator, and the channel emulator outputs the processed multipath signal for throughput testing.

The RTS method has the characteristics of low implementation cost, easy operation, simple calibration steps, and high measurement repeatability. However, it has also experienced multiple challenges. The key technologies to overcome these challenges contribute to the effectiveness and reliability of the RTS method in MIMO OTA testing, enabling comprehensive evaluation and optimization of MIMO system performance in practical wireless communication scenarios.

9.3.2 Solution Method of Inverse Matrix

In the RTS method, after obtaining the antenna pattern and calculating the measured signal, it is necessary to load the inverse matrix of the propagation matrix in the RF module, so as to realize the OTA direct connect and feed the measurement signal to the MIMO terminal receiver. In addition, the ill-conditioned degree of the propagation matrix contributes to the measurement error. If the error cannot be controlled and quantified, the accuracy of the measurement result becomes unevaluable. Therefore, methods to search for a suitable propagation matrix and find its inverse are the core technology of the RTS method. In fact, the solution of inverse matrix is also one of the basic technologies for RTS

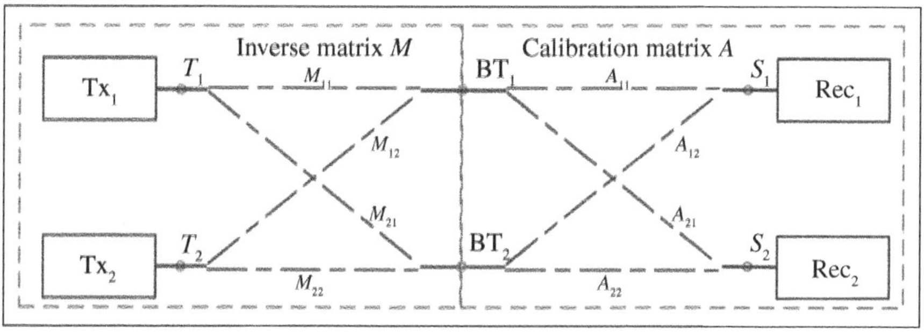

Figure 9.15 Schematic diagram of introducing inverse matrix. Source: Reproduced with permission from Ref. [49]/IEEE.

engineering implementation. Here we take a 2×2 MIMO system as an example to summarize the solution process of the inverse matrix. As shown in Figure 9.15, the relationship between the received signals S_1 and S_2 of the DUT and the signals T_1 and T_2 output by the channel emulator can be expressed as:

$$\begin{bmatrix} S_1 \\ S_2 \end{bmatrix} = \begin{bmatrix} A_{11} & A_{12} \\ A_{21} & A_{22} \end{bmatrix} \cdot \begin{bmatrix} M_{11} & M_{12} \\ M_{21} & M_{22} \end{bmatrix} \cdot \begin{bmatrix} T_1 \\ T_2 \end{bmatrix} \tag{9.59}$$

When the DUT is fixed at a certain position, the inverse matrix solution method at this position can be divided into three steps, as shown in Figure 9.16.

Step 1: Only open port BT_1, set the transmission signal $T_2 = 0$, and output the transmission signal T_1. As shown in Figure 9.16a, the inverse matrix can be expressed as:

$$\boldsymbol{M} = \begin{bmatrix} 1 & 0 \\ 0 & 0 \end{bmatrix} \tag{9.60}$$

Then the received signal can be expressed as:

$$\begin{bmatrix} S_1 \\ S_2 \end{bmatrix} = \boldsymbol{A} \cdot \boldsymbol{M} \cdot \begin{bmatrix} T_1 \\ 0 \end{bmatrix} = \begin{bmatrix} |A_{11}| e^{j\varphi_{11}} T_1 \\ |A_{21}| e^{j\varphi_{21}} T_1 \end{bmatrix} \tag{9.61}$$

Since the power values of T_1, R_1, and R_2 are known, the values of $|A_{11}|$ and $|A_{21}|$ can be obtained as:

$$|A_{11}| = \frac{|R_1|}{|T_1|}, |A_{21}| = \frac{|R_2|}{|T_1|} \tag{9.62}$$

Similarly, the values of $|A_{12}|$ and $|A_{22}|$ can also be obtained.

Step 2: Open ports BT1 and BT2, and set the transmission signal $T_2 = 0$. In order to achieve the effect of OTA direct connect, it is required to realize that the transmitted signal T_1 can only be received by the R_1 port after passing through the inverse matrix, and the R_2 port cannot receive the signal radiated from port 1 after

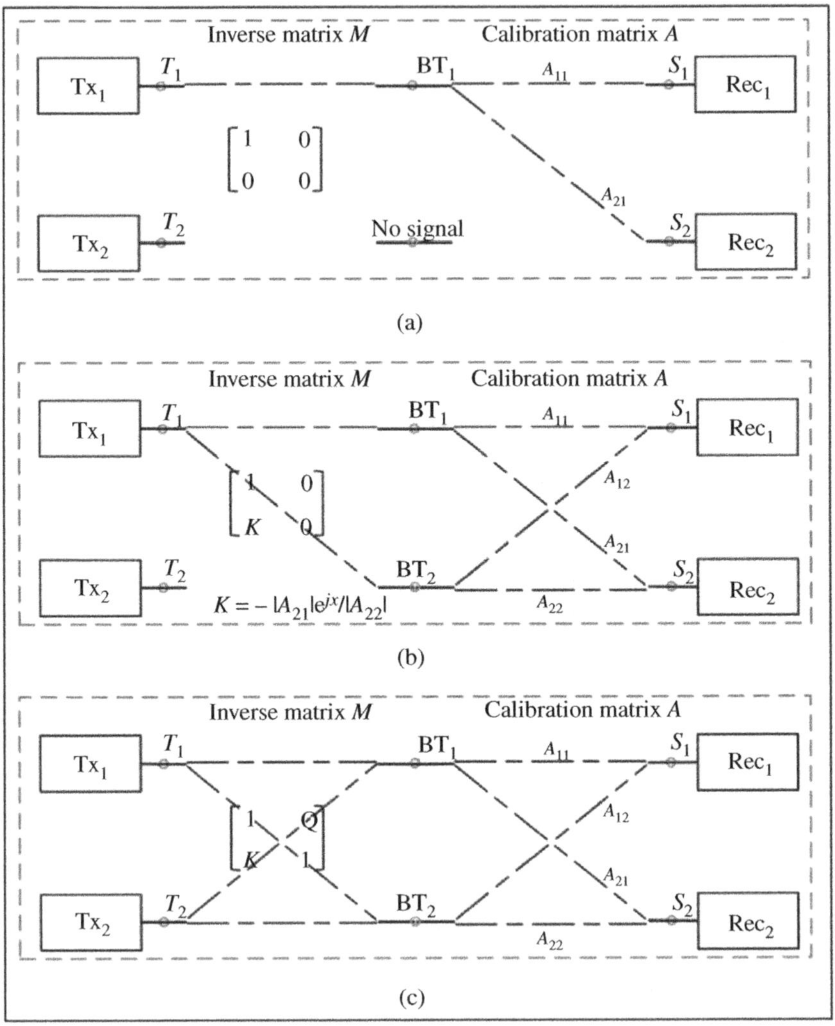

Figure 9.16 Steps of solving the inverse matrix at a fixed position. (a) Step 1. (b) Step 2. (c) Step 3. Source: Reproduced with permission from Ref. [49]/IEEE.

the signal cancellation. Therefore, the inverse matrix can be set to:

$$\boldsymbol{M} = \begin{cases} \begin{bmatrix} 1 & 0 \\ -\dfrac{|A_{21}|}{|A_{22}|}e^{jx} & 0 \end{bmatrix}, & (\text{if}|A_{22}| \geq |A_{21}|) \\ \begin{bmatrix} -\dfrac{|A_{22}|}{|A_{21}|}e^{jx} & 0 \\ 1 & 0 \end{bmatrix}, & (\text{if}|A_{22}| < |A_{21}|) \end{cases} \tag{9.63}$$

where x is an adjustable phase. Without loss of generality, we assume that $|A_{22}| \geq |A_{21}|$ (since the derivation process of $|A_{22}| < |A_{21}|$ is similar). Then the received signal can be expressed as:

$$\begin{bmatrix} S_1 \\ S_2 \end{bmatrix} = \begin{bmatrix} \left(A_{11} - A_{12} \dfrac{|A_{21}|}{|A_{22}|} e^{jx} \right) T_1 \\ \left(A_{21} - A_{22} \dfrac{|A_{21}|}{|A_{22}|} e^{jx} \right) T_1 \end{bmatrix} \tag{9.64}$$

By adjusting phase x to make $R_2 = 0$, thereby the signal received by R_2 port from T_1 can be eliminated. Let $x = \alpha$ at this time, then:

$$A_{21} = A_{22} \frac{|A_{21}|}{|A_{22}|} e^{j\alpha} \tag{9.65}$$

where $\alpha = \varphi_{21} - \varphi_{22}$.

Similarly, another phase β can be obtained following the same processes:

$$A_{12} = A_{11} \frac{|A_{12}|}{|A_{11}|} e^{j\beta} \tag{9.66}$$

where $\beta = \varphi_{12} - \varphi_{11}$.

Step 3: For a complete RTS test, set the inverse matrix to:

$$\boldsymbol{M} = \begin{bmatrix} 1 & -\dfrac{|A_{12}|}{|A_{11}|} e^{j\beta} \\ -\dfrac{|A_{21}|}{|A_{22}|} e^{j\alpha} & 1 \end{bmatrix} \tag{9.67}$$

Then the received signal can be expressed as:

$$\begin{aligned} \begin{bmatrix} S_1 \\ S_2 \end{bmatrix} &= \begin{bmatrix} \left(A_{11} - A_{12} \dfrac{|A_{21}|}{|A_{22}|} e^{j\alpha} \right) T_1 + \left(A_{12} - A_{11} \dfrac{|A_{12}|}{|A_{11}|} e^{j\beta} \right) T_2 \\ \left(A_{21} - A_{22} \dfrac{|A_{21}|}{|A_{22}|} e^{j\alpha} \right) T_1 + \left(A_{22} - A_{21} \dfrac{|A_{12}|}{|A_{11}|} e^{j\beta} \right) T_2 \end{bmatrix} \\ &= \begin{bmatrix} \left(A_{11} - A_{12} \dfrac{|A_{21}|}{|A_{22}|} e^{j\alpha} \right) T_1 \\ \left(A_{22} - A_{21} \dfrac{|A_{12}|}{|A_{11}|} e^{j\beta} \right) T_2 \end{bmatrix} \end{aligned} \tag{9.68}$$

It can be seen from the above formula that after applying the inverse matrix $\boldsymbol{M}$, the first receiving antenna of DUT only receives the signal from port T_1, and the second receiving antenna only receives the signal from port T_2. There is no cross interference of signals, and the purpose of OTA direct connect is realized.

In theory, if the inverse matrix is applied in the throughput test, the crosstalk between transmitted signals does not exist. However, in actual testing, due to the limitations of the testing instruments and the system, the crosstalk caused by reflection and coupling cannot be completely eliminated. Therefore, in order to evaluate the crosstalk between signals after applying the inverse matrix, the concept of virtual cable isolation is introduced. The results in [49] show that the isolation of the antenna at each test angle can be obtained based on the antenna pattern information measured in the first stage. Therefore, after measuring the antenna pattern, the isolation of the DUT at each angle should be calculated first. Then the position with the highest isolation should be selected as the best test position. Adjust the DUT to this position by using the turntable, and then calculate the inverse matrix of the DUT at this position.

9.4 Other Testing Methods

The RC-based methods, MPAC methods, and RTS methods have been standardized by CTIA and 3GPP and are widely used for SISO/MIMO OTA testing. Besides those OTA methods mentioned above, there is another promising testing method named as wireless cable method.

The proposal of the wireless cable method was inspired by the conducted testing, which is widely adopted in industry for RF performance evaluations in the early stage of RF front-end circuitry and baseband development. In the conducted testing, the DUT is directly connected to the measurement instruments with RF coaxial cables via antenna ports, i.e. with antennas bypassed. Nevertheless, in the later stage of device development, ex-factory inspection, for example, extra modifications are needed for the DUT to maintain support for conducted testing, making this method disruptive and inconvenient. Therefore, it is highly desirable to have a substitution without cable connection for the conducted testing.

The wireless cable method was first proposed by Wei Fan from Aalborg University in 2017 [54], further investigated by Fengchun Zhang from Aalborg University [55–57], and implemented by Feilong Wang from the China Academy of Information and Communications Technology [58]. Essentially, this testing method can be regarded as an OTA replacement of the conventional conducted testing method, which directed the signals via virtual wireless cable rather than actual RF cable. Unlike the aforementioned method conducted in RC or anechoic chamber setup, the wireless cable method can be implemented inside an RF-shielded box, which is an economical solution for OTA testing.

Principles and critical technical problems of the wireless cable method are summarized as follows and the advantages and disadvantages of this method are briefly described.

9.4.1 Principle of Wireless Cable Method

9.4.1.1 Signal Model

To begin with, let us have a look at the schematic diagram and the principles of the conducted method and the wireless cable method. It should be noted that the main goal of the wireless cable method is to provide a "wireless substitution" for the conventional conducted method. The signal model for a MIMO system with M transmitters and N receivers can be described as

$$\boldsymbol{y}(f,t) = \boldsymbol{H}(f,t)\boldsymbol{x}(f,t) + \boldsymbol{n}(f,t) \tag{9.69}$$

where $\boldsymbol{y}(f, t) \in \mathbb{C}^{N\times 1}$ represents the receive signal vector at the receive antenna ports, $\boldsymbol{x}(f, t) \in \mathbb{C}^{M\times 1}$ denotes the transmit signal vector at the transmitter antenna ports, and $\boldsymbol{n}(f, t) \in \mathbb{C}^{M\times 1}$ denotes the noise vector at the receiver antenna ports, which consist of the environment noise as well as the noise caused by the receiver. For simplicity, the noise vector is neglected and 2×2 MIMO system is assumed in the following discussion. The system diagrams of the conducted method and wireless cable method are shown in Figures 9.17 and 9.18, respectively.

1. Conducted Testing Method

As for the conducted testing shown in Figure 9.17, the testing setup includes a base station emulator to establish a cellular connection such as GSM, LTE, or other standard protocols, a channel emulator to synthesize the target channel and perform the signal convolution, and a DUT which connected to the channel emulator with RF cables. The test signal is directly conducted to the DUT antenna ports via the RF cables, with the DUT antennas bypassed. So, we can represent the received signal as

$$\boldsymbol{y}(f,t) = \boldsymbol{A}\boldsymbol{H}(f,t)\boldsymbol{x}(f,t) \tag{9.70}$$

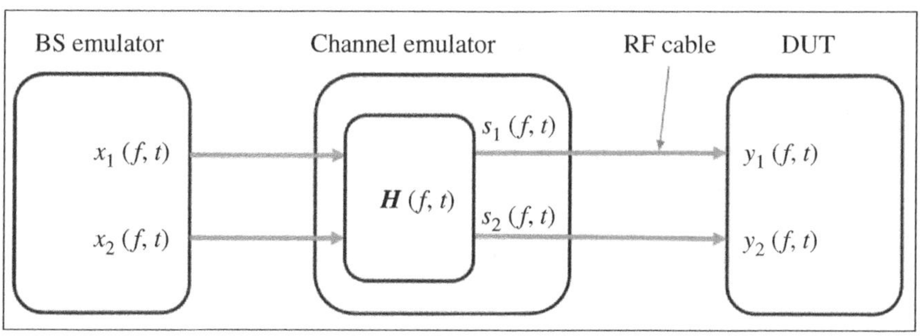

Figure 9.17 System diagram of conducted testing for 2×2 MIMO setup. Source: Reproduced with permission from Ref. [54]/IEEE.

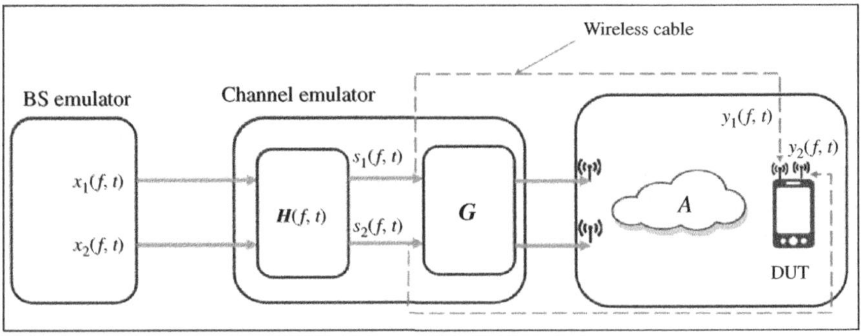

Figure 9.18 System diagram of the wireless cable method. ***A*** represents the transfer relations of signals between the CE output ports and receive antenna ports. Dotted lines: equivalent wireless cable connections. Source: Reproduced with permission from Ref. [54]/IEEE.

where $\boldsymbol{A} \in \mathbb{C}^{2\times 2}$ is the transfer matrix, with $a_{i,j}$ the complex coefficient between the jth channel emulator output port and the ith DUT antenna port, $i, j \in [1, 2]$. Since the complex loss caused by the RF cables can be measured by VNAs or other instruments and compensated easily in practical situations, the transfer matrix $\boldsymbol{A}$ can be regarded as an identity matrix after calibration. Therefore, it seems that there is no crosstalk between different cables, no cable loss, and no phase delay introduced by RF cable connection. Namely, as depicted in Figure 9.17, the signal $s_1(f, t)$ is only received by the DUT antenna port 1 and $s_2(f, t)$ is only received by the DUT antenna port 2 without any alterations.

Although the conducted testing method is easy to understand and widely accepted in the industry, the performance evaluation via conducted testing faces the following limitations.

1. Conducted testing inevitably requires the DUT to be equipped with antenna connectors for RF cable connection. Furthermore, the DUT needs to be opened for access to the antenna connectors. As for GSM or CDMA terminals, the batteries are usually detachable and the back cover of the terminal can be easily removed, which facilitates the conducted testing. However, for LTE or 5G terminals, the antennas may be integrated with the RF front-end circuitry for miniaturization and integration and the whole devices are hard to disassemble, which would hinder the RF cable connection.
2. The antenna connectors are usually delicate and easily damaged. For the DUTs without antenna connectors, temporary RF connectors can be soldered to the mainboard for RF cable connection, which is unstable and easily broken. This modification may also increase the measurement uncertainty and the error probability.

3. The RF cables and connectors might introduce inconsistency due to impedance mismatch between cables and connectors and affect the device's overall performance. Besides, as mentioned in Section 9.3, the self-generated noise and interference cannot be considered properly determined as well.
4. For future massive MIMO antenna systems with hundreds of antennas, hundreds of manual cable connections would be required. Given that, the conducted method may be prohibitive due to the complexity, cost, and the potential level of error.
5. The conducted testing method may affect the integrity of the DUTs and the overall OTA performance. Optimization based on the conducted testing may give unrealistic performance evaluations.

2. Wireless Cable Method

The system diagram of the wireless cable method is shown in Figure 9.18. This system consists of a BS emulator, a CE, an RF-shielded anechoic box to reduce the influence of the external environment, and a DUT. Note that the aim of the wireless cable method is to calibrate out the transfer matrix $\boldsymbol{A} \in \mathbb{C}^{N\times K}$ and direct the signals to the corresponding receive antenna ports without crosstalk, similar to having RF cables in a wireless way, where K denotes the number of OTA antennas inside the RF anechoic chamber. In the wireless cable method, the received signal vector can be expressed as

$$\boldsymbol{y}(f,t) = \boldsymbol{AGH}(f,t)\boldsymbol{x}(f,t) + \boldsymbol{n}(f,t) \tag{9.71}$$

where $\boldsymbol{G}$ is the calibration matrix, which satisfies $\boldsymbol{AG} = \boldsymbol{I}_N \in \mathbb{R}^{N\times N}$. For simplicity, matrix $\boldsymbol{G}$ is implemented in the channel emulator with the channel model $\boldsymbol{H}(f, t)$ and the2×2MIMO case is discussed. With the transfer matrix calibrated out ($\boldsymbol{AG} = \boldsymbol{I}_2$), $s_1(f, t)$ would be transmitted only to the DUT antenna's port 1 and $s_2(f, t)$ only to the DUT antenna's port 2 in a radiated way without actual RF cables, which is equivalent to conducted testing.

The principle of wireless cable method is similar to that of the RTS method in the second stage. The main distinction for those two methods is whether to execute the measurement of DUT's antenna patterns or not. When the RTS method is first proposed in 2014, the transfer matrix $\boldsymbol{A}$ was estimated through the multiplication of the following terms: the measurement antenna gains, the path loss, and the DUT antenna gains [46]. However, this estimation method relies on the measurement of the DUT antenna pattern and the pre-knowledge of the relative positions between the measurement antennas and the DUT antennas, which is not supported by all DUTs. Also, it only holds under far-field conditions, which require an anechoic chamber setup with sufficiently large dimensions. Therefore, although the aim of RTS second stage is also to calibrate the transfer matrix $\boldsymbol{A}$, this estimation method is not appropriate for the wireless cable method.

A new calibration method appropriate for the wireless cable method was first proposed in [1]. The implementation of tuning matrix $\boldsymbol{G}$ is further illustrated in Figure 9.19. Amplitude and phase control at each output port are required to adjust the complex gain for each output signal. And two RF combiners are needed for the case in which $K = 2$ OTA antennas are placed inside the shielded box. As for the cases with $K = 4$ OTA antennas as shown in Figure 9.20, the number of OTA antennas is equal to the number of CE output ports that are needed for calibration matrix implementation to avoid the need for the RF combiners. The average received power level, denoted as a reference signal received power (RSRP) value in LTE (for each DUT antenna) can be collected via a universal serial bus cable by a dedicated tool installed on a computer with the DUT connected to it. The signals at the CE output ports 1–4 can be expressed as $c_1 s_1(f,t)$, $c_1\omega_1 s_1(f,t)$, $c_2 s_2(f,t)$, and $c_2\omega_2 s_2(f,t)$, respectively, with

$$\begin{bmatrix} s_1(f,t) \\ s_2(f,t) \end{bmatrix} = \boldsymbol{H}(f,t)\boldsymbol{x}(f,t) \tag{9.72}$$

where ω_1 is the complex weight to adjust signal $s_1(f,t)$ at port 1 or port 2 and ω_2 is the complex weight to adjust signal $s_2(f,t)$ at port 3 or port 4. c_1 and c_2 are gain weight in real numbers. This is to adjust the power for the first and second virtual cable transmission, respectively. After the output signal adjustment, two RF combiners are implemented to direct the mixed signal to the OTA antennas, as shown in Figure 9.19. The signal model for the proposed method can be expressed as

$$\boldsymbol{y}(f,t) = \boldsymbol{A}\left[c_1\begin{pmatrix}1\\ \omega_1\end{pmatrix} \quad c_2\begin{pmatrix}1\\ \omega_2\end{pmatrix}\right]\boldsymbol{H}(f,t)\boldsymbol{x}(f,t) \tag{9.73}$$

For notational clarity, we name

$$\boldsymbol{B} = \boldsymbol{A}\left[c_1\begin{pmatrix}1\\ \omega_1\end{pmatrix} \quad c_2\begin{pmatrix}1\\ \omega_2\end{pmatrix}\right] = \begin{bmatrix} b_{11} & b_{12} \\ b_{21} & b_{22} \end{bmatrix} \tag{9.74}$$

According to the discussion about the principle, we expect that the matrix $\boldsymbol{B}$ is equal to an identity matrix and the received signal vector is $\boldsymbol{y}(f,t) = \boldsymbol{H}(f,t)\boldsymbol{x}(f,t)$. Therefore, the objective function can be written as

$$\min_{c_1,c_2,\omega_1,\omega_2}\left\|\boldsymbol{A}\left[c_1\begin{pmatrix}1\\ \omega_1\end{pmatrix} \quad c_2\begin{pmatrix}1\\ \omega_2\end{pmatrix}\right] - \begin{bmatrix}1 & 0\\ 0 & 1\end{bmatrix}\right\| \tag{9.75}$$

As (9.7) shows, the entry b_{11} and b_{22} that denote the propagation coefficients for the direct links are expected to be equal in amplitude; the entry b_{12} and b_{21} that denote the propagation coefficient for the crosstalk are expected to be close to zero. In other words, the wireless cable connection can be fulfilled with two conditions. (1) Crosstalk coefficients b_{12} and b_{21} are minimized. (2) The powers of the desired links are determined by b_{11} and b_{22} are equal.

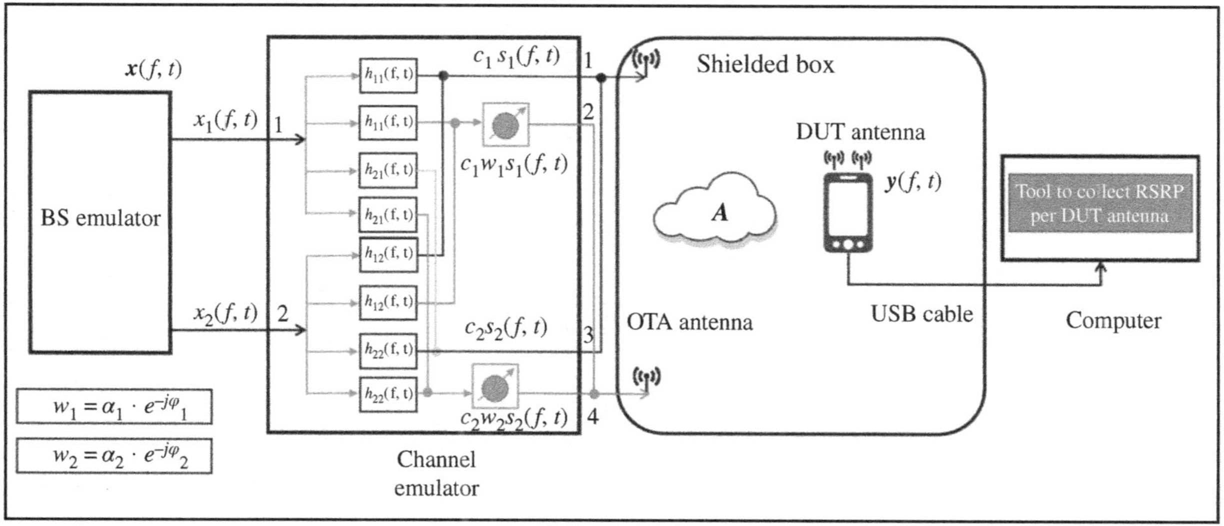

Figure 9.19 Illustration of the proposed calibration method for a 2×2 MIMO system with $K = 2$ OTA antennas. Source: Reproduced with permission from Ref. [54]/IEEE.

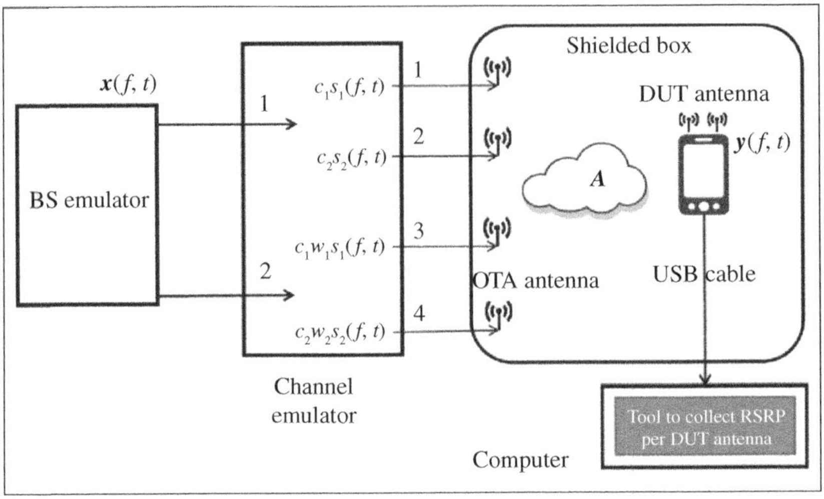

Figure 9.20 Illustration of the proposed calibration method for a 2×2 MIMO system with $K = 4$ OTA antennas. Source: Reproduced with permission from Ref. [54]/IEEE.

The proposed calibration method is detailed as follows.

1. Place the DUT in an RF-shielded anechoic box.
2. Bypass the channel model in the CE (i.e. $\boldsymbol{H} = \boldsymbol{I}_2$) and establish a cellular connection.
3. Find optimal complex weight ω_1 to ensure the first wireless cable connection from s_1 to y_1.
4. Find optimal complex weight ω_2 to ensure the second wireless cable connection from s_2 to y_2.

Here, we take the complex weight ω_1 for example, to illustrate the optimization procedure. Firstly, CE output ports transmitting s_1 and $\omega_1 s_1$ are enabled. Gain weight c_1 is adjusted to offer a good signal quality at the DUT. The RSRP values at the two receive antenna ports are

$$\begin{aligned} \text{RSRP}_1(\omega_1) &= |c_1|^2 |b_{11}|^2 P(x_1) \\ \text{RSRP}_2(\omega_1) &= |c_1|^2 |b_{21}|^2 P(x_1) \end{aligned} \tag{9.76}$$

where $P(x_1)$ denotes the average power level for the transmit signal x_1. According to (9.74), the desired link is $b_{11} = c_1(a_{11} + \omega_1 a_{12})$ and the crosstalk is $b_{21} = c_1(a_{21} + \omega_1 a_{22})$. Since the RSRP values can be obtained through dedicated tools, the optimal complex weight ω_1 can be acquired through a rotation-based method. In terms of the $\text{RSRP}_2(\omega_1)$, the gain term of the complex weight ω_1 is fixed with the phase term sweeping within [0°, 360°]. Obviously, the optimal phase $\hat{\varphi}_1$ is found when $\omega_1 a_{22}$ and a_{21} are out of phase (i.e. with a phase difference of 180°). After that, the phase term is fixed to $\hat{\varphi}_1$ with the gain term sweeping within the feasible region. Also, the optimal gain $\hat{\alpha}_1$ is found when $\omega_1 a_{22}$ and a_{21} are equal in amplitude and out of phase. Therefore, the optimal complex weight is $\omega_1 = \hat{\alpha}_1 e^{j\hat{\varphi}_1}$. Similar procedure can be executed to obtain the optimal complex weight ω_2. Thc gain weight c_1 and c_2 are adjusted to maintain power balance between the two wireless cable connections. Finally, the wireless cable connection is established with target in (9.75) satisfied approximately.

The quality of each established wireless cable connection can be expressed by the isolation levels defined as

$$I_1 = \frac{\text{RSRP}_1(\omega_1)}{\text{RSRP}_2(\omega_1)} \tag{9.77}$$

$$I_2 = \frac{\text{RSRP}_2(\omega_2)}{\text{RSRP}_1(\omega_2)} \tag{9.78}$$

And the minimum between I_1 and I_2 are regarded as the isolation level for the 2×2 wireless cable connection.

3. Wireless Cable Method with $K > N$ OTA Antennas

When it comes to the cases where the number of OTA antennas is more than that of the DUT antennas, i.e. $K > N$, more degree of freedom is provided to further improve the performance of this novel calibration method compared to the $K = N$ cases.

It should be noted that the wireless cable connection cannot be established with enough quality in some special conditions due to the characteristics of the transfer matrix $\boldsymbol{A}$. Clearly, the number of OTA antennas should be no less than the number of DUT antennas in the system setup. The target that $\boldsymbol{AG}$ can approximate $\boldsymbol{I}_N$ would not be achieved with a rank deficient matrix $\boldsymbol{A}$. Nevertheless, a full-rank matrix $\boldsymbol{A}$ is just a prerequisite for the existence of an ideal solution, with the matrix $\boldsymbol{G}$ implemented in any specified resolution level. This is not practical in real system and therefore a small condition number of matrix $\boldsymbol{A}$ is also crucial to achieve a good wireless cable connection, which confines the influence of the implementation errors. For $K = N$, an appropriate transfer matrix $\boldsymbol{A}$ can be found via alterations of the relative positions between the DUT and OTA antennas, such as replacing or rotating the DUT or OTA antennas or by positioning metallic reflectors in the RF shielded box. This trial-and-error approach would complicate the measurement procedure. For $K > N$, extra flexibility is provided to search for the best routing matrix that can achieve the best wireless cable transmission. Routing matrix specifies the rules of how signals are directed to OTA antennas. For a 2×2

MIMO system with $K = 4$ OTA antennas, as shown in Figure 9.20, the received signal vector at DUT side is

$$\boldsymbol{y}(f,t) = \boldsymbol{ARs}(f,t) \tag{9.79}$$

where $\boldsymbol{A} \in \mathbb{C}^{2\times4}$ is the transfer matrix from CE output ports to DUT antenna ports and $\boldsymbol{R} \in \mathbb{C}^{4\times2}$ is the routing matrix set in CE. Since we need to divide four CE output ports to two groups, we have three possible routing matrixes $\boldsymbol{R}$,

$$\boldsymbol{R} = \left\{ \begin{bmatrix} 1 & 0 \\ \omega_1 & 0 \\ 0 & 1 \\ 0 & \omega_2 \end{bmatrix}, \begin{bmatrix} 1 & 0 \\ 0 & 1 \\ \omega_1 & 0 \\ 0 & \omega_2 \end{bmatrix}, \begin{bmatrix} 1 & 0 \\ 0 & 1 \\ 0 & \omega_2 \\ \omega_1 & 0 \end{bmatrix} \right\} \tag{9.80}$$

Note that the second routing matrix $\boldsymbol{R}$ is used in Figure 9.20, i.e. $s_1(f, t)$ and $\omega_1 s_1(f,t)$ are directed via port 1 and port 3, while $s_2(f,t)$ and $\omega_2 s_2(f,t)$ are directed via port 2 and port 4. The achievable quality of wireless cable connection (usually evaluated by the minimum isolation level calculated by (9.77) and (9.78)) can be further improved with the degree of freedom provided by routing matrix.

4. Wireless Cable Method in High-Order MIMO Cases

The aforementioned method is mainly considering the 2×2 MIMO cases. However, the algorithm is not directly applicable for high-order MIMO DUTs. The work in [55] provides a fresh insight for the search of the optimal calibration matrix using heuristic stochastic optimization algorithm (e.g. PSO algorithm in this paper). On the other hand, [56] and [58] convert the research focus from the search of the optimal calibration matrix within the feasible region to the estimation of the transfer matrix. The amplitude of the entries in the transfer matrix can be easily obtained through received power measurement (e.g. RSRP). However, the phase differences cannot be directly obtained from amplitude-only measurement results. In [56], a novel closed-form calibration method is proposed based on solving linear equations to obtain the unknown phase difference between the entries in the transfer matrix. Finally, the work in [58] determined the unknown phase difference between the transfer matrix entries via a rotation-based method. Those methods extend the application range of wireless cable method to the high-order MIMO cases.

9.4.2 Pros and Cons of Wireless Cable Method

The wireless cable method is a non-intrusive and cost-effective OTA testing method with adequate precision, which requires the CE output ports that only equal to or slightly more than the number of receive antennas, regardless of the

DUT size and complexity of channel models (2D or 3D-channel model). Moreover, compared to the wireless cable connection established in [46], this method can be executed in any environment with frequency flat channels, without requiring sufficiently large anechoic chambers to ensure the far-field conditions. Since the main problem faced by the wireless cable method and the RTS method in the second stage is almost identical, those two OTA testing solutions establish the wireless cable connection in a pretty similar way [49, 54, 59].

However, as an OTA replacement of the conventional conducted method, the wireless cable method is subjected to some deficiency compared to the conducted testing. The antenna characteristics of the DUT are ignored during the wireless cable measurement, so much so that this method could not evaluate the actual end-to-end performance (i.e. DUT antennas included) promoted by CTIA and 3GPP standardization organizations. For example, the influence of human body loss can be measured by attaching the DUT to a specific anthropomorphic mannequin head and or hand phantom when performing the antenna pattern measurements in the RTS methods. The antenna pattern measurement results that include the body loss influence are subsequently reported back to the channel emulator for calculation of the test signal that truly reproduce the real environment. However, since the wireless cable method do not acquire the DUT's antenna patterns but calibrate the transfer matrix like the RTS method, those impact on the RF performance cannot be properly evaluated through the wireless cable method. Also, the influence of different DUT orientations with given channel models cannot be assessed via the wireless cable method either. Despite these imperfections of the wireless cable method, it does perform well as an OTA replacement of the conducted testing method and have great potential in industry applications.

References

1 (2016). *Test Plan for Wireless Large-Form-Factor Device Over-the-Air Performance.* Washington, DC: CTIA Certification.

2 Hill, D.A. (2009). *Electromagnetic Fields in Cavities: Deterministic and Statistical Theories.* Hoboken, NJ: Wiley.

3 Gifuni, A., Bastianelli, L., Moglie, F. et al. (2018). Base-case model for measurement uncertainty in a reverberation chamber including frequency stirring. *IEEE Transactions on Electromagnetic Compatibility* 60 (6): 1695–1703.

4 Remley, K.A., Wang, C.-M.-J., Williams, D.F. et al. (2016). A significance test for reverberation-chamber measurement uncertainty in total radiated power of wireless devices. *IEEE Transactions on Electromagnetic Compatibility* 58 (1): 207–219.

5 Li, F., Huang, S., Xue, W. et al. (2022). Signal and coherence bandwidth effects on total radiated power measurements of LTE devices in reverberation chambers. *IEEE Transactions on Instrumentation and Measurement* 71: 1–3.

6 Kildal, P.-S., Chen, X., Orlenius, C. et al. (2012). Characterization of reverberation chambers for OTA measurements of wireless devices: Physical formulations of channel matrix and new uncertainty formula. *IEEE Transactions on Antennas and Propagation* 60 (8): 3875–3891.

7 Xu, Q., Huang, Y., Xing, L., and Tian, Z. (2016). Extract the decay constant of a reverberation chamber without satisfying nyquist criterion. *IEEE Microwave and Wireless Components Letters* 26 (3): 153–155.

8 Grimmett, G. and Stirzaker, D. (2001). *Probability and Random Processes*, 3rde. New York, NY: Oxford University Press.

9 Xue, W., Li, F., and Chen, X. (2021). Effects of signal bandwidth on total isotropic sensitivity measurements in reverberation chamber. *IEEE Transactions on Instrumentation and Measurement* 70: 1–8.

10 Skårbratt, A., Åsberg, J., and Orlenius, C. (2011). Over-the-air performance testing of wireless terminals by data throughput measurements in reverberation chamber. *European Conference on Antennas and Propagation (EUCAP)*, 11–15 April 2011, Rome, Italy, pp. 615–619.

11 Kildal, P.-S., Hussain, A., Chen, X.M. et al. (2011). Threshold Receiver Model for Throughput of wireless devices with MIMO and frequency diversity measured in reverberation chamber. *IEEE Antennas and Wireless Propagation Letters* 10: 1201–1204.

12 Kildal, P.-S., Hussain, A., Durisi, G., Orlenius, C., and Skårbratt, A. (2012). LTE MIMO multiplexing performance measured in reverberation chamber and accurate simple theory. *2012 6th European Conference on Antennas and Propagation (EUCAP)*, 26–30 March 2012, Prague, Czech Republic, pp. 2299–2302.

13 Chen, X., Kildal, P.-S., and Gustafsson, M. (2013). Characterization of implemented algorithm for MIMO spatial multiplexing in reverberation chamber. *IEEE Transactions on Antennas and Propagation* 61 (8): 4400–4404.

14 Chen, X. (2014). Throughput modeling and measurement in an isotropic-scattering reverberation chamber. *IEEE Transactions on Antennas and Propagation* 62 (4): 2130–2139.

15 Chen, X. (2013). Throughput multiplexing efficiency for MIMO antenna characterization. *IEEE Antennas and Wireless Propagation Letters* 12: 1208–1211.

16 Chen, X. (2014). Experimental investigation and modeling of the throughput of a 2 × 2 closed-loop MIMO system in a reverberation chamber. *IEEE Transactions on Antennas and Propagation* 62 (9): 4832–4835.

17 Fan, W., Carreño, X., Sun, F. et al. (2013). Emulating spatial characteristics of MIMO channels for OTA testing. *IEEE Transactions on Antennas and Propagation* 61 (8): 4306–4314.

18 Kyösti, P., Jämsä, T., and Nuutinen, J.-P. (2012). Channel modelling for multiprobe over-the-air MIMO testing. *International Journal of Antennas and Propagation* 2012.

19 Baum, D.S., Hansen, J., Salo, J., Del Galdo, G., Milojevic, M., and Kyösti, P. (2005). An interim channel model for beyond-3G systems: extending the 3GPP spatial channel model (SCM). *2005 IEEE 61st Vehicular Technology Conference*, 30 May–01 June, 2005, Stockholm, Sweden, pp. 3132–3136.

20 Kyösti, P., Juha Meinilä, Hentilä, L., et al. (2007). IST-4-027756 WINNER II D1.1.2 V1.0 WINNER II Channel Models. IST-WINNER II Technical Report.

21 3GPP (2017). Study on cannel model for frequencies from 0.5 to 100 GHz. *3GPP TR* 38: 901.

22 S. Jaeckel, L. Raschkowski, K. Börner, L. Thiele (2017). Quasi deterministic radio channel generator user manual and documentation, Document Revision: v2.0.0.

23 Pei, H.L., Chen, X.M., Huang, X.Y. et al. (2022). Key issues and algorithms of multiple-input-multiple-output over-the-air testing in the multi-probe anechoic chamber setup. *Science China Information Sciences* 65 (3): 1–27.

24 Boyd, S. and Vandenberghe, L. (2004). *Convex optimization*. Cambridge University Press.

25 Ji, Y., Fan, W., Pedersen, G.F., and Wu, X. (2018). On channel emulation methods in multi-probe anechoic chamber setups for over-the-air testing. *IEEE Transactions on Vehicular Technology* 67 (8): 6740–6751.

26 Huang, X.Y., Chen, X.M., Pei, H.L., and Li, Y.S. (2021). Accuracy comparisons of channel emulation methods for two-dimensional uniform and three-dimensional sectored multi-probe anechoic chamber. *Applied Computational Electromagnetics Society Journal* 36 (9): 1139–1144.

27 Fan, W., Szini, I., Foegelle, M.D., Nielsen, J.Ø., and Pedersen, G.F. (2014). Measurement uncertainty investigation in the multi-probe OTA setups. *2014 8th European Conference on Antennas and Propagation (EuCAP)*, 06–11 April 2014, The Hague, Netherlands, pp. 1068–1072.

28 Fan, W., Hentilä, L., Kyösti, P., and Pedersen, G.F. (2017). Test zone size characterization with measured MIMO throughput for simulated MPAC configurations in conductive setups. *IEEE Transactions on Vehicular Technology* 66 (11): 10532–10536.

29 Fan, W., Kyösti, P., Nielsen, J.Ø., and Pedersen, G.F. (2016). Wideband MIMO channel capacity analysis in multiprobe anechoic chamber setups. *IEEE Transactions on Vehicular Technology* 65 (5): 2861–2871.

30 Imai, T., Okano, Y., Koshiro, K., Saito, K., and Miura, S. (2010). Theoretical analysis of adequate number of probe antennas in spatial channel emulator for

MIMO performance evaluation of mobile terminals. *2010 4th European Conference on Antennas and Propagation (EuCAP)*, 12–16 April 2010, Barcelona, Spain, pp. 1–5.

31 Khatun, A., Laitinen, T., Kolmonen, V.M., and Vainikainen, P. (2012). Dependence of error level on the number of probes in over-the-air multiprobe test systems. *International Journal of Antennas and Propagation* 2012: 1–6.

32 Fan, W., Kyösti, P., Ji, Y. et al. (2017). Experimental evaluation of user influence on test zone size in multi-probe anechoic chamber setups. *IEEE Access* 5: 18545–18556.

33 Wang, H., Wang, W.M., Gao, H.Q. et al. (2019). Plane wave compensation technique for multiple-input multiple-output over-the-air testing in small multi-probe anechoic chamber. *IET Microwaves, Antennas and Propagation* 13 (9): 2625–2631.

34 Wang, W.M., Li, M.Y., Liu, Y.A. et al. (2017). Novel physical probe configurations in a multi-probe based 3D MIMO OTA setup. *Journal of China Universities of Posts and Telecommunications* 24 (1): 60–66.

35 Zhang, Z., Tian, L., Zhang, J., Zhang, J., Yu, L., and Liu, Z. (2019). Analysis of test volume size in 3D MIMO OTA for 5G. *IEEE International Conference on Communications Workshops (ICC Workshops)*, 20–24 May 2019, Shanghai, China, pp. 1–6.

36 Fan, W., Sun, F., Nielsen, J.Ø. et al. (2014). Probe selection in multiprobe OTA setups. *IEEE Transactions on Antennas and Propagation* 62 (4): 2109–2120.

37 Yang, X., Zhang, P., Chen, J., Ma, N., and Liu, B. (2018). Probe subset selection in 3D multiprobe OTA setup. *IEEE 29th Annual International Symposium on Personal, Indoor and Mobile Radio Communications (PIMRC)*, 09–12 September, Bologna, Italy, pp. 1–6.

38 Fan, W., Szini, I., Nielsen, J.Ø., and Pedersen, G.F. (2013). Channel spatial correlation reconstruction in flexible multiprobe setups. *IEEE Antennas and Wireless Propagation Letters* 12: 1724–1727.

39 Kyösti, P., Hentilä, L., Fan, W. et al. (2018). On radiated performance evaluation of massive MIMO devices in multiprobe anechoic chamber OTA setups. *IEEE Transactions on Antennas and Propagation* 66 (10): 5485–5497.

40 Wang, H., Wang, W., Wu, Y. et al. (2020). Probe selection for 5G massive MIMO base station over-the-air testing. *IEEE Antennas and Wireless Propagation Letters* 19 (11): 1998–2002.

41 Stoica, P. and Moses, R. (2005). *Spectral Analysis of Signals*. Morgan & Claypool.

42 Zhang, X., Qiao, S., Peng, M., and Li, Y. (2019). Probe selection for over-the-air test in 5G base stations with massive multiple-input multiple-output. *China Communications* 16 (7): 1–12.

43 Kyösti, P., Fan, W., Pedersen, G.F., and Latva-Aho, M. (2016). On dimensions of OTA setups for massive MIMO base stations radiated testing. *IEEE Access* 4: 5971–5981.

44 Kyösti, P., Fan, W., and Kyröläinen, J. (2017). Assessing measurement distances for OTA testing of massive MIMO base station at 28 GHz. *2017 11th European Conference on Antennas and Propagation (EuCAP)*, 19–24 March 2017, Paris, France, pp. 3679–3683.

45 Jing, Y., Wen, Z., Kong, H., Duffy, S., and Rumney, M. (2011). Two-stage over the air (OTA) test method for MIMO device performance evaluation. *2011 IEEE International Symposium on Antennas and Propagation (APSURSI)*, 03–08 July 2011, Spokane, WA, USA, pp. 71–74.

46 Yu, W., Qi, Y., Liu, K. et al. (2014). Radiated two-stage method for LTE MIMO user equipment performance evaluation. *IEEE Transactions on Electromagnetic Compatibility* 56 (6): 1691–1696.

47 Shen, P., Qi, Y., Yu, W. et al. (2018). A decomposition method for MIMO OTA performance evaluation. *IEEE Transactions on Vehicular Technology* 67 (9): 8184–8191.

48 Shen, P., Qi, Y., Fan, J., and Wang, Y. (2017). The advantages of the RTS method in MIMO OTA measurements. *2017 IEEE International Symposium on Electromagnetic Compatibility & Signal/Power Integrity (EMCSI)*, 07–11 August 2017, Washington, DC, USA, pp. 470–473.

49 Shen, P., Qi, Y., Yu, W., and Li, F. (2021). Inverse matrix autosearch technique for the RTS MIMO OTA test. *IEEE Transactions on Electromagnetic Compatibility* 63 (4): 962–969.

50 Shen, P., Yu, Q., Beetner, D.G., and Qi, Y. (2023). A throughput fast measurement method for two-antenna equipped wireless MIMO terminals. *IEEE Transactions on Instrumentation and Measurement* 72: 1–8.

51 (2016). Test plan for 2 × 2 downlink MIMO and transmit diversity over-the-air performance. Revision 1.1, CTIA, Washington, DC, USA.

52 Jing, Y., Kong, H., and Rumney, M. (2016). MIMO OTA test for a mobile station performance evaluation. *IEEE Instrumentation and Measurement Magazine* 19 (3): 43–50.

53 Jing, Y., Kong, H., and Rumney, M. (2016). Radiated two-stage MIMO OTA test method progress for antenna performance evaluation. *2016 Asia-Pacific International Symposium on Electromagnetic Compatibility (APEMC)*, 17–21 May 2016, Shenzhen, China, pp. 729–731.

54 Fan, W., Kyösti, P., Hentilä, L., and Pedersen, G.F. (2017). MIMO terminal performance evaluation with a novel wireless cable method. *IEEE Transactions on Antennas and Propagation* 65 (9): 4803–4814.

55 Fan, W., Zhang, F., Kyösti, P. et al. (2018). Wireless cable method for high-order MIMO terminals based on particle swarm optimization algorithm. *IEEE Transactions on Antennas and Propagation* 66 (10): 5536–5545.

56 Zhang, F., Fan, W., and Wang, Z. (2021). Achieving wireless cable testing of high-order MIMO devices with a novel closed-form calibration method. *IEEE Transactions on Antennas and Propagation* 69 (1): 478–487.

57 Zhang, F., Li, M., Zhang, X. et al. (2022). Achieving wireless cable testing for MIMO terminals based on maximum RSRP measurement. *IEEE Transactions on Antennas and Propagation* 70 (8): 7084–7093.

58 Wang, F., Wang, Z., Qiao, S., Guo, Y., and Zhang, X. (2022). High-order MIMO terminal testing with the reduced-order wireless cable method. *2022 IEEE 95th Vehicular Technology Conference: (VTC2022-Spring)*, 19–22 June 2022, Helsinki, Finland, pp. 1–5.

59 Shen, P., Qi, Y., Yu, W. et al. (2019). OTA measurement for IoT wireless device performance evaluation: challenges and solutions. *IEEE Internet of Things Journal* 6 (1): 1223–1237.

Index

a
Active feed networks 56, 59
Active reconfigurable MIMO antenna systems 147
Air interface access delay 174, 188
Antenna unit 176, 181–185, 267

b
Base station 2, 4, 6, 11, 85–86, 173, 175–176, 179–181, 185–187, 245
Beam manipulation 218–219

c
Channel emulation 333, 338, 341, 344–346
Circularly polarized array 311
Correlation coefficient 9, 13, 18–20, 36, 130, 157, 331
Cross-band 176, 180–181, 188, 308

d
Decoupling techniques 11, 285, 298–299, 305, 311, 316
Diversity gain 13, 23, 25, 29–30, 36, 131, 174, 200, 217
Dual polarized array 305

e
Energy efficiency 6, 85, 174, 188–189
Equal-gain combining 27

f
FR1 band 88, 96, 102
FR2 band 75, 96, 99, 109, 113, 119
Frequency flat MIMO channel 16
Frequency-reconfigurable 147–152, 166
Frequency-selective MIMO channels 16–17

g
Generalized scattering matrix 21–22

i
Impedance-matching networks 41–42
In-band 176, 180, 188, 302, 308
Isolation 4, 9, 13, 17–18, 43, 49–50, 54–56, 60–61, 63–64, 73, 78–79, 95–98, 109–113, 116, 124–127, 131, 150, 152, 157, 159, 163, 165, 168, 177–185, 202, 216, 264, 287, 289, 299–307

l
Linearly polarized arrays 299, 316

MIMO Antenna Systems for 5G and Beyond, First Edition.
Xiaoming Chen, Ping Jack Soh, and Mohammad S. Sharawi.

m

Massive-MIMO (m-MIMO) antenna 11, 176, 187–189
Metamaterial 178, 202, 235, 248, 302–304, 312, 316
Microstrip line 41, 50–53, 80, 95, 100, 107, 114, 117
Millimeter wave band 141
MRC diversity gain 33–36
MTS 191, 217–232, 235–239, 241, 243, 248–261
Multi-band MIMO antennas 102
Multiple-input-multiple-output (MIMO) Technology 4, 6, 9–11, 13, 85, 150, 173, 186, 225, 271, 289, 290, 343
  antenna systems 3, 11, 15, 41, 59, 60, 65, 85, 147, 157, 168, 191–192, 199, 285
  capacity 13, 16, 30–31, 33, 36, 296
  channel 15–19, 30–32, 36, 240, 330
Mutual coupling 11, 18, 41, 49, 53, 63, 118–119, 121, 123, 134, 152, 179, 285–301, 304–309, 311–312, 314, 316

o

OAM 11, 191, 253, 258–263
On-chip antennas 192–196, 198–199
Over-the-air tests 323

p

Parasitic structures 177
Passive feeding networks 50, 56
Passive MIMO antenna systems 85
Pattern reconfigurable 63, 155–161
PIN diode 60–62, 148–149, 157, 159–163, 244
Plane wave synthesis (PWS) 333, 336–341, 343, 345–346
Polarization reconfigurable 161–164

r

Radiated two-stage (RTS) method 348–352, 355–356, 359, 365
Reflective intelligent surfaces (RISs) 11, 191, 239–240, 243, 247

s

Signal processing complexity 175, 188
Single-band 10, 86, 88, 136–137, 142
Space applications 11, 263
Spectrum efficiency 173, 188, 230
Sub-6 GHz band 4, 85, 114, 140, 142
Sub-THz antennas 191, 204–205
System coverage 175, 188
System reliability 176, 188

t

Target channel model 333, 343, 345–346
Throughput testing 11, 329, 348, 352
Total active reflection coefficient 9, 13, 20

u

Ultra-wide band (UWB) MIMO antennas 113

w

Wireless cable method 356–360, 363–365

Printed and bound by CPI Group (UK) Ltd, Croydon, CR0 4YY
17/09/2026
14964255-0003